FIELD GUIDE TO

SHALLOW WATER

SEASTARS

OF AUSTRALIA

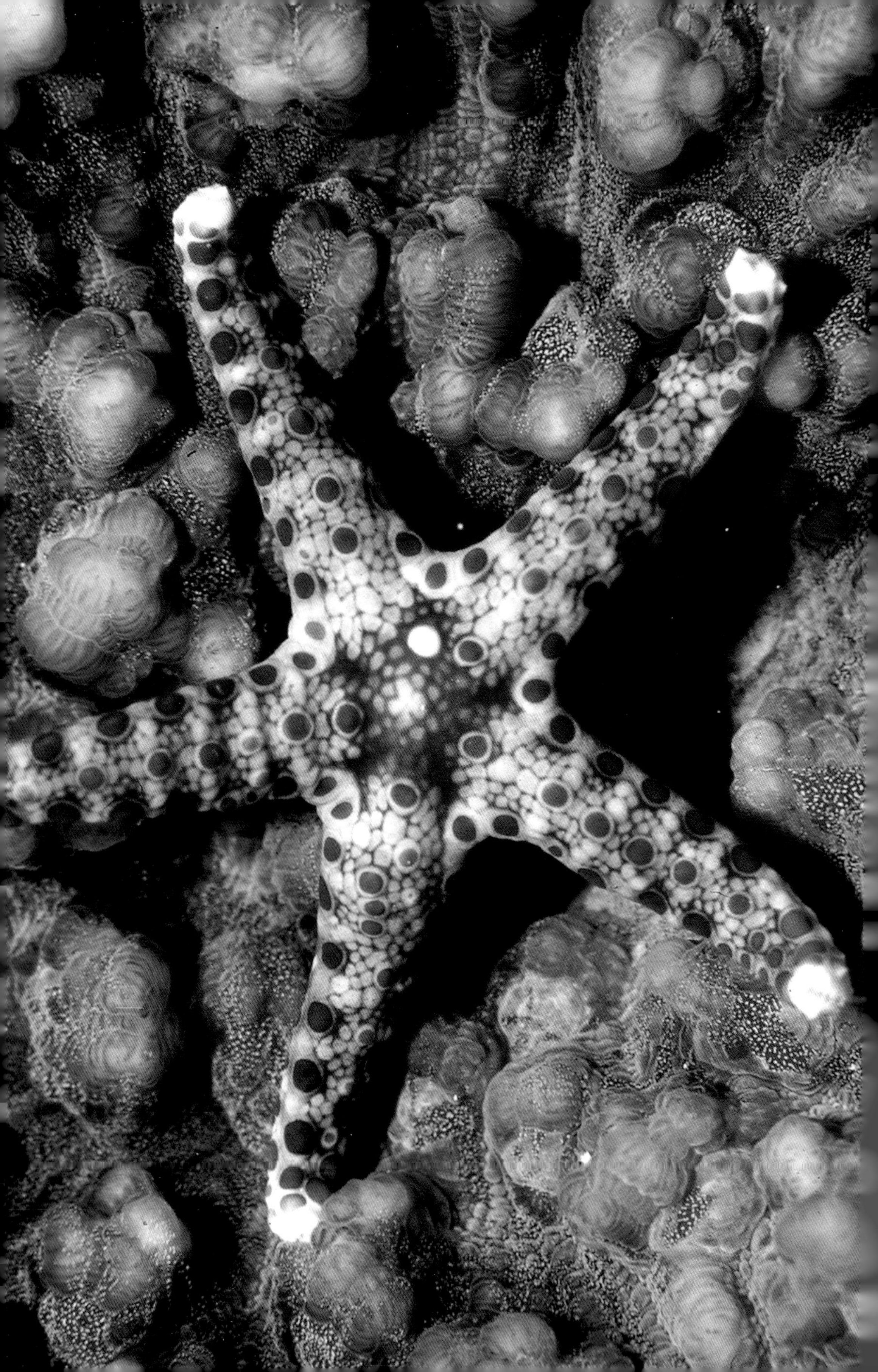

FIELD GUIDE TO SHALLOW WATER SEASTARS OF AUSTRALIA

LOISETTE M. MARSH AND JANE FROMONT

THE AUTHORS

LOISETTE MARSH. A childhood encounter with a purple seastar off Canada's west coast stimulated Loisette's love of marine life. Her special interest in the taxonomy of seastars was sparked by her postgraduate work on rock platforms near Perth, where echinoderms are some of the most dominant animals. Loisette obtained her Bachelor and Master of Science in Zoology from the University of Western Australia, and became an echinoderm specialist under the guidance of Dr Ernest Hodgkin. In 1970, Loisette joined the Western Australian Museum Department of Aquatic Zoology and became the Curator of Marine Invertebrates, a position she held until her retirement at the end of 1993.

Scuba diving became an essential skill for marine fieldwork from the 1970s, making it possible to sample previously inaccessible coral reef slopes. From 1958 until her retirement, Loisette was a scuba diver which, on numerous expeditions enabled her to record many coral reef seastar species formerly not known to occur in Australia. Loisette has written two books, published numerous scientific papers and reports and named 10 species of echinoderms. In acknowledgement of her contributions to science, 13 species have been named after her. Loisette is now Emeritus Curator at the Western Australian Museum and continues to research echinoderms to this day.

JANE FROMONT. The wonder that unfolded on her first scuba dive in northern New Zealand led Jane to complete a Bachelor of Science in Zoology at the University of Auckland in 1978. Jane first met Loisette in 1981 when she worked on the sponge collections in the Western Australian Museum under Loisette's guidance. This mentoring of her as a young graduate instilled in Jane a lifelong passion for marine invertebrates. Jane went on to complete a Master of Science at Auckland University (1985) and a PhD at James Cook University (1990), both on sponge taxonomy and biology. Jane became Curator of Marine Invertebrates at the Western Australian Museum in 1996, and went on to become Head of Department and Senior Curator in Aquatic Zoology. She retired at the end of 2019. Jane has published in excess of 130 scientific papers on subjects ranging from sponge taxonomy and biology to marine ecology and natural products chemistry. She is honoured to have assisted Loisette in the publication of this book.

Echinaster varicolor occurs in many colour forms throughout sheltered temperate and tropical waters along the western coastline of Australia (C. Wood).

CONTENTS

Fromia indica is commonly found on coral reefs in tropical waters around Australia and through the Indo-West Pacific. Great Barrier Reef, Queensland (N. Wu).

ACKNOWLEDGEMENTS

Numerous people have contributed to bringing this book to publication. It could not have been achieved without the dedication of Tim Cumming (Publications Coordinator, Western Australian Museum), who discussed interpretations and suggested many modifications to better highlight species characteristics and ensure as much clarity for the reader as possible. We are deeply indebted to Tim as his input has greatly improved the book, and we cannot thank him enough for his time and efforts.

We thank Dr David Sutton, who in a voluntary capacity read every word and made countless suggestions to improve the consistency and readability of the text. His close attention to detail has greatly improved the book, and we are deeply grateful.

We are very grateful to Alec Coles (Chief Executive Officer, Western Australian Museum), Diana Jones (Executive Director, Collections and Research), and Jason Fair (Director Fremantle Museums and Business Development) for their unwavering support for the publication of this book.

We thank Dr Chris Mah (National Museum of Natural History, Smithsonian, Washington, D.C. and Seastar Editor for the World Register of Marine Species) for numerous discussions about the taxonomic placement of seastars.

For access to specimens, and for information and images related to their seastar collections, we gratefully acknowledge: Dr Tim O'Hara, P. Mark O'Loughlin and Melanie McKenzie (Museums Victoria); Gavin Dally and Suzanne Horner (Northern Territory Museum of Arts and Sciences); Dr Stephen Keable and Helen Stoddart (Australian Museum); Dr Andrea Crowther (South Australian Museum); Dr Peter Davie and Darryl Potter (Queensland Museum); Dr Dave Pawson (Smithsonian Institution, Washington, D.C.); Dr Robert Woollacott (Museum of Comparative Zoology, Harvard); Dr Nadia Ameziane-Cominardi (Muséum National d'Histoire Naturelle, Paris); the late Ailsa Clark (formerly of the Natural History Museum, London); and Dr Frank Rowe (formerly of the Australian Museum).

Those who provided underwater photographs are credited with each image. Without their skills there would be no book. It was also critical to obtain images of specimens held in museum collections, and their distinguishing characteristics, for which we gratefully acknowledge the efforts of the following photographers: William Wong (Australian Museum specimens); J. Ben Boonen,

Lupita Bribiesca, P. Mark O'Loughlin (Museums Victoria specimens); Michael Hammer (Northern Territory Museum of Arts and Sciences specimens); Janet Atkinson, Angela Bannon, Andrea Crowther, Laura Fazzalari, James Hartley, Geoff Lemmey, Amy Pfitzner, Paul Stokes, Elizabeth Thornett (South Australian Museum specimens); Clay Bryce, Aaron Cosgrove-Wilke, Andrew Hosie, Mark Salotti (Western Australian Museum specimens); Peter Marsh and David Sutton photographed specimens from the Australian Museum and Western Australian Museum collections.

For their technical assistance and support we thank staff of the Aquatic Zoology Department at the Western Australian Museum, especially Oliver Gomez, Ana Hara, Jenelle Ritchie and Mark Salotti. The Western Australian Museum librarians Bobbie Bruce and Wendy Crawford helped us to obtain important references, and were always ready to assist.

We thank Anne Nevin, who typed an early version of the manuscript, and Jill Ruse and Tim Cumming for their illustrations and line drawings.

Our thanks go to current and former Western Australian Museum colleagues for collecting seastars incidentally to their own work: Dr Gerry Allen, Dr Paddy Berry, Clay Bryce, Dr Ray George, Andrew Hosie, Dr Barry Hutchins, Dr Gary Morgan, Sue Morrison, Alison Sampey, Shirley Slack-Smith, Dr Fred Wells, and the late Dr Barry Wilson. We similarly thank Dr J.E.N. (Charlie) Veron of the Australian Institute of Marine Science (AIMS), Townsville, and Dr John Keesing of the Commonwealth Scientific Industrial Research Organisation (CSIRO), Perth.

For translations from German we thank the late Ada Neumann and Volker Framenau. Pam Beesley, Australian Biological Resources Study (ABRS) assisted with grant management.

Loisette particularly acknowledges the late Dr E.P. Hodgkin of the University of Western Australia for his mentorship, initiating this project, and for the extensive collection of south-western Australian seastars that formed the basis of the Western Australian Museum asteroid collection.

The task would not have been possible without our families and the years of support they gave in sustaining our belief that it would be published one day. To Jenni and Peter Marsh, and Jane's family David, Zollie and Hanna Sutton, we thank you for believing in us and the importance of this seastar book.

Funding for this book was provided by the Western Australian Museum, the Australian Biological Resources Study and Woodside Energy.

INTRODUCTION

Seastars (Asteroidea: asteroids) comprise one of the five classes of living Echinodermata (spiny-skinned animals: echinoderms). The other classes are the Ophiuroidea (brittle and basket stars: ophiuroids), the Echinoidea (sea urchins, heart urchins and sand dollars: echinoids), the Crinoidea (feather stars and sea lilies: crinoids) and the Holothuroidea (sea cucumbers or bêche-de-mer: holothurians). Echinoderms have a long fossil record, although the record is more extensive for the echinoids than the asteroids. The skeletons, or tests, of the former are more robust than those of asteroids, which usually separate into multiple ossicles (small calcareous elements embedded in the dermis of the body wall) after death. The five classes of echinoderms arose during the Palaeozoic Era. Asteroids are recognisable in the Ordovician fossil record, but were probably present in the Cambrian Period (500+ million years ago), and diversified widely during the Mesozoic Era (approximately 100–200 million years ago).

The present-day world fauna includes in the order of 1,900 species of asteroids, approximately 280 of which have been recorded from Australia. Among these, 202 described species (72%) occur at less than 30 m depth, and form the subject of this book. Some of the species (86; 43% of the total) are endemic to Australia (found only in Australian waters). Among the endemic Australian species that occur at shallow depths, some are ubiquitous around the coastline, while others have ranges of a few tens or hundreds of kilometres. Some species that are widely distributed in the tropical Indo–Malayan region (including Indonesia) only reach the atolls and bank reefs (the Rowley Shoals, Scott and Ashmore Reefs) off north-western Australia, or the northernmost reefs of the Great Barrier Reef, Queensland.

The aim of this book is to facilitate identification of the shallow water seastars of Australia, based on the illustrations, keys and descriptions that are provided. While many seastars can be readily identified by careful examination of the illustrations, others are difficult to identify, even by specialists.

'New' species continue to be found as a result of more detailed collecting along well studied shores, and by detection of cryptic genera through the use of genetic analysis, which can facilitate the separation of morphologically similar species.

Iconaster longimanus is the only species of the genus found in Australian waters (C. Bryce).

For example, several new genera and species have been discovered in the family Asterinidae. Genera that would benefit from this type of detailed investigation include *Goniodiscaster* and *Anthenea* in the family Oreasteridae, and *Ophidiaster, Tamaria* and *Hacelia* in the family Ophidiasteridae.

Each species description provided is accompanied by a distribution map, and distribution patterns are discussed below in the subsection *Distribution Patterns of Australian Seastars.*

The basic morphology of seastars and the unique terminology used to describe them is explained in diagrams associated with the species descriptions and in the glossary. Species descriptions are based on original descriptions of the holotype, where available, or on a modern revision.

The list of species is based on *The Zoological Catalogue of Australia, volume 33, Echinodermata* (Rowe and Gates, 1995), with the addition of species described from 1993 (the last date for species included by Rowe and Gates, 1995) to the end of 2019.

HISTORY OF SEASTAR COLLECTING IN AUSTRALIA

The first major exploration and collecting expedition to Western Australia was the French expedition led by Captain Nicolas Baudin (1801–1803), the intent of which was to survey parts of the coast of Australia not charted during the 17th and 18th centuries. The expedition included botanists, zoologists and other scientists, who sought to learn as much as possible about Australia. Because of illness, and disagreements with the captain, many scientists left the ship at Mauritius. François Péron remained as the sole zoologist/anthropologist on the expedition, but was ably assisted by the artist Charles Lesueur. The expedition surveyed and collected off Western Australia (WA) between Geographe Bay and the Kimberley, prior to restocking in Timor (Péron 1807–1816). The expedition returned to Australia in 1802 and surveyed and collected in Tasmania and off the south coast of Australia; it returned to France via Mauritius, where Baudin died. Lesueur made exquisite watercolour paintings of marine specimens collected during the expeditions. These are held at the Natural History Museum of Le Havre, but few have been published. Jangoux (1984) catalogued the seastar paintings but illustrated only one. From these he identified 36 seastars to species and a further six to genus. Péron completed the first volume of the narrative of the expedition (1807), but died in 1810. The second volume was completed by Louis de Freycinet and published between 1807 and 1816. The marine invertebrates were described by Lamarck (1816), but as he did not have access to Péron's field notes the localities are vague. Lamarck described 14 species of asteroids from Australia, including four endemic species.

Seastars were collected from south-western Australia by the German botanical collector Ludwig Preiss, whose specimens were described by Müller and Troschel (1840, 1842, 1844). Australian seastars were also described by J.E. Gray of the British Museum (Natural History) in 1840 and 1847, and an illustrated version of his 1840 paper was published in 1866. He described 18 new species of asteroids, but 10 of the names are now regarded as synonyms of previously identified species.

The British Admiralty *Challenger* Expedition (1872–1876) was equipped to explore the deep sea, worldwide. Sampling during the *Challenger* Expedition was only brief in Australian waters, including in Bass Strait, New South Wales (NSW), the Torres Strait and the Arafura Sea, but no sampling occurred off the western half of the continent. Sladen (1882, 1883, 1889) described 12 new species of asteroids from the *Challenger* collections made in Australia, most of which were from deep water.

Originally described by Lamarck as *Asterias calcar* in 1816, *Meridiastra calcar* is one of the few endemic species that were collected by Péron and Lesueur on the first exploration and collecting expedition to Australia (D. Hobern).

Several sites off north-western Australia were dredged during worldwide collecting as part of the German *Gazelle* Expedition (1874–1876). These included off Shark Bay, in the Dampier Archipelago and on the North West Shelf. Based on this expedition, Studer (1883, 1884) listed eight species of asteroids from Australia, including three new species from WA and one from Queensland.

In the course of a four year cruise around the world, in 1881 HMS *Alert* spent six months surveying the Queensland coast and the Torres Strait. The surgeon R.W. Coppinger dredged and collected from the shore at many islands and ports, and obtained specimens from pearl divers. From amongst these, Bell (1884) described three new species of asteroids, of which only one is now regarded as a valid species. In all, he listed 28 species from the HMS *Alert* collections, including from Queensland and NSW.

During the German Hamburg Expedition of 1905, Michaelsen and Hartmeyer sampled at Shark Bay, Geraldton, Fremantle, Bunbury and Albany. Asteroids were not included in *Die Fauna Sud-West Australiens*, a five volume report of the expedition (1907–1930), but some were subsequently described by Döderlein in the Siboga reports (1917, 1920, 1935, 1936). Some of the new species have since fallen into synonymy, but the Hamburg Expedition made the most significant contribution to knowledge of the echinoderms of Western Australia since the Baudin Expedition.

The pre-eminent contributor to knowledge of Australian echinoderms was Hubert Lyman Clark (Museum of Comparative Zoology, Harvard), who published on Australian echinoderms from 1909–1946. He made two extended visits to Australia, in 1929 and 1932, visiting all parts of the Australian coast except the western half of the south coast and the Gulf of Carpentaria. He collected from the shore, in shallow water (by dredging), and also obtained specimens from pearl divers during two lengthy stays in Broome. He spent time at Lord Howe Island, and on the Murray Islands in the Torres Strait (Clark, 1921, 1938). Clark collected in excess of 11,000 specimens of echinoderms encompassing 422 species in 184 genera. He described 52 new seastars from among the total of 216 species of echinoderms recognised by Rowe and Gates (1995). He also examined museum collections, particularly those in the Australian Museum (Sydney) and the Western Australian Museum (Perth) (Clark, 1914). From these he described a further 30 species of seastars, which have since been synonymised with previously known species. He also described seven new genera, of which three are still recognised as valid (see Rowe and Gates, 1995). Clark (1946) summarised his studies in *The Echinoderm Fauna of Australia: its Composition and its Origin.*

In the late 19th and early 20th centuries, state and national museums began to appoint Australian taxonomists. Amongst these was Arthur Livingstone (Australian Museum), who had assisted Clark at Lord Howe Island and Broome, and with Clark took part in the British Great Barrier Reef Expedition of 1928–29 (Livingstone, 1932a). Livingstone was at the Australian Museum from 1920 to 1941, and described three new genera and 16 currently recognised species of seastars (Livingstone, 1930, 1936).

Aided by the widespread use of SCUBA, which enabled collecting in previously inaccessible habitats, from the 1950s zoologists in Australian museums and universities contributed many revisions of families and genera, including descriptions of new taxa. More recently, the combination of genetic analysis and detailed morphological studies has revealed the occurrence of cryptic species, particularly in the family Asterinidae. During the past 50 years Australian Asteroidea were described by a number of authors including: A.M. Clark (1966), Shepherd (1967–68), Dartnall (1967–80), Gibbs et al. (1976), Marsh (1976–2009), Rowe (1976–1989), Rowe and Albertson (1987–1988), Rowe and Marsh (1982), Byrne (1991), Naughton and O'Hara (2009) and O'Loughlin et al. (1990–2015).

In 1995 the Australian Biological Resources Study published a catalogue of Australian Echinoderms (Rowe and Gates, 1995), which provided the first compilation of species of seastars in Australia, and formed the basis of this present book. Subsequently, new species have been described and many species newly recorded following fieldwork on coral reefs off north-western Australia (for example Marsh, 2009; O'Loughlin and Bribiesca-Contreras, 2015), whilst temperate species have been described from southern Australia (Dartnall et al., 2003; Naughton and O'Hara, 2009; O'Loughlin, 2002). Genetic and morphological studies of Asterinidae (O'Loughlin and Rowe, 2005, 2006; O'Loughlin and Waters, 2004) have resulted in additional new species descriptions, while other studies have led to taxonomic changes to genera and species (Mah and Foltz, 2011a, 2011b). Updates on new species and changes to higher level taxonomy have been recorded on the World Register of Marine Species: World Asteroidea database: http://www.marinespecies.org (Mah, 2019).

At the time of writing the most recent publications on Australian echinoderms are by Byrne and O'Hara (2017), and Edgar (2019).

BIOLOGY OF SEASTARS

FORM AND FUNCTION

Seastars, like other echinoderms, have several features unique to the phylum including: five-fold radial symmetry (pentamerism) in the adult stage and larvae having bilateral symmetry; a water vascular system; and the presence of pincer-like organs on the body surface (pedicellariae). Echinoderms are solely marine organisms, as they have no mechanism to control cellular influx of fresh water by osmosis under low salinity conditions. Nevertheless, asteroids in the Baltic Sea have adapted to salinities as low as 8 parts per thousand (ppt) (Stickle and Diehl, 1987), and as high as 60 ppt in the Western Arabian Gulf (Price, 1982). At least four of these Arabian Gulf species are found in tropical Australian waters at normal seawater salinity (35 ppt).

Seastars are found in all marine habitats, including on sand, mud or rubble, and on rocks and coral reefs, from intertidal shores to the deepest oceans. Each species typically occurs in a relatively narrow depth range (from less than one metre to several hundred metres), which may be related to factors including access to a preferred food, predation and temperature. For instance, on the continental shelf in Western Australia (which is warmed by the Leeuwin Current in winter), the tropical seastar *Euretaster insignis* extends south to the temperate Cape Leeuwin area, while in shallow coastal waters its southern limit is in Cockburn Sound, near Fremantle (Marsh, personal observation).

Seastars have a large or small central disc, from which five (rarely fewer) or more arms radiate (Figure **1A**). These are barely evident in pentagonal or circular species, where the arms can only be seen on the actinal surface, where the ambulacral furrows radiate from the mouth; the furrows extend from the mouth to the end of each arm, and support rows of tube feet that can be extended or withdrawn for locomotion (Figures **1B**, **2**, **5**).

The anus and madreporite are on the abactinal surface, with the anus located centrally, while the madreporite, often evident as a white striated lump, is located between the anus and the margin (Figure **1A**). The madreporite, or sieve plate, is the external entry point of the water vascular system (Figure **2**). This involves a stone canal (a tube strengthened by calcareous deposits) that leads to a ring canal circling the central disc, from which a radial canal extends along each arm to the arm tip. Small lateral branches off each radial canal end in ampullae (bulb-like sacs) in the body cavity; these are connected to the locomotory tube feet that extend through the skeleton along the ambulacral furrow (Figures **1B**, **2**, **5**).

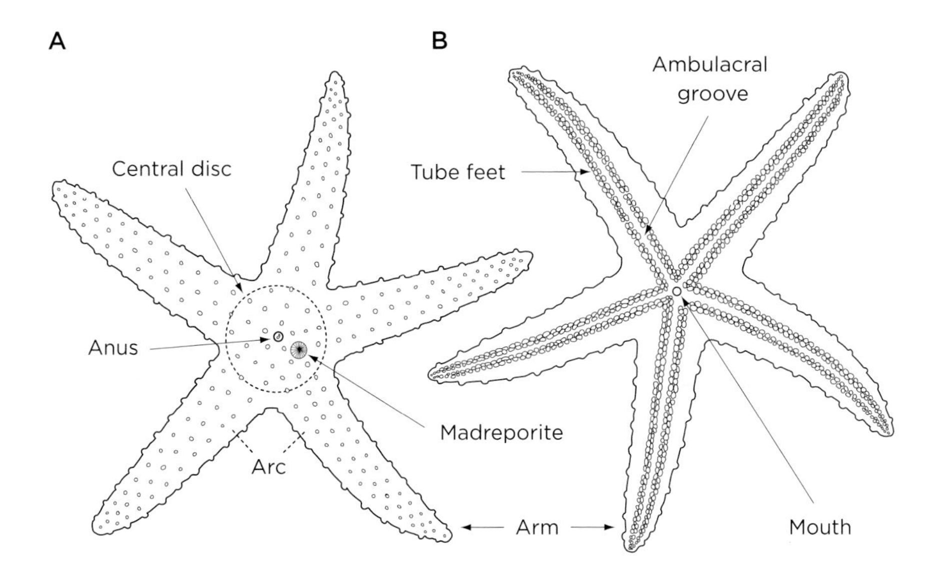

Figure 1: Basic external features of a typical seastar.
A) The abactinal (aboral) view. **B)** The actinal (oral) view.

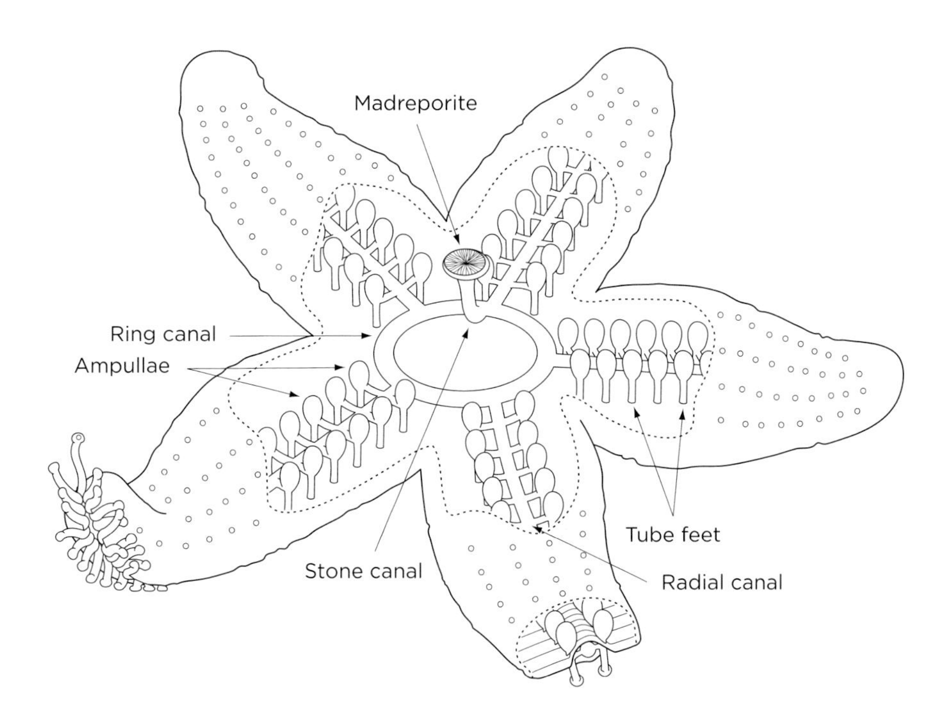

Figure 2: The water vascular system of a typical seastar.

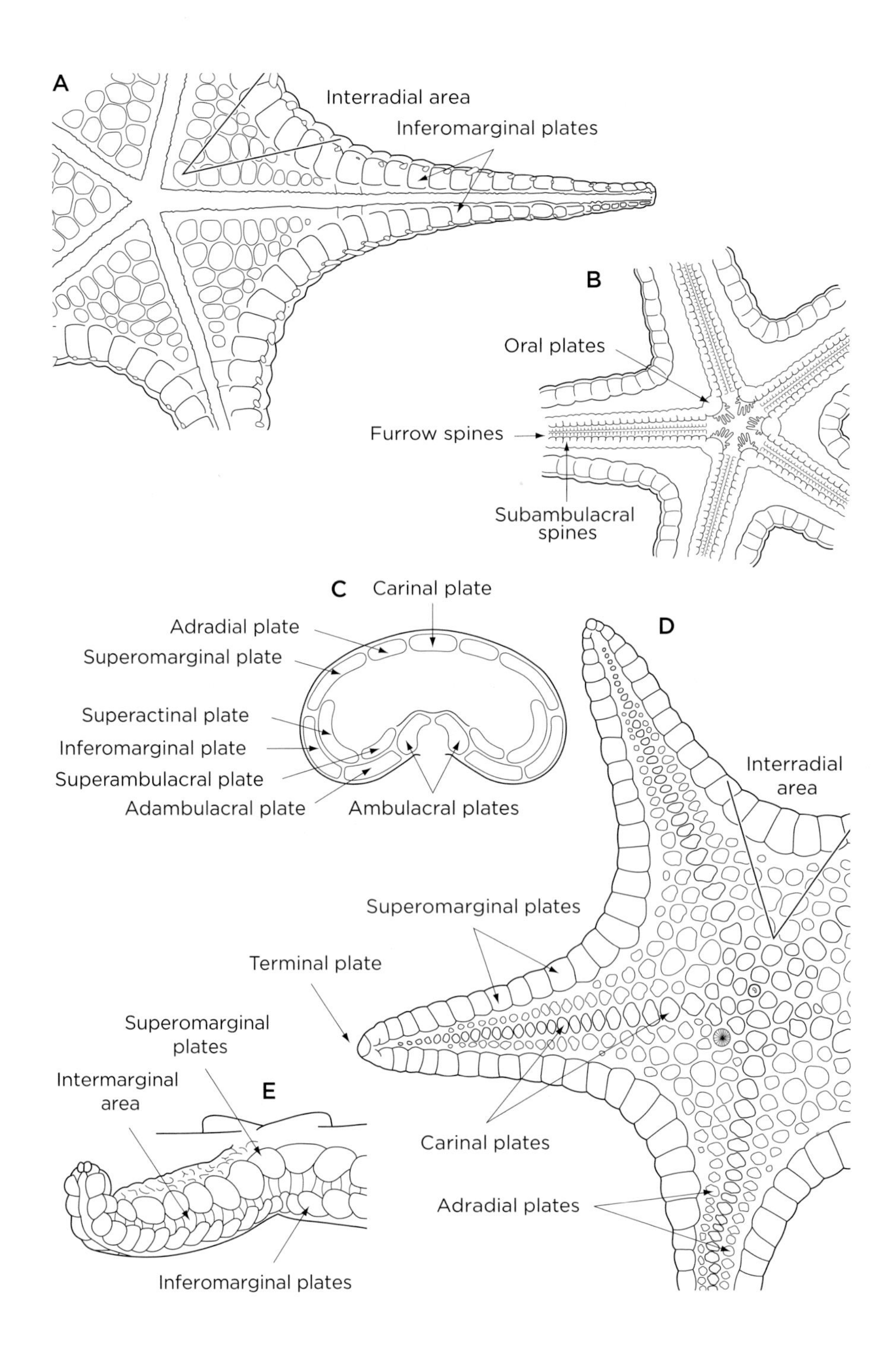

Figure 3: The skeletal structure of a typical seastar.
A) The actinal surface. B) The ambulacral groove. C) Cross section of the arm plate structure. D) The abactinal surface. E) The marginal plates.

The tube feet usually occur in two rows, although some members of the family Asteriidae have four rows. To extend a tube foot, a valve in the lateral branch of the radial canal closes, isolating the tube foot. The ampulla contracts and fluid pressure extends the muscular tube foot, which goes through a 'stepping' motion. The integrated movement of hundreds of tube feet, coordinated in waves, moves the seastar forward. Locomotion is slow, up to 2.7 m per hour for *Linckia laevigata*, and 20 m per hour for the Crown-of-thorns seastar (*Acanthaster planci*) moving across sand to regain the reef. Most seastars have a sucker at the tips of each tube foot, which enables it to adhere tightly to a hard surface; the tube feet tear if they are forcibly removed rather than releasing their grip. Members of two families, Luidiidae and Astropectinidae (order Paxillosida), have pointed or knobbed tube feet, which may aid locomotion in their sandy habitat. The terminal tube foot of each arm is sensory, and has a red eyespot at its base.

The skeleton of seastars is composed of many separate calcareous ossicles embedded in the body wall (Figure 3). This includes connective tissue that has the property of being very stiff and rigid or, when relaxed, allowing the seastar to wrap around a prey item. Externally, the skeletal plates may be concealed in the dermis, covered by a very thin epidermis, or the plates may be completely bare. Pedicellariae, or pincer organs, are embedded in the skin, or in some cases in pits in the skeleton (Figure 4). These are used to keep the body surface clean and clear of parasites. Pedicellariae in seastars are not venomous, in contrast to those of some echinoids.

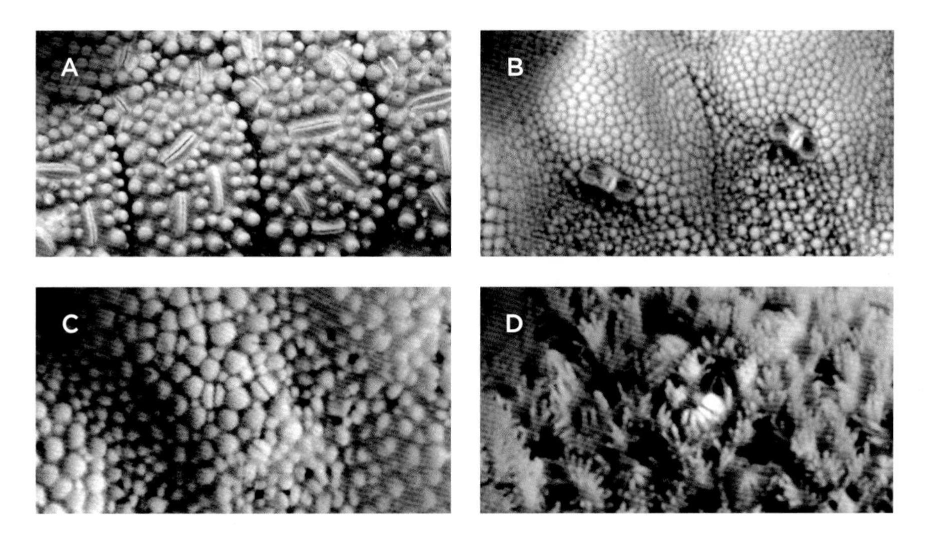

Figure 4: Typical forms of seastar pedicellariae.
A) Bivalved (e.g. *Anthenea conjungens*). B) Sugar-tongs (e.g. *Hacelia helicosticha*).
C) Split-granule (e.g. *Goniodiscaster seriatus*). D) Fasciculate (e.g. *Nepanthia crassa*).

Papulae fill the spaces between the unique 'warty' pustules of *Echinaster callosus* (R. Steene).

Numerous papulae and pincer-like pedicellariae cover the abactinal surface of *Acanthaster planci* between venomous spines (C. Bryce).

Respiration (gas exchange) in seastars takes place mainly through tiny finger-like outgrowths (termed papulae) of the coelom (Figure **5A**). These usually project from the abactinal surface between the skeletal plates in large or small areas termed papular areas or papular spaces. In some seastars the papulae cover most of the surface, giving it a velvety texture. A living Crown-of-thorns seastar (*Acanthaster planci*) may appear reddish because of the extended papulae, whereas the skin is in fact grey when the papulae are retracted.

The mouth is located in the centre of the disc on the actinal (lower) surface (Figures **1B**, **3B**, **5B**) and is surrounded by oral plates bearing spines. In many species the digestive system consists of a two-chambered stomach, intestine, rectal caecae and paired pyloric caecae. The latter two are both secretory absorptive organs that fill the body cavity within each arm along with the paired gonads (Figure **3C**).

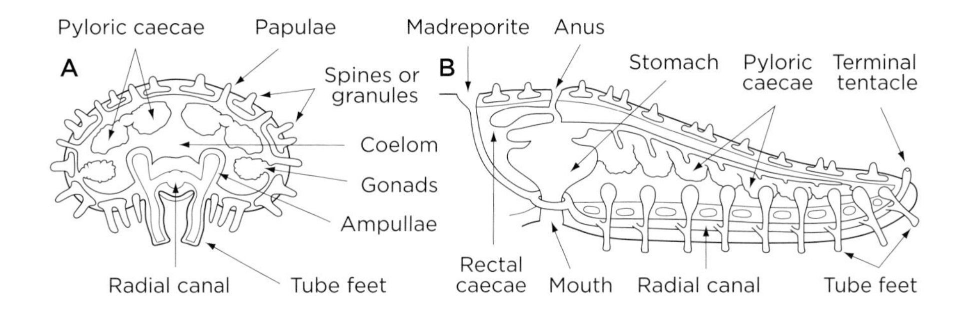

Figure 5: Cross sections of a typical seastar arm showing respiratory, water vascular and digestive features.

FOOD AND FEEDING

As for the other classes of echinoderms, seastar feeding methods are very different from those in any other phylum. Members of the Luidiidae and Astropectinidae ingest relatively small prey, which include bivalves, gastropods and sea urchins, but as they have no rectum or anus they eject the indigestible parts through the mouth. In these two families, captured prey are pushed into the mouth by the pointed tube feet. *Luidia* species sometimes ingest entire heart urchins (e.g. *Echinocardium*), which results in the abactinal body wall distending and gives the appearance of a large hump in the disc. *Astropecten granulatus* has been recorded to consume 57 different prey species (Martin, 1970); it takes 10–30 minutes to

Astropecten sumbawanus feeding on a tusk shell (*Dentallium*) discovered in the substrate (C. Bryce)

ingest a prey item and 26–48 hours to digest it, depending largely on the size of the prey. This form of feeding is termed intraoral because digestion takes place internally.

Asteroids other than Luidiidae and Astropectinidae feed extraorally, by everting the stomach through the mouth and spreading it over the prey, which is partially digested externally. *Acanthaster planci* has a large disc and short arms and everts its stomach over living coral covering an area approximately the size of the disc. Digestion of the coral tissues leaves a clean white coral skeleton after as little as three hours.

Ciliated cells lining the floor of the stomach create a current that carries the partly digested food and mucus into the stomach and upwards to the pyloric caeca, where the food is further digested by enzymes and stored as glycogen and lipids. Undigested particles are passed to the rectal caecae, where further digestion and storage takes place (Figure **5B**). Any undigested material remaining is expelled through the anus.

Many seastars feed on encrusting sponges, bryozoans and compound ascidians, which commonly occur in dense communities covering reef walls and shaded jetty piles.

Many species evert the stomach to feed on detritus and small organisms among the sand and algal biofilm forming the substrate. *Stellaster*, a goniasterid, fills its stomach with mud as a means of capturing and digesting small infaunal animals. Some asterinids have been found with their stomachs full of red or green algae, which is presumably digested along with the attached biofilms of microorganisms. Others feed on encrusting algae, and juvenile *Acanthaster* eat the encrusting coralline alga *Porolithon*. Many seastars are opportunistic scavengers on dead animals.

The introduced asteroid *Asterias amurensis* is an example of species that feed on large well protected prey. For example, they are able to pull the shells of oysters and mussels slightly apart with their tube feet, insert the stomach through the very small opening created, and digest the soft parts of these animals. Some seastars seek any tiny natural opening (as small as 100 µm) near the hinge of a bivalve, through which they can insert a fold of the stomach. Ciliary feeding is another method of food gathering. The everted stomach has a ciliated mucus-coated surface, which enables surface detritus and microorganisms to be trapped and transferred to the digestive tract by ciliary currents, along the ambulacral groove.

Where known, the prey items for each species described in the text are noted.

A large group of *Asterias amurensis* feeding on a bed of mussels (*Mytilus*) (M. Norman).

SEASTAR REPRODUCTION AND DEVELOPMENT

Seastars have a diversity of reproductive strategies. These range from regeneration of the entire animal from a single detached arm, to the release of small eggs that grow into a plankton-feeding swimming larvae. Other seastars have large yolk-filled eggs that quickly settle to the bottom producing non-feeding (lecithotrophic) larvae that live on the yolk until metamorphosis. Brooding the young either externally or internally is another reproductive method.

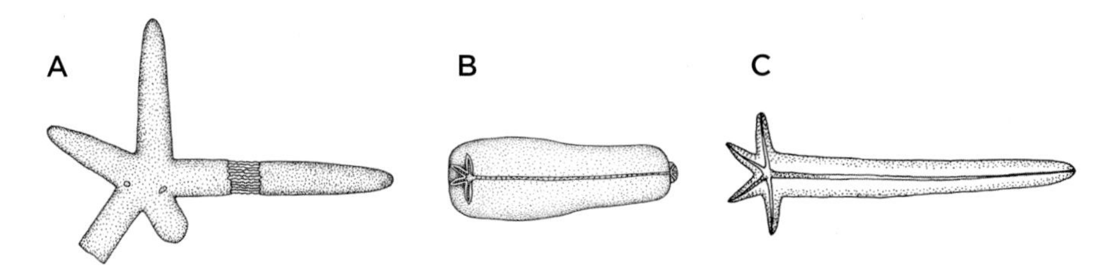

Figure 6: Asexual development stages of a 'comet' from an arm tip (e.g. *Linckia multifora*). **A)** An arm is stretched and broken from the seastar. **B)** A new mouth and disc are developed by the arm. **C)** New arms are generated from the disc.

ASEXUAL REPRODUCTION

All seastars have the ability to replace parts of their bodies following either external or internal injury. This capacity is exploited by a few species that can reproduce by casting off an arm which regenerates a new body (autotomy); the source seastar simply grows a new arm. This ability is most highly developed in *Linckia multifora* (family Ophidiasteridae), which can regenerate a new seastar from part of an arm (Figure **6**), even parts that lack a tip or any part of the disc (Edmondson, 1935). Several other ophidiasterid species can regenerate the body from an entire arm. An arm that regenerates a body and other arms is termed a 'comet'. This form of reproduction enables species to increase their population asexually, which is an advantage on remote island reefs.

Another form of asexual reproduction is termed fission. This involves one-half of the animal deliberately pulling away from the other half, with each half

Linckia multifora almost completely regenerated from a single arm [top]; *Mithrodia clavigera* producing replacement arms [bottom] (C. Bryce).

then regenerating an entire seastar. Stichasterids, such as *Allostichaster polyplax*, use this method of reproduction and are often found with four short and four longer arms. The fission process takes 1–24 hours to complete, depending on the species.

PARTHENOGENESIS

(development of an egg without fertilisation)

Parthenogenesis is a type of asexual reproduction only known to occur in one species of seastar, *Ophidiaster granifer* (Yamaguchi and Lucas, 1984), which has large (600 μm diameter) yolky eggs. The majority of these float and develop into pelagic lecithotrophic larvae, but others sink to the bottom and adhere to the substrate, where they start to metamorphose nine days after spawning. Metamorphosis can be delayed up to 10 weeks if no suitable substrate is found. This species is widely distributed across the Indo-West Pacific, and no males have ever been found.

SEXUAL REPRODUCTION

The gonads lie in the body cavity in the arms. They consist of a series of tufts connected by a shared gonoduct opening in the angle formed by the arms, or open independently through a series of gonoducts on the interradial edge of the arm (e.g. *Linckia guildingi*). On coral reefs several species of seastars have been reported to spawn synchronously with the mass spawning of hard corals.

For species that spawn by broadcasting eggs and sperm, the sexes are usually separate; however, hermaphroditic individuals have occasionally been found (e.g. in *Fromia indica*; Marsh, 1988). There is no sexual dimorphism in seastars, although in *Archaster* species the males are smaller and mount the females such that the arms of the male and female alternate. The female spawns first, followed by the male. This is termed pseudo-copulation, as the gametes are broadcast and no true copulation takes place. However, the chance of fertilisation is increased, and consequently *Archaster* individuals are abundant in suitable habitats (Komatsu, 1983).

The fertilised eggs develop into larvae that pass through either a bipinnaria or a brachiolaria larval stage, or both (Figure **7**). For species that develop in the plankton, the embryos hatch from a small (120–150 μm) egg at the blastula stage (a hollow ball of cells) and develop into a bipinnaria larva that is bilaterally

Multiple groups of *Archaster angulatus* engaging in pseudo-copulation (P. Baker).

symmetrical, has swimming lobes and feeds on smaller planktonic particles; this mode of feeding is termed planktotrophic. The larva swims with the aid of two bands of cilia, and has a mouth, stomach and anus. After a period of weeks or months the larva develops three short blunt brachiolar arms (tipped with adhesive cells) with a sucker between the three arms. The larva is now termed a brachiolaria, and it is this stage that settles on a suitable substrate. Following settlement the brachiolaria larva metamorphoses into a juvenile seastar, absorbs the brachiolar arms and starts to feed on benthic organisms.

In some genera (e.g. *Luidia* and *Astropecten*) there is no brachiolaria stage, and metamorphosis (which occurs after 2–3 months) takes place in the plankton directly from the bipinnaria, then settlement occurs.

In other genera (e.g. *Fromia* and some asterinids) the eggs are large (approximately 400 µm diameter) and full of yolk. The embryo hatches into a ciliated oval larva that swims at the water surface for a few days, then becomes pear-shaped and develops brachiolar arms and a sucker. The larva swims near the bottom and the brachiolar arms make contact with the substrate. The larva moves about, probably assessing the surface, until the central sucker attaches and secretes a substance that cements it in place, approximately 10 days after fertilisation. Metamorphosis (in *Meridiastra gunnii*) commences by 12–13 days after fertilisation. The larval arms are resorbed, and the major parts of the water vascular system (ring canal, radial canals and tube feet) develop from the coelom on the left side of the larva, which becomes the actinal side of the adult. The larval mouth and anus form as new openings arising from the larval oesophagus and intestine.

The Asterinidae exhibit a diverse range of reproductive strategies, from free spawning with planktotrophic development (*Patiriella regularis*) to lecithotrophic brachiolaria (*Meridiastra gunnii*; Figure **7**). In *Parvulastra exigua*, masses of large eggs (400 µm diameter; Figure **7**) are deposited and cemented to the substrate; the eggs develop into encapsulated brachiolariae, which are not brooded.

The simplest form of brooding is that which occurs beneath the adult body (external brooding). This is presumed to occur in *Tosia neossia* and *Tegulaster alba*, where the adult remains with the juveniles during their development.

Intraovarian brooding of small eggs (120 µm diameter) associated with direct development, involving accelerated acquisition of adult characters and

Fromia indica broadcasting eggs (A. Hogget).

Tosia neossia brooding juveniles during their development (N. Coleman).

release of crawl-away juveniles, occurs in *Parvulastra vivipara* and *P. parvivipara* (Figure 7). Despite the small size of the eggs the development is lecithotrophic, with brachiolaria larvae swimming in the gonadal fluid. Nutrition is enhanced by larval cannibalism. These species are hermaphrodites, and self fertilisation can occur. The juveniles are born by rupture of the abactinal surface of the adult (no gonoducts are present). The resulting wound heals, and the seastars can reproduce several times a year for several years (Chia, 1976).

The method of reproduction, where known, is noted in the text describing each species. It can also be inferred from the size of the eggs, if this is known.

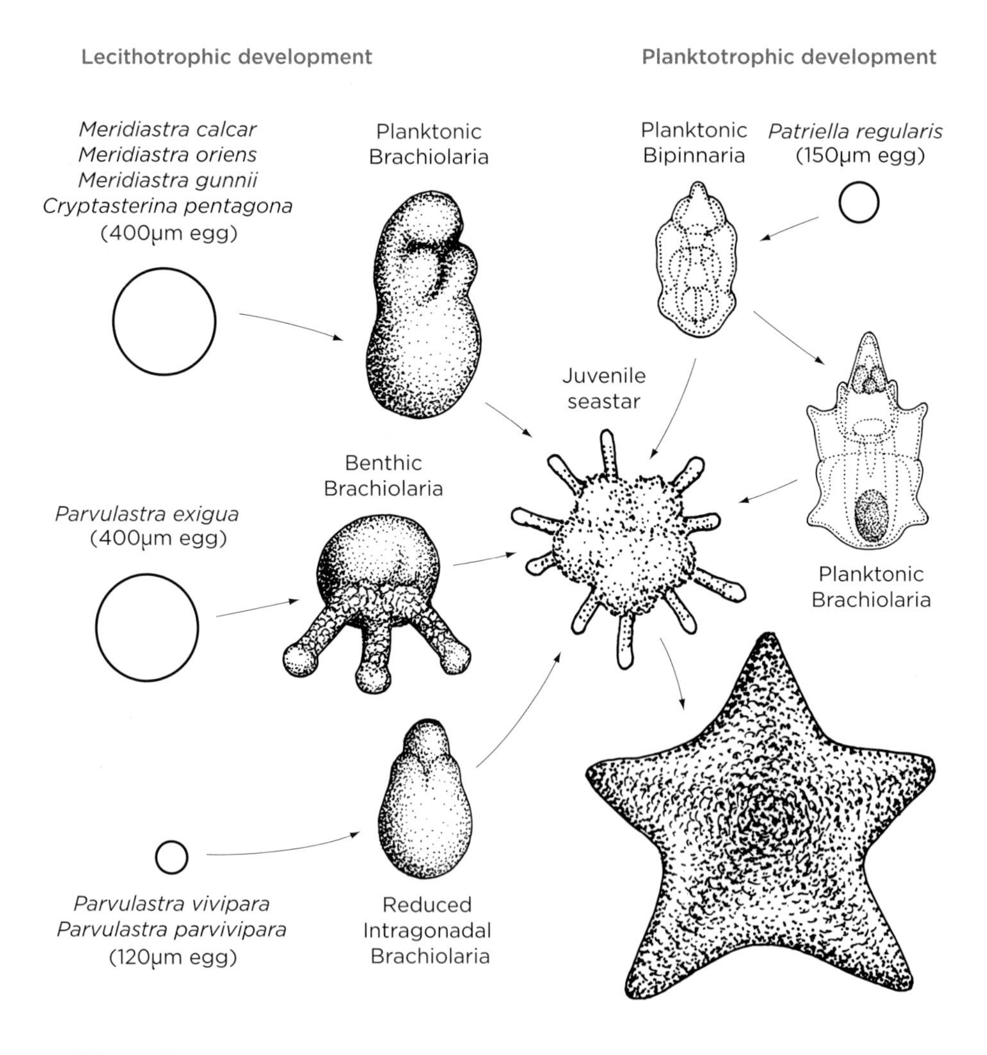

Figure 7: The developmental patterns in various species of seastars.

DISTRIBUTION PATTERNS OF AUSTRALIAN SEASTARS

Australian seastars include many South-East Asian and Indo-Pacific coral reef species, but of greater interest is the distribution within Australia of species found nowhere else. Factors to be considered include the geological history of the region, the fossil history of various taxa, the patterns of warm and cold currents around Australia and the availability of suitable habitats and food.

GEOLOGICAL HISTORY

In the late Jurassic Period (160 million years ago) Australia was joined to that landmasses that are now New Zealand, New Guinea, South Africa and South America, forming part of the supercontinent Gondwana. At this time only the west coast of WA was open to the Tethys Sea, after India had drifted away. During the early Cretaceous Period, rifting began to separate Australia from Gondwana; seafloor spreading commenced on the north-western margin and progressed anticlockwise around the continent. By the late Cretaceous (70–66 million years ago) Australia had moved away from Antarctica, with Tasmania still joined to Victoria, and northern Australia and the Sahul Shelf forming a land bridge to New Guinea. The coastline of southern Australia was colonised from the west by warm water species, and the new shore and shelf habitat provided unoccupied space for the evolution of new genera and species.

The eastern coast of Australia supported a cool water fauna that extended around Antarctica to what is now South America (O'Hara and Poore, 2000).

At the end of the Cretaceous Period, Australia was affected by a worldwide mass extinction event that destroyed both terrestrial (e.g. dinosaurs) and marine (e.g. ammonites) fauna. However, many seastar families in the Tethys Sea survived the extinction, and went on to colonise the new coastline of southern Australia.

The separation of Tasmania from Antarctica enabled the cool water eastern Australian fauna to mingle with the warm water western fauna. The evolution of endemic species in southern Australia was supplemented by further invasions of tropical species, and a smaller number of Southern Ocean species that arrived on circumpolar Subantarctic currents (O'Hara and Poore, 2000).

During the Pliocene (5.3–1.8 million years ago) Australia drifted north, to within 1–2° of its present position, and the area corresponding to the present southern Great Barrier Reef was then in waters warm enough for the initiation of reef growth.

During the Pleistocene, from 1.8 million to 10,000 years ago, the sea level fluctuated markedly during the course of 60 glacial–interglacial cycles (Teidemann et al., 1994). The glacial periods were characterised by sea levels that were as much as 110 m below present (Wells and Okado, 1996), and water temperatures were cooler than at present. During the last interglacial (125,000 years ago) the sea was 1–2°C warmer than at present, and tropical corals occurred as far south as Newcastle on the east coast of Australia, and along the west coast to Albany, on the south-west coast of WA (Kendrick et al., 1991).

The sea stabilised at its present level 6,000 years ago, and coral reefs expanded to form the Great Barrier Reef and the reefs off the WA coast; in the case of Ningaloo Reef and the Houtman Abrolhos, corals colonised a pre-existing Pleistocene reef that formed during the last interglacial (120,000 years ago) (Wyrwoll et al., 2009).

Among the many marine organisms found in tropical Australia, some species are endemic while others are widespread throughout tropical oceans. Approximately 42% of tropical Australian seastars are endemic species, while the remaining 58% are widely distributed in the Indo-West Pacific (east Africa to the mid Pacific Ocean), the East Pacific, the Indian Ocean, South-East Asia, and the West Pacific.

FOSSIL HISTORY

The phylogeny of the class Asteroidea has proved to be contentious, with little consensus on relationships; this is reflected in the occurrence of a number of different classification schemes. Cladograms generated for post-Palaeozoic asteroids by Gale (1987) and Blake (1987) differ in major respects. Gale considered the Paxillosida to be the basal neoasteroid group and the Forcipulatida to be the most advanced, whereas Blake (1987, 1988) argued that the Paxillosida were highly specialised, and that the Forcipulatida were basal.

The major groups of seastars (the orders Valvatida, Forcipulatida, Spinulosida and Paxillosida) differentiated in the Ordovician Period, 500 million years ago. Spencer and Wright (1966) list the ages of the first orders, families and genera of seastars currently recognised. While the orders are recognised from the Ordovician, there is little evidence for the presence of most present day families prior to the Jurassic Period. However, well preserved seastar ossicles dated to

A well preserved fossil of *Australaster giganteus* from the extinct family Palaeasteridae. Discovered in the Humevale Formation of Victoria, it is dated from the late Ordovician Period (R. Nistri).

the late Triassic Period demonstrate that much of the neoasteroid radiation had taken place by the late Triassic. Gale (2011) concluded that the sudden appearance of diverse taxa having close affinities to modern families in the Early and mid Jurassic is an artefact of the extensive record of marine sediments from these ages, and the lack of Permian and Triassic sedimentary rocks (286–248 million years ago).

Modern seastar families first appeared in the Early Jurassic, and genera including *Mediaster* and *Astropecten* appeared in the late Miocene (10–5 million years ago) (Spencer and Wright, 1966; Gale, 2011). No records are available from Australian fossil sites.

Mah and Blake (2012) summarised the divergent arguments for a basal position for the Forcipulatida versus the Paxillosida, and concluded that despite comprehensive phylogenetic efforts (Mah and Foltz, 2011b), basal relationships among major lineages of seastars remain unclear. The transfer of two Ophidiasterid genera (*Fromia* and *Neoferdina*) and the genus *Nectria* from the Oreasteridae family to the family Goniasteridae (Mah and Foltz, 2011a) is followed here (Table 1).

Table 1: Classification of the Asteroidea recorded from depths of 30 m or less in the Australian region (following Byrne and O'Hara, 2017).

ORDER/FAMILY	GENERA
Order: Paxillosida	
Astropectinidae	*Astropecten*
	Bollonaster
	Craspidaster
Order: Valvatida	
Luidiidae	*Luidia*
Acanthasteridae	*Acanthaster*
Archasteridae	*Archaster*
Asterinidae	*Ailsastra*
	Anseropoda
	Aquilonastra
	Cryptasterina
	Disasterina
	Indianastra
	Meridiastra
	Nepanthia
	Paranepanthia
	Parvulastra
	Patiriella
	Pseudonepanthia
	Tegulaster
Asterodiscididae	*Asterodiscides*
Asteropseidae	*Asteropsis*
	Petricia
	Valvaster
Chaetasteridae	*Chaetaster*
Goniasteridae	*Anchitosia*
	Anthenoides
	Fromia
	Iconaster
	Mediaster
	Nectria
	Neoferdina
	Pentagonaster
	Pseudogoniodiscaster
	Stellaster
	Styphlaster
	Tosia
Mithrodiidae	*Mithrodia*
	Thromidia
Ophidiasteridae	*Bunaster*
	Cistina
	Dactylosaster
	Gomophia
	Hacelia
	Leiaster
	Linckia
	Nardoa
	Ophidiaster
	Tamaria
Oreasteridae	*Anthaster*
	Anthenea
	Bothriaster
	Choriaster
	Culcita
	Goniodiscaster
	Gymnanthenea
	Halityle
	Pentaceraster
	Poraster
	Protoreaster
	Pseudoreaster
Order: Velatida	
Pterasteridae	*Euretaster*
Order: Spinulosida	
Echinasteridae	*Echinaster*
	Metrodira
	Plectaster
Order: Forcipulatida	
Asteriidae	*Asterias*
	Astrostole
	Coscinasterias
	Sclerasterias
	Stolasterias
Stichasteridae	*Allostichaster*
	Smilasterias
	Uniophora

CURRENTS

A dominate influence on the ocean adjacent to the western half of Australia is the Leeuwin Current, defined by Cresswell and Golding (1980) as a poleward flowing eastern boundary current that carries warm low salinity water southward off the Western Australian coast. The warm, low salinity water flowing from the Pacific Ocean through the Indonesian Archipelago to the Indian Ocean (the Indonesian Throughflow) is lower in density than the cooler, saltier ocean water off south-west Australia. When these water masses meet the resulting density difference sets up a sea level gradient of approximately 0.5 m along the Western Australian coast, with the result that the Leeuwin Current flows 'downhill', and strengthens as it flows south. The flow rate and volume of transport is greatest between April and September (Godfrey and Ridgway, 1985).

The Leeuwin Current is fed from several sources. South Pacific Ocean water passes through the species rich East Indies Triangle of Biodiversity (also known as the Coral Triangle), where it can pick up planktonic larvae and carry them to Ashmore Reef and beyond. Much of the Indonesian Throughflow joins the westward-flowing south equatorial current in the Indian Ocean, while part of the current branches off southward and joins the shelf edge Holloway Current and Indian Ocean water to form the shelf edge Leeuwin Current in the region of North West Cape (WA) (D'Adamo et al., 2009). The Leeuwin Current swings eastward at Cape Leeuwin (south-west tip of WA), and weakens as it flows along the south coast shelf.

During its passage south along the west coast shelf margin, the Leeuwin Current produces cold core and warm core eddies that can entrap larvae and carry them offshore. While the Leeuwin Current flows most strongly in autumn and winter, in summer the coast is bathed by the cool north flowing Capes Current (Pearce and Pattiarachi, 1999). This originates at about 34°S and follows the coast, sometimes as far north as Shark Bay. Its source upwells from the bottom of the Leeuwin Current, and it is driven by the strong southerly winds that occur during summer. The Capes Current flows inshore of the Leeuwin Current, which follows the 200 metre isobath.

The presence of the Capes Current accounts for the pronounced difference between the largely tropical fauna of the Houtman Abrolhos on the shelf edge, and the temperate fauna of the adjacent coast. This difference was first observed by Saville-Kent, who postulated 'that an ocean current setting in from the equatorial area of the Indian Ocean penetrates as far south as this island group without impinging on the adjacent mainland…' (Saville-Kent, 1897).

North of the Capes Current the cool Ningaloo Current flows northwards between the Leeuwin Current and the coast north of Shark Bay to north of North West Cape. It is strongest during summer, being driven by strong southerly winds, while the Leeuwin Current is weak (Taylor and Pearce, 1999). The waters off the long east–west orientated coastline of southern Australia are influenced by the east flowing but weakening Leeuwin Current in winter, and the year-round Flinders Current flowing west between Tasmania and Cape Leeuwin. The Flinders Current is stronger in summer and is the source of the Leeuwin Undercurrent which flows northwards on the continental slope under the Leeuwin Current (Pattiarachi and Woo, 2009).

On the east coast of Australia the East Australian Current, fed by the west flowing Pacific South Equatorial Current, follows the Great Barrier Reef southwards and swings offshore from NSW, carrying larvae of tropical species to Lord Howe Island, and to Middleton and Elizabeth Reefs. At times the East Australian Current continues south, taking warm water species to the east coast of Tasmania.

TIDES

While most of the species covered in this book live subtidally, the extreme differences in the tidal ranges on different parts of the Australian coast influence the depths at which individual species are found. Most of the east and south coasts have a moderate semi-diurnal tidal range of 2–4 m, which decreases to 1–2 m on the Victorian and South Australian coasts. In south-western Australia the range is approximately 1 metre or less, but the predicted tide level is heavily influenced by high or low atmospheric pressure, strong southerly or south-westerly winds, strong easterly winds, and the seasonal (autumn and winter) influence of the Leeuwin Current, which raises the sea level by nearly 0.5 m. The tides here are diurnal (one high and one low each day), so in early summer rock platforms can be exposed for many hours each day, providing a very stressful environment for fauna inhabiting shallow rock pools.

The predicted tides on the west coast sometimes remain at the same level through the 24 hour cycle, indicating that wind and atmospheric pressure are more important than the tide in determining the water level.

The tides to the north are semidiurnal and the amplitude increases, from 1.3 m at Carnarvon to 2.3 m at Exmouth, 4.2 m at Dampier, 6.2 m at Port Hedland and 9 m at Broome. This macrotidal regime drives strong currents between islands, including the Montebello Islands and Dampier Archipelago, mobilising fine sediments during the spring tide cycle; the resulting extreme turbidity is harmful to most species of seastars. Exposure of shallow water species to extremely high temperatures at low tide is often mitigated by the occurrence of low tides at dawn and dusk.

The maximum spring tide range occurs off the Kimberley coast, with an amplitude of almost 10 m at Yampi Sound (WA), from where it gradually decreases eastward, to 6.7 m at Cape Voltaire (WA) and Darwin (NT), and 2.7 m at Weipa, on the Cape York Peninsula (Queensland). The tidal ranges given here are the difference between the highest astronomical tide and the mean lower low water (MLLW) (Australian National Tide Tables, 2015).

SEASTAR HABITATS

Echinoderms are exclusively marine and, with few exceptions, are found in waters of 34–35 ppt salinity, from intertidal areas to the deep sea. Several widespread tropical species occur in the Persian Gulf at a salinity of 60 ppt (Price, 1982). In WA the seastars enter the lower reaches of the Swan River in summer when the salinity is that of seawater.

The coastline of Australia encompasses a diverse range of climatic and habitat types, within which a multitude of microhabitats occur.

Southern Australia lies in the temperate zone, and has sea surface temperatures ranging from 15.5 to 25°C in summer and 11.5 to 20°C in winter (O'Hara and Poore, 2000). The waters of northern Australian are tropical, with summer temperatures of 28–29°C on the Great Barrier Reef. Off WA the seawater temperatures can be more variable. For example, in the Dampier Archipelago the inshore temperatures can range from 18°C in July to 34°C in February.

Broadly, habitats range from shores with surf beaches and rocky headlands (often with intertidal platforms at their base) that are exposed to the open sea and surf, to those in bays sheltered from the surf, and estuaries, or open coasts, sheltered by offshore islands or reefs. In addition to many small bays, large embayments occur on all coasts, and include locations including Moreton Bay (Queensland); Port Jackson, Lake Macquarie, Botany Bay and Jervis Bay (NSW); Port Phillip and Westernport Bays (Victoria); the St Vincent and Spencer Gulfs (SA); King George Sound, Cockburn Sound, Shark Bay and Exmouth Gulf (WA); and the Gulf of Carpentaria (NT and Queensland).

Linckia laevigata on a reef flat among coral and rubble (S. Morrison).

Coral reefs offer diverse habitat for echinoderms, providing a solid or honey-combed, more or less flat platform having a cover of living and dead coral, and mainly algal covered reef flats. Reef crevices provide habitat for crinoids, ophiuroids and seastars, many of which emerge at dusk to feed. Back reefs usually consist of areas of living and dead coral interspersed with patches of sand, and these commonly deepen into a sand- or rubble-floored lagoon that provides habitat for luidiids, astropectinids and archasterids.

On the east and west coasts of Australia coral reefs occur on or near the coast north of the Tropic of Capricorn (23°26'22"S). In Western Australia, corals occur on offshore reefs much further south because of the Leeuwin Current, which flows near the edge of the continental shelf, coming close to the coast near Cape Naturaliste and Cape Leeuwin at the south-western corner of Western Australia. Cool north flowing currents close to the coast restrict the shallow water seastars to temperate species on most of the west coast.

Protoreaster lincki on an algae covered reef flat [top] (S. Morrison);
Fromia polypora in a reef crevice among sponges [bottom] (Z. Richards).

Sand, mudflats and seagrass beds provide important habitats for seastars in sheltered waters on both tropical and temperate shores, and each habitat type supports a characteristic suite of species.

In northern Australia sponge gardens occur subtidally. These are dense communities of sponges, soft corals and gorgonians, which provide suitable substrate for many seastars. Shaded jetty piles mimic deeper water substrates that receive less light, and attract a rich growth of encrusting sponges, tunicates, bryozoans and hydroids, which provide food for a diversity of small seastars otherwise found on deeper rocky shores.

Many seastars are masters of camouflage, their bright colours blending with those of sponges and tunicates on which they feed. However, others contrast with their background and are probably ignored by predators, their appearance perhaps indicating they are poisonous.

The bright colouration of many seastars, such as *Nectria ocellata*, could give predators the impression that they are poisonous (G. Edgar).

The colour patterns of *Culcita novaeguineae* [top] and *Ophidiaster granifer* [bottom] blend exceptionally well into sediment-rich soft coral and sponge gardens (C. Bryce).

ZOOGEOGRAPHY

Since the breakup of Gondwana the Australian continent has continued its northern drift towards Indonesia. It is still moving north at a rate of approximately 7 cm each year (Geocentric Datum of Australia 2020).

Australia lies broadly between latitudes 10°40' (Cape York) in the north and 43°39'S (South East Cape, Tasmania) in the south, and longitudes 112°59'E (Cape Inscription, WA) in the west and 153°38'E (Cape Byron, NSW) in the east. The extent of shallow water (to 50 m depth) is estimated to be 1,298,000 km^2 (Bunt, 1987); the area to a depth of 30 m, the depth limit of this book, is a little less.

Australia is well suited to zoogeographical analysis as it has eastern and western coasts trending north–south, and a long southern coastline; this facilitates the study of both latitudinal and longitudinal gradients, particularly on the temperate coasts (O'Hara and Poore, 2000).

O'Hara and Poore (2000) analysed the distribution patterns for echinoderms and decapod crustaceans from Australia south of 30°S to 1,000 m depth. They used statistical analysis to describe patterns of species composition, richness and endemism, and related these patterns to distance and temperature gradients. They found that species richness was relatively constant from east to west, but varied with latitude from high in the warmer regions to low in cool temperate southern Tasmania. They concluded that as Australia split from Gondwana and drifted northwards, the continued invasion and divergence of species of tropical origin caused the progressive extinction of some cool temperate Gondwanan species that were at the limits of their temperature range. There is a low level of immigration of cool temperate species and some in situ endemic speciation.

Wilson and Allen (1987) separated the marine fauna into those having a general southern Australian distribution, and groups endemic to particular regions (the west, south-western, south, south-eastern and east coast regions).

O'Hara and Poore (2000) used cluster analysis to identify five primary subregions: the subtropical west coast; the south-west coast from Perth to the South Australian Gulfs; the South Australian Gulfs to central Victoria (including Tasmania); Wilsons Promontory to Sydney; and the warm temperate region north of Sydney to 30°S (with some exceptions extending to 24°S).

Although their analysis covered a wider depth range than considered in this book, and included all classes of echinoderms, it applies well to the temperate seastar fauna. The northern half of Australia adds a large tropical component to the seastar fauna; this includes species distributed beyond Australia, which make up 55% of the fauna. Widespread coral reef seastars in Australia are mainly known from offshore reefs including the Great Barrier Reef (Queensland), Ashmore Reef (Timor Sea), the Rowley Shoals and Scott Reef (off northern WA), the Houtman Abrolhos (off mid-western WA) and clear water fringing reefs such as Ningaloo Reef (north-western Australia) (Marsh, personal observation).

Of the 202 species in 72 genera of seastars known from shallow water depths (< 30 m) around Australia, 86 species (43%) are not found elsewhere. These indigenous or endemic species are predominantly shallow shelf or coastal species from both tropical and temperate coasts. Several of these have very short distributional ranges, typically a few hundred kilometres or less.

Three species have been excluded from this zoogeographic overview. These are: *Goniodiscaster foraminatus*, which was described from Shark Bay but despite

One of the colour variations of *Nepanthia crassa*, a species endemic to the central Western Australian coast (C. Bryce).

extensive collecting has not been found there since the Baudin Expedition; *Patiriella inornata*, for which no locality or distribution data are recorded; and *Asterias amurensis*, an introduced North Pacific species. Of the 86 endemic species, 31 (36%) are found in shallow shelf and coastal waters of northern Australia and 55 (64%) are found in southern Australia. Among the endemic seastars in northern Australia, approximately one third occur throughout the entire area, one third only occur in the north-east and one third only occur in the north-west.

The scope of this book is limited to seastars occurring to depths of 30 m. This is partly for practical reasons, but also because the deeper shelf fauna, which is very sparse but very species rich, is largely unstudied, particularly in the vast North West Shelf area of Western Australia.

It is difficult to differentiate zoogeographical provinces clearly because of the variety of individual distributions, particularly on the south coast of Australia.

AUSTRALIAN ENDEMIC GENERA FROM DEPTHS OF LESS THAN 30 M

There are 86 (43%) endemic species in Australia, but few endemic genera. Currently nine endemic genera are recognised: five from the south, three from the north and one on the north-east coast (Table 2).

Table 2: Australian endemic genera of seastars and the number of species reported from less than 30 m depth. Five of the genera have a southern distribution (S), three have a northern distribution (N), and one occurs in the north-east (N-E).

GENERA	SPECIES	DISTRIBUTION
Anthaster	1	S
Nectria	6	S
Plectaster	1	S
Pseudogoniodiscaster	1	N-E
Pseudoreaster	1	N
Styphlaster	1	N
Tegulaster	4	N
Tosia	3	S
Uniophora	3	S

The most speciose endemic genus is *Nectria*, a goniasterid. It is believed to have originated in the south-western corner of WA, and to have spread eastwards (Zeidler and Rowe, 1986). The genus is confined to southern Australia and contains eight species, two of which only occur in waters deeper than 45 m (*Nectria humilis,* from northern Tasmania, and *N. ocellifera*, from near Albany to Geraldton, WA). All eight species are to some degree sympatric across southern Australia on reef or rocky coasts having moderate exposure to wave action. Four species have their eastern limit in Bass Strait, one extends to southern NSW, and one extends to the mid coast of NSW. Port Gregory (28°12'S) is the northern limit of the genus in WA.

Historical records of *Nectria* from Fiji, New Zealand and Mauritius are believed to be erroneous, as no further specimens of this genus have been found despite intensive collecting in these areas.

A number of the genera are monotypic (contain a single species). The oreasterid genus *Anthaster* contains a single species found in south-western Australia; the single *Pseudoreaster* species is from north-western Australia; the northern endemic monotypic goniasterid genera *Pseudogoniodiscaster* and *Styphlaster* are from Queensland and north-western Australia, respectively; and *Plectaster* is a southern monotypic echinasterid genus. *Uniophora* is a southern stichasteriid genus that will need to be subjected to molecular studies to separate the species satisfactorily. Ten species of this genus were originally described from southern Australia and south-west WA, but this number was reduced to three by Shepherd (1967b).

ENDANGERED SPECIES

Parvulastra vivipara and *Smilasterias tasmaniae* are both very rare in Tasmania. *P. vivipara* is listed as vulnerable under the *Commonwealth Environmental Protection and Biodiversity Conservation Act 1999*, and as endangered under the *Tasmanian Threatened Species Protection Act 1995*, which also lists *S. tasmaniae* as Threatened. *Patiriella littoralis* has not been found for many years and is now considered to be extinct (O'Hara et al., 2019). The cool temperate species from south-eastern Tasmania are at the limit of available habitat, and will have nowhere to migrate to if ocean warming continues.

HOW TO IDENTIFY A SEASTAR

A number of key features contribute to identifying seastar species, including:

Overall shape: for example, whether the specimen is flattened or cushion-like, has short or long arms, or a wide or narrow angle between the arms.

Texture of the external surface: for example, the presence and location of granules or spines.

Body measurements: in particular from the centre of the disc to the tip of the arm (R), from the centre of the disc to the margin between the arms (r), and the breadth of the arm (br) (Figure **8**).

Presence and sizes of calcareous skeletal structures: these are often plate-like and can give the surface a mosaic appearance or may border the arms (marginal plates) (Figure **3C–E**). Some species have unique plates not found in other species.

Other body features: for example, the presence of multiple madreporites, more than five arms, and body colour.

In the taxonomic section a key to each seastar order is provided first. Then, for each order, a diagnosis is provided, followed by a key and diagnosis for each family and genus. A species key is provided where more than three species occur in Australia. Where only two species are found in Australia, no key is provided as species descriptions can be directly compared. The characteristics of each species are described including information on their food preferences and reproduction, where these are known.

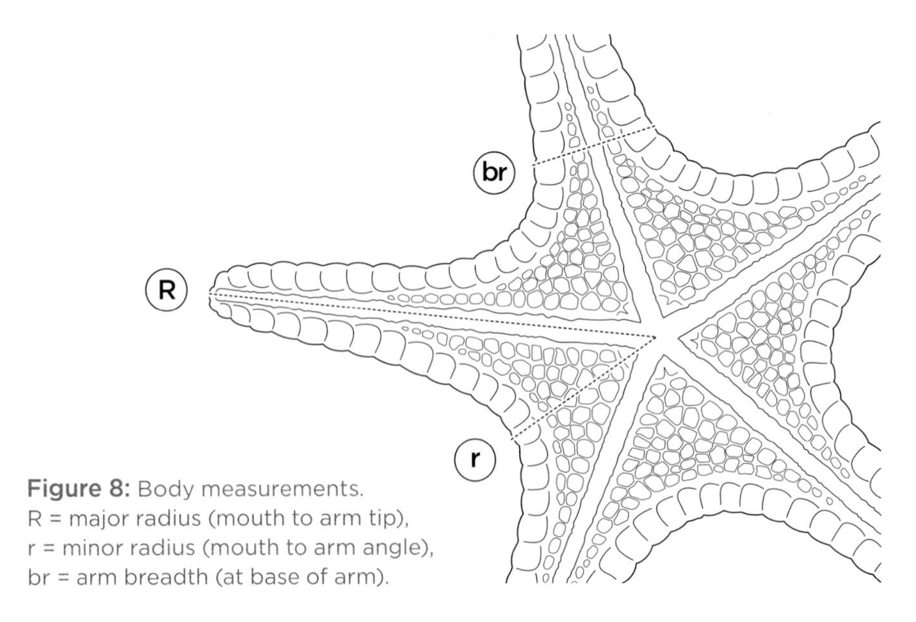

Figure 8: Body measurements.
R = major radius (mouth to arm tip),
r = minor radius (mouth to arm angle),
br = arm breadth (at base of arm).

CLASSIFICATION OF SEASTARS

KEY TO THE ORDERS OF ASTEROIDEA IN SHALLOW AUSTRALIAN WATERS

1	Abactinal surface covered by paxillae; tube feet pointed, without sucking discs; rectum and anus lacking.	**Paxillosida** page 48
	Tube feet ending in a sucking disc; rectum and anus present.	**2**
2	Marginal plates more or less conspicuous; body form highly variable from pentagonal or stellate to those with a small disc and long arms; pedicellariae, when present, valvate or sometimes modified spiniform.	**Valvatida** page 90
	Marginal plates inconspicuous, often set below the ambitus; pedicellariae, when present, modified spiniform, often with integrated basal piece.	**3**
3	Complex pedicellariae with basal piece and either crossed or straight valves, often both present; tube feet in four rows in shallow water species.	**Forcipulatida** page 466
	Pedicellariae absent; tube feet in two rows.	**4**
4	Body large usually thickened, stellate; marginal, abactinal and actinal plates paxilliform, supporting a supradorsal membrane in Pterasteridae; skeleton reticulate; marginal and actinal plates may be absent; adambulacral spines transversely aligned, usually webbed.	**Velatida** page 442
	Body small, arms long, usually cylindrical; skeleton reticulate (except in small *Metrodira*) with abactinal, marginal and actinal plates; intermarginal plates present; spines or spinelets small, often skin-covered.	**Spinulosida** page 446

PAXILLOSIDA

Astropecten sumbawanus, Dampier Archipelago, Western Australia (S. Morrison).

Species of the Paxillosida have tube feet in two rows, pointed or with knob-like tips, without sucking discs. The body shape is varied, though most shallow water species have a small disc and long, tapering arms. The abactinal plates are in the form of paxillae. Pedicellariae, when present, are simple, composed of spinelets or granules. The digestive system lacks a rectum or anus.

KEY TO THE FAMILIES OF PAXILLOSIDA IN SHALLOW AUSTRALIAN WATERS

1	The edge of the body is defined by the two series of marginal plates; the superomarginals are sometimes smaller than the inferomarginals but are always conspicuously different from the paxillae.	**Astropectinidae** page 49
	The edge of the body is defined by the large inferomarginal plates; the superomarginals are indistinguishable from the paxillae.	**Luidiidae** page 81

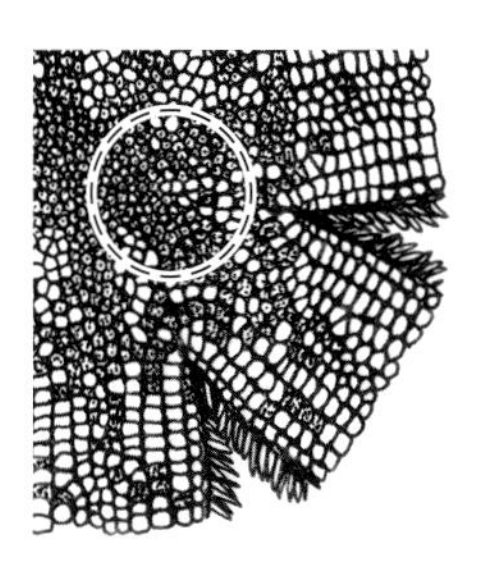

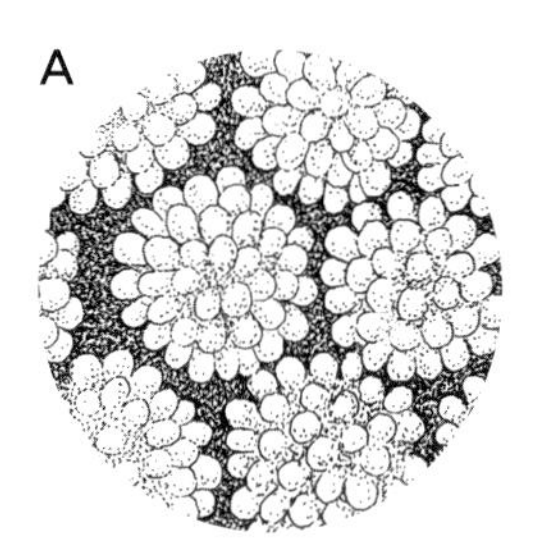

A

Paxillae composed of a cluster of central spinelets or granules

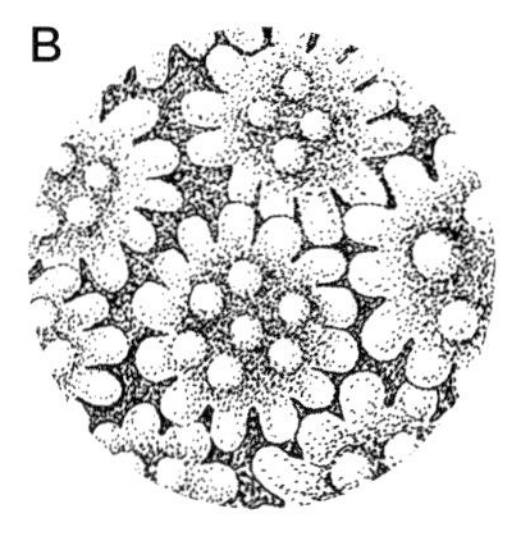

B

Paxillae composed of a group of central spinelets or granules and an array of peripheral spinelets

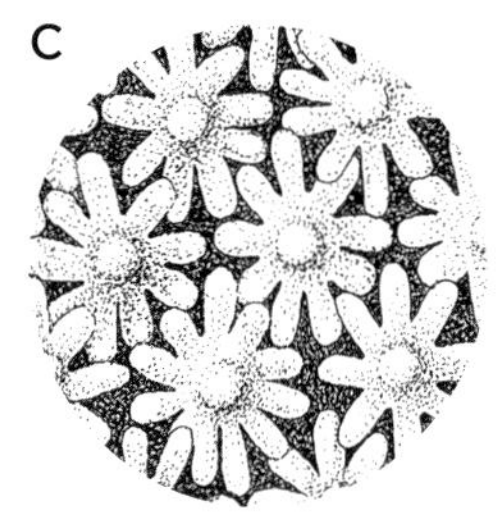

C

Paxillae composed of a single central spinelet or granule and an array of peripheral spinelets

Figure 9: Typical compositions of paxillae in species of Astropectinidae and Luidiidae. A) *Bollonaster pectinatus*. B) *Astropecten zebra*. C) *Astropecten monacanthus*.

ASTROPECTINIDAE

Most shallow water Astropectinids have small discs and fairly long arms with acute arm angles. The body is covered by paxillae and **is outlined by prominent superomarginal plates and the inferomarginals bear conspicuous spines**. Actinal interradial areas are reduced to a few plates in most species. The tube feet are pointed or with rounded tips, without terminal discs.

The family is cosmopolitan with 26 genera, eight of which, comprising 23 species, are found in Australian waters. Of these, fourteen species in three genera are found in depths of less than 30 m.

KEY TO THE GENERA OF ASTROPECTINIDAE IN SHALLOW AUSTRALIAN WATERS

1	Inferomarginal spines small appressed, inconspicuous; interradial areas with large plates; body goniasterid in shape; interradial arcs rounded with very broad marginal plates; adambulacral plates with about six radiating furrow spines and a fringe of small subambulacral spines.	***Craspidaster*** page 76
	Inferomarginal spines large, conspicuous; interradial areas small, with up to three rows of small actinal plates; arms triangular to long, narrow; interradial arcs usually angular; three furrow spines, subambulacral spines elongate.	2
2	Arms long, flat, narrow, with acute arm angles; actinal interradial areas usually with only two plates either side of the midline; inferomarginal plates with a single large spine, with or without smaller spines below it.	***Astropecten*** page 51
	Arms fairly short, broad at base tapering to a sharp tip; actinal interradial areas with three rows of actinal plates in the arm angle, on each side of the midline; inferomarginal plates with up to six thin marginal spines.	***Bollonaster*** page 76

Astropecten polyacanthus, Red Sea (C. Cooper).

GENUS *ASTROPECTEN* Gray, 1840

Species of this genus generally have five arms. The arms and disc are well defined by the marginal plates. Both superomarginal and inferomarginal plates are well developed, often one or both with conspicuously enlarged spines. The inferomarginal plates are often large and more obvious. Generally the actinal plates are few and restricted. The madreporite is often hidden by encroaching paxillae; abactinal paxillae are present, and sometimes also pedicellariae. Often an epiproctal cone occurs centrally on the disc. The papulae are restricted to the disc and dorsolateral parts of arms. Internally, the ampullae of tube feet are large and double, and strong superambulacral plates are present.

Type species: *Astropecten aranciacus* (Linnaeus, 1758)
(north-eastern Atlantic, Mediterranean Sea)

Fourteen species occur in Australian waters, 11 are found in depths of less than 30 m.

Astropecten granulatus, Dampier Archipelago, Western Australia (C. Bryce).

KEY TO THE SPECIES OF *ASTROPECTEN* IN SHALLOW AUSTRALIAN WATERS

1	No spines on any superomarginal plates.	**2**
	Spines present on some or all superomarginal plates.	**6**
2	Spinelets cover inferomarginal plates; spines along the distal edge; SA and south-western Australia.	***A. preissi*** page 63
	Scales cover inferomarginals; a few spines may be present; tropical.	**3**
3	Scales on inferomarginal plates elongate, upright with some short spines; inferomarginal spine slender, pointed, longer than two inferomarginal plates; $R \leq 55$ mm.	***A. sumbawanus*** page 66
	Scales on inferomarginals truncate or rounded, more or less appressed.	**4**
4	Abactinal paxillae crowded, with about eight central and 16 peripheral spinelets; 3–4 smaller spines below large inferomarginal spine; groups of superomarginal plates on each side of arm are bright rose red, there are irregular groups of blackish spots in the interradial area and on arms; northern WA and NT.	***A. pulcherrimus*** page 64
	Abactinal paxillae not as above.	**5**
5	Abactinal paxillae with one or a few central spinelets and 7–10 marginal ones; 1–2 small spines or elongate scales below main inferomarginal spine.	***A. monacanthus*** page 58
	Abactinal paxillae with 20–30 central granules or spinelets; one small spine below main inferomarginal spine.	***A. granulatus*** page 54
6	A large spine on the first superomarginal plate then usually two small compressed spineless plates followed by smaller spines on the rest of the marginal plates.	***A. polyacanthus*** page 60
	Spines on one or more superomarginals.	**7**
7	Inferomarginal plates with small spines or spinelets covering the plate; spines only on the first 1–2 superomarginals.	***A. velitaris*** page 72
	Inferomarginals with scales covering the plates; spines on more than two superomarginals.	**8**

Astropecten vappa, Singapore (R. Tan).

8	Spines on all superomarginals.	**9**
	Spines on some superomarginals.	**10**
9	One spine (occasionally two spines) on each superomarginal.	***A. vappa*** page 70
	Two or three spines on each superomarginal.	***A. triseriatus*** page 68
10	Spines usually only on the first 4–5 superomarginals, sometimes more; $R \leq 125$ mm.	***A. zebra*** page 74
	Spine usually on first superomarginal and then on all or only distal superomarginals; $R \leq 45$ mm.	***A. indicus*** page 56

GRANULATED SAND STAR

Astropecten granulatus

Müller and Troschel, 1842

Western Australian Museum (WAMZ6195).

DESCRIPTION

Arms fairly narrow, tapering evenly, margin thick, at least basally, vertical; proximal superomarginal plates appear broader than long, distally as long as broad from above; no spines on superomarginals, which are covered with close-spaced, flat-topped granules on the upper surface becoming more scale-like on the lateral face; paxillar area less than half arm width, at the fifth marginal plate up to 22 central paxillar spines, up to 30 on disc paxillae; inferomarginals with a large flattened spine at the upper edge of the plate and a smaller one

Roebuck Bay, Western Australia (L. Marsh).

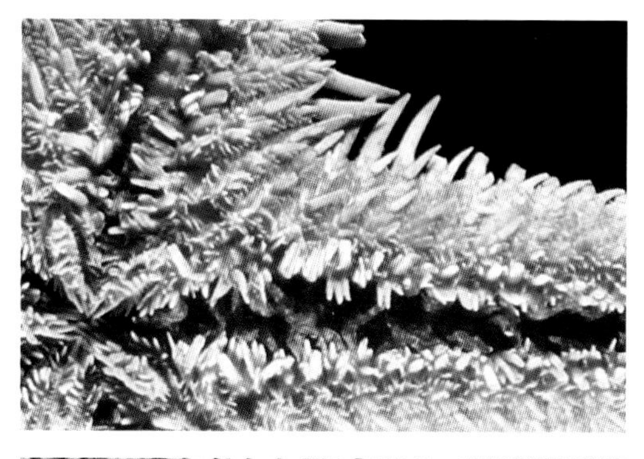

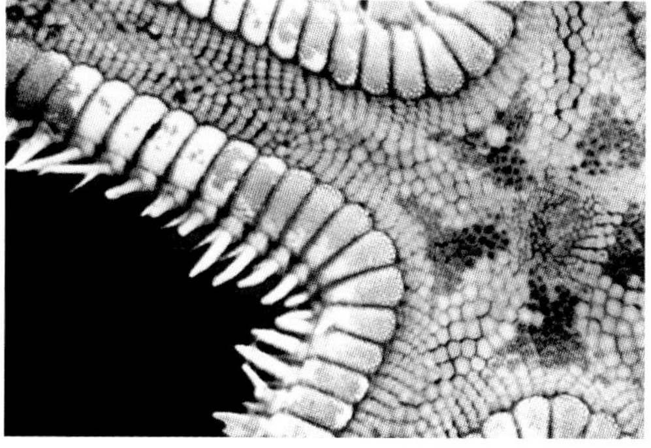

Furrow and actinal plate characteristics [top], South Australian Museum (SAMAK709); superomarginal characteristics [bottom], Western Australian Museum (WAMZ6195).

(if any) below it, 3–5 small to medium sized flat, pointed spines along distal edge of the plate; **inferomarginals covered with flat, truncate to round-ended spinelets (larger than scales)**; furrow spines in a fan of three; two subambulacral spines, one large, one small, 3–4 spinelets on outer part of plate.

MAXIMUM SIZE

R/r/br = 130/20/25 mm (R = 2.3–6.5r).

COLOUR

Light brown, blue-grey or yellowish grey (matching the substrate), there is a dark brown triangular spot in each interradius of the disc; the first 3–4 superomarginal plates and some distal ones dark, forming 1–2 bands in small specimens.

FOOD

Gastropod molluscs, which are engulfed whole.

HABITAT, DEPTH AND AUSTRALIAN DISTRIBUTION

At low tide buried in silty sand, sandy mud or muddy reef, 0–130 m. From Geraldton, WA around north-western, northern and north-eastern Australia to Tin Can Bay (~26°S), Queensland.

TYPE LOCALITY

Not recorded.

FURTHER DISTRIBUTION

From east Africa and the western Indian Ocean to the Philippines.

REFERENCES

Clark, H.L., 1946: 76.

John, 1948: 486–489; pl. 1 figs 1–2.

Livingstone, 1932a: 242; pl. 8 figs 2–3.

Sladen, 1889: 215–216; pl. 35 figs 3–4; pl. 39 figs 4–6.

INDIAN SAND STAR
Astropecten indicus
Döderlein, 1889

DESCRIPTION
Arms more or less blunt at the tip, paxillar areas ending abruptly; only one large inferomarginal spine; distal superomarginal plates or whole series appearing broader than long from above; **distal superomarginals each with a spine on the outer part of the plate, spine on first superomarginal not larger than the others**; paxillae rarely with more than 12 central granules or spinelets, often one central granule ringed by about seven others; mid-radial arm paxillae often reduced to a single granule or spinelet; ventral side of inferomarginals with few spines among small rounded scales; spines may be present only on interradial plates or absent when R < 30 mm.

Superomarginal plate characterics [top], Singapore (R. Tan); inferomarginal plate characteristics [bottom], Western Australian Museum (WAMZ6203).

MAXIMUM SIZE
R = 45 mm.

COLOUR
Dark grey.

FOOD
Bivalve molluscs.

HABITAT, DEPTH AND AUSTRALIAN DISTRIBUTION
On sand flats, shallow water, 0–57 m. Tropical Queensland coast south from Townsville to Ball Bay, north of Mackay, Queensland.

TYPE LOCALITY
Sri Lanka.

FURTHER DISTRIBUTION
From the Persian Gulf to southern India and Indonesia.

REFERENCES
Döderlein, 1889: 828; pl. 31 fig. 2a–d.
John, 1948: 489–490.

Western Australian Museum (WAMZ6203).

Singapore (R. Tan).

SINGLE-SPINED SAND STAR

Astropecten monacanthus

Sladen, 1883

Western Australian Museum (WAMZ6211).

DESCRIPTION

Five arms tapering to a point towards the end; paxillar area of arms taking up nearly three-quarters of the width of the arm; paxillae are large and well spaced, with one or more large central granule-like spinelets and 7–10 very short club-shaped spinelets around the margin of the paxilla; **superomarginal plates narrow, without spines**, covered with large, flat-topped, spaced granules; **inferomarginals with a single lateral spine, the remainder of the plate is covered with moderately spaced rounded to slightly pointed scales; just below the marginal spine are two elongated scales** or small spines; there are three short, blunt-tipped furrow spines, the middle one slightly larger than the other two, followed by three rather flat,

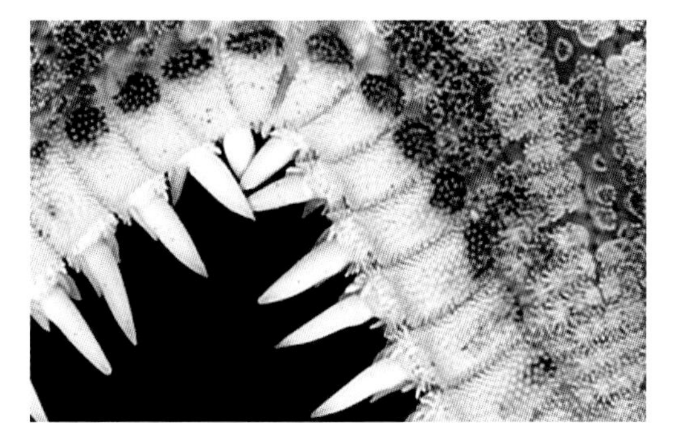

Superomarginal plate characteristics [top], Western Australian Museum (WAMZ6211); inferomarginal plate characteristics [bottom], Western Australian Museum (WAMZ6210).

truncate subambulacral spines, the middle one of which is broader and more robust than the other two; there may be one or two small additional spines between the furrow and subambulacral spines.

MAXIMUM SIZE

R/r/br = 38/11/12 mm (R = 3.4r).

COLOUR

Greyish, blotched with dark grey or brown, particularly on the mid-length of the arms.

FOOD

Carnivorous.

HABITAT, DEPTH AND AUSTRALIAN DISTRIBUTION

On sand or mud, 0–50 m. Ningaloo, WA to Abbot Point, Bowen, Queensland.

TYPE LOCALITY

Philippines.

FURTHER DISTRIBUTION

From east Africa and the Red Sea to Indonesia and the Philippines, north to southern China including Hainan Island.

REFERENCES

Clark, H.L., 1946: 76.

Clark and Rowe, 1971: 46; pl. 5 figs 1–2.

John, 1948: 491–492.

Sladen, 1889: 216–217; pl. 33 figs 7–8; pl. 37 figs 10–12.

COMB SAND STAR

Astropecten polyacanthus

Müller and Troschel, 1842

Houtman Abrolhos, Western Australia (C. Bryce).

DESCRIPTION

The arms are fairly broad and have pointed tips; the superomarginal plates appear narrower than long from above and are high vertically on the sides of the arms; **the first superomarginal bears a long stout spine, the second to fourth (or third) plates are reduced in size and spineless**, thereafter each plate bears a large spine; the paxillar area occupies two-thirds of the arm width at the fifth marginal plate; the thick, flat inferomarginal spine can be as long as three marginal plates, there is a shorter spine below it and a row of spines along the distal edge of the plate; the rest of the **plate is covered by long, flattened, usually pointed scales**; this is a distinctive species easily recognised by the distribution of the large superomarginal spines.

MAXIMUM SIZE

R/r = 105/21 mm (R = 5r).

COLOUR

Variable, for example light brown with darker median stripe or brown blotches on arm; pinkish beige, purple, cream, red or orange; occasionally deep purple paxillae; spines can be partially coloured.

FOOD

Mainly gastropod and bivalve molluscs, which are engulfed whole.

Port Jackson, New South Wales [top] (J. Turnbull); Amakusa, Japan [bottom] (G. Edgar).

REPRODUCTION

In the Red Sea gonads mature during May. Females release small eggs (164 µm) and the bipinnaria larva metamorphoses to a juvenile three days after fertilisation without going through a brachiolaria stage. The larvae swim near the surface.

HABITAT, DEPTH AND AUSTRALIAN DISTRIBUTION

On sand, sometimes on coral reefs, 0–185 m. From Dunsborough, WA around northern and eastern Australia to Narooma, NSW. In Queensland it is found in coastal and Great Barrier Reef habitats, also Norfolk and Lord Howe Islands, Tasman Sea.

TYPE LOCALITY

Red Sea.

FURTHER DISTRIBUTION

Eastern Persian Gulf, tropical Indo-West Pacific, from east Africa to Hawaii, and Pitcairn Island north to southern China, Japan and Taiwan; south to southern New Zealand and Stewart Island.

REFERENCES

Chiu et al., 1990b: 1033–1038.

Clark, H.L., 1946: 74–75.

Clark, H.E.S. and McKnight, 2000: 46–52; pl. 12.

Mortensen, 1937: 42–43; pl. 10 figs 2, 5 (reproduction).

Superomarginal plate characteristics at the arm angle, Timor-Leste (N. Hobgood).

REMARKS

Astropecten polyacanthus is attracted to tetrodotoxin (Saito and Kishimoto, 2003); its tissues contain toxic saponins as well as the puffer fish toxin tetrodotoxin (Migazawa et al., 1987). *Astropecten phragmorus* is very similar to *A. polyacanthus* but has spines on the second and third superomarginal; it is usually found in water deeper than 30 m.

PREISS' SAND STAR
Astropecten preissi
Müller and Troschel, 1843

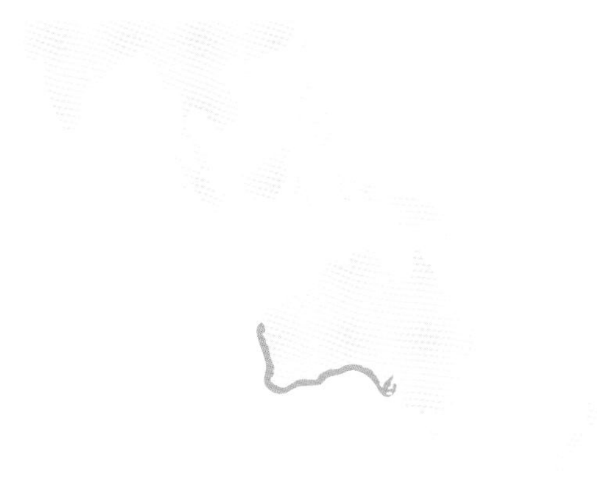

DESCRIPTION
The arms are very long with narrow ends, not sharply pointed; the superomarginal plates appear slightly wider than long from above, **there are no superomarginal spines; the plates are thickly covered by small papillae**; the paxillar area covers about two-thirds the width of the arm at the fifth marginal plate; the inferomarginal plates have a pointed, thin, somewhat flattened large marginal spine with two or three shorter spines below it; the rest of the plate is covered by spinelets with a row of long thin pointed spines along the distal edge.

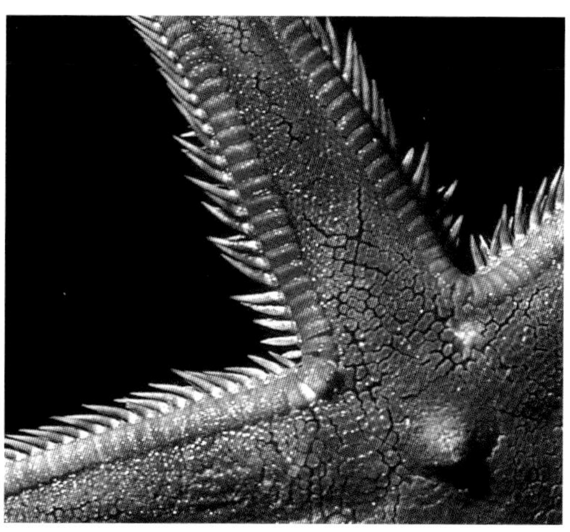

Western Australian Museum (WAMZ6329), superomarginal plate characteristics [bottom].

MAXIMUM SIZE
R/r = 155/22 mm (R = 7r).

COLOUR
Uniform purple or brown to brownish orange. Tube feet and furrow spines yellow.

FOOD
Bivalves and other sand dwelling animals which are engulfed whole.

HABITAT, DEPTH AND DISTRIBUTION
On sand or mud, in sheltered waters, 0–140 m. Gulf St Vincent, SA to Shark Bay, WA. Locally abundant in Cockburn Sound and in the Swan River estuary. Southern Australian endemic.

TYPE LOCALITY
Fremantle, WA.

FURTHER DISTRIBUTION
Australian endemic.

REFERENCES
Clark, H.L., 1946: 77.
Döderlein, 1917: 162.
Shepherd, 1968: 735 (food).

BEAUTIFUL SAND STAR

Astropecten pulcherrimus

H.L. Clark, 1938

Mary Anne Group, Western Australia (J. Keesing).

DESCRIPTION

Disc small, arms taper gradually to a rather wide and rounded tip, terminal plate has a distinct median depression; superomarginals slightly wider than long, viewed from above; **superomarginal and terminal plates covered with granules, no superomarginal spines**; paxillae on the disc and arms numerous and densely crowded; in the interradial areas the larger paxillae have about eight granules on top and 16 marginal papillae; **inferomarginal plates closely covered with short, flat, scale-like spinelets, truncate or rounded at tip**; at the outer end of each plate is a single stout spine, truncate near the disc, pointed on distal plates; on the distal margin of the inferomarginal plates are 3–4 (up to 5) similar but smaller spines; distally two of these next to the marginal spine are enlarged, and near the arm tip the smaller spines disappear; there are three furrow spines,

Superomarginal plate characteristics [top], Mary Anne Group, Western Australia (J. Keesing); inferomarginal plate characteristics [bottom], Western Australian Museum (WAMZ90876).

the middle one curved, the other two set back slightly; there are two subambulacral spines with widened tips followed by 2–3 more with narrower tips, then several spinelets.

MAXIMUM SIZE

R/r = 135/23 mm (R = 3.5–5.8r).

COLOUR

Brownish orange with irregular groups of blackish spots in the interradial areas and scattered on arms. Bright rose red superomarginal plates (usually 6–9) in groups on each side of arms. Actinal surface white.

FOOD

Probably carnivorous.

HABITAT, DEPTH AND DISTRIBUTION

On sand, ~9 m. Mary Anne Group, WA to Bountiful Island, Gulf of Carpentaria, NT. Northern Australian endemic.

TYPE LOCALITY

Lagrange Bay, WA.

FURTHER DISTRIBUTION

Australian endemic.

REFERENCES

Clark, H.L., 1938: 68; pl. 1 fig. 2.
Clark, H.L., 1946: 76.

REMARKS

Astropecten pulcherrimus is distinguished from *A. granulatus* and *A. monocanthus* by the armature of the inferomarginal plates and the abactinal spinelets.

SUMBAWA SAND STAR
Astropecten sumbawanus
Döderlein, 1917

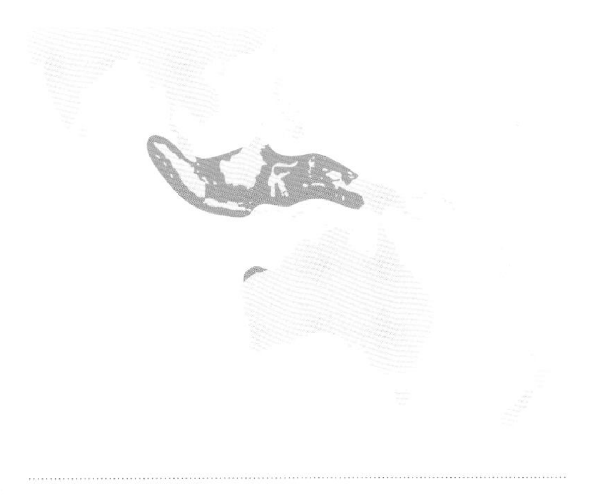

DESCRIPTION
The arms of this small species taper quickly to an acute tip; the superomarginal plates are broader than long from above except in very small specimens; paxillae have up to three central spines, which are not enlarged; **the superomarginal plates are covered with coarse, well spaced granules but no spines**; the big inferomarginal spines are slender and pointed, longer than two marginal plates; **the inferomarginal plates are covered by upright, slender, long scales with short spines among them** and below the big marginal spine is a shorter similar spine and several smaller spines.

MAXIMUM SIZE
R/r = 65/15 mm (R = 4.3r).

COLOUR
Sand coloured.

FOOD
Mainly small molluscs (bivalves, gastropods, scaphopods) and crustaceans. Also scavenges dead animals.

HABITAT, DEPTH AND AUSTRALIAN DISTRIBUTION
On silty sand, 0–36 m. In Australia it is only known from Exmouth Gulf to Dampier Archipelago, WA. Common on intertidal sand flats in Exmouth Gulf and near Dampier.

TYPE LOCALITY
Sumbawa Island, Indonesia.

FURTHER DISTRIBUTION
Indonesia.

REFERENCES
Döderlein, 1917: 54, 159; pl. 4 fig. 11; pl. 6 fig. 9; pl. 12 figs 10, 10a; pl. 15 figs 2, 2a.
Wells and Lalli, 2003: 209–216 (food).

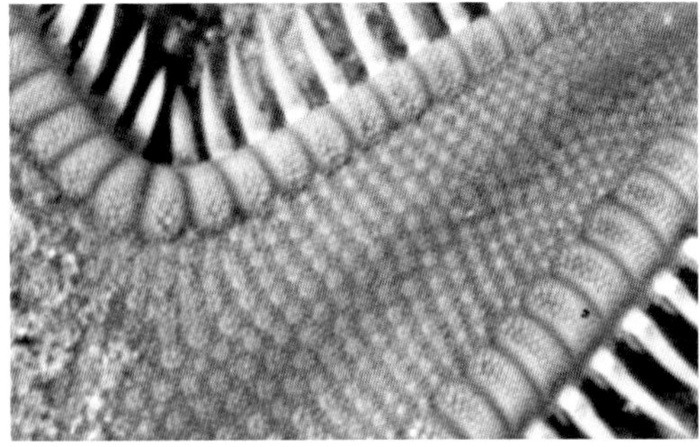

Feeding on a tusk shell (*Dentallium*) [left], superomarginal plate characteristics [above], Dampier Archipelago, Western Australia (C. Bryce).

Dampier Archipelago, Western Australia (S. Morrison).

Dampier Archipelago, Western Australia (C. Bryce).

THREE-SPINED SAND STAR
Astropecten triseriatus
Müller and Troschel, 1843

Swan River estuary, Western Australia (L. Marsh).

Swan River estuary, Western Australia (B. Wilson

DESCRIPTION

The arms are broad and rounded at the ends; the paxillar area occupies about three-quarters of the arm width at the fifth marginal plate; there are up to 10 central paxillar spinelets, the central one of which is often enlarged, surrounded by 15 almost capitate spinelets; the superomarginals are narrow, only slightly broader than long, covered on their lateral surface with flat granules becoming spine-like near the base of the large spines; **most of the superomarginal plates bear**

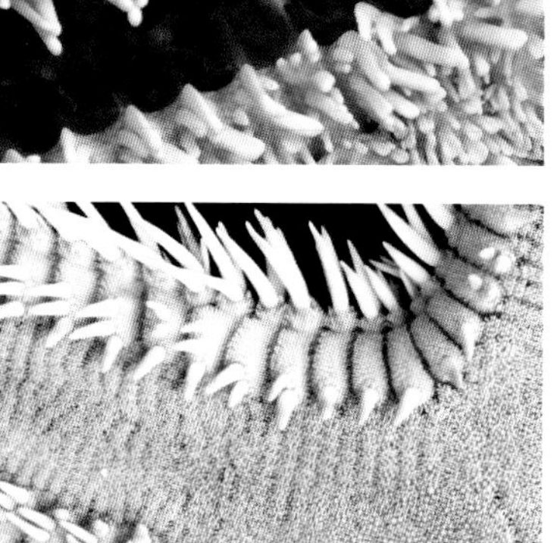

Inferomarginal [top] and superomarginal plate characteristics [bottom], Western Australian Museum (WAMZ6435).

three strong spines but only one on the first few plates; the inferomarginal plates are covered by spatulate scales with a row of spines along the distal edge and a large marginal spine with two smaller ones below it; there are three furrow spines, the middle one longest, curved; one large distal and a small proximal subambulacral spine with 3–4 smaller spines on the outer part of the plate.

MAXIMUM SIZE

R/r/br = 114/22/24 mm (R = ~5r).

COLOUR

Variable, typically pale or dark grey-brown, sometimes deep purple with inferomarginal spines reddish at base shading to white at the tips. Actinal surface cream.

FOOD

Carnivorous.

HABITAT, DEPTH AND AUSTRALIAN DISTRIBUTION

On sand or mud, 0–46 m. From Seacliff, SA to Ningaloo, WA, including the Swan River estuary.

TYPE LOCALITY

South-western Australia.

FURTHER DISTRIBUTION

Fiji and Hawaii.

REFERENCES

Clark, A.M. and Rowe, 1971: 45.

Döderlein, 1917: 125; pl. 5 figs 2–3; pl. 11 figs 5–6.

John, 1948: 500–502; pl. 2 fig. 4.

REMARKS

Astropecten triseriatus is closely related to *A. vappa*, which sometimes has more than one superomarginal spine on some plates.

SPINY SAND STAR
Astropecten vappa
Müller and Troschel, 1843

Western Australian Museum (WAMZ88503).

DESCRIPTION

The arms are moderately broad or fairly narrow tapering to pointed arm tips; the paxillar area occupies about two-thirds of the arm width at the fifth marginal plate; there are up to 15 central paxillar spinelets; the superomarginal plates appear slightly broader than long from above, they are covered with scale-like papillae and all have a **cylindrical spine, fairly large on the first plate then decreasing in size; the spine is on the inner edge of the first superomarginal but by the fifth plate it is near the outer edge**; sometimes there are two outer spines and sometimes an inner spine next to the outer spine; **the inferomarginal plates are covered with short, wide, spatulate scales with rounded ends** and a row of short, blunt spines along the distal edge; the large slightly flattened marginal spine is longer than three inferomarginal plates and has a small spine below it, and a smaller one behind that is part of the distal row; three furrow spines, middle one largest; two subambulacral spines followed by smaller spines on the outer part of the plate.

MAXIMUM SIZE

R/r = 105/18 mm (R = 5.8r),
br at fifth superomarginal = 16 mm.

COLOUR

Variable, often dark brown or grey, variegated light and dark grey or blue. Small specimens (R < 50 mm) are light yellowish or cream, mottled with grey or with a dark V in the arm angle and irregular dark blotches on arms, sometimes forming up to four indistinct bands. Actinal surface cream.

FOOD

Small bivalves and gastropods, which are ingested whole.

HABITAT, DEPTH AND AUSTRALIAN DISTRIBUTION

On silty sand, weed and sand, or sand and shells, 0–128 m; moderately common. All coasts of Australia.

TYPE LOCALITY

South-western Australia.

FURTHER DISTRIBUTION

From Sri Lanka to Indonesia, the Philippines, southern China, Taiwan and Japan.

REFERENCES

Clark, H.L., 1946: 75–76 (synonymy of *Astropecten hartmeyeri* Döderlein, 1917 with *A. vappa*).

Döderlein, 1917: 54, 156; pl. 5 fig. 8; pl. 14 figs 6, 6c.

Liao and Clark, A.M., 1995: 76–80; pl. 4 fig. 1.

Mah, 2019: as *Astropecten carcharicus.*

Superomarginal characteristics at the arm angle, Singapore (R. Tan).

REMARKS

Liao and Clark (1995) after examining over 70 specimens of *Astropecten vappa* and *A. carcharicus* from the Western Australian Museum collection, many from Shark Bay (the type locality of *A. carcharicus* Döderlein, 1917), concluded that the latter fell within the variability of *A. vappa* (type locality south-western Australia) and relegated *A. carcharicus* into synonymy with *A. vappa*. Mah (2019) records *A. carcharicus* as a valid species.

VELITAR SAND STAR

Astropecten velitaris

von Martens, 1865

Onslow, Western Australia (J. Keesing).

DESCRIPTION

The arms are slightly constricted at the base in large animals then widen and taper to a fairly acute tip; **the superomarginals lack spines except for a conspicuous erect spine on the first superomarginal** and on the second plate in large specimens; the largest abactinal paxillae have about 20 round-topped central granules surrounded by about the same number of similar radiating round-ended spinelets; lateral paxillae on the arms small with 2–4 central granules and up to 12 marginal granules; superomarginal plates wider than long covered with low rounded granules, thinner and longer on the sides adjacent to plates; inferomarginals with a long, pointed marginal spine equal to 2.5–3 plates in length, below this are three smaller curved spines in a row, then a row of 4–5 slightly smaller spines on the distal edge of the plate, the rest of the plate is covered in **spinelets**; there is one large curved furrow spine and 4–5 smaller subambulacral spines followed by about four spinelets.

MAXIMUM SIZE

R/r/br = 115/16/18 mm (R = ~7r).

COLOUR

Greenish grey to beige, irregularly blotched with dark grey, forming irregular cross-bands near the middle of each arm. Superomarginal plates, terminal plate and an interradial line purple or violet; superomarginal spines may be orange, tipped with purple.

FOOD

Gastropod and bivalve molluscs, crustaceans and other infauna, which are engulfed whole.

REPRODUCTION

In the Red Sea reproduction takes place during May. The planktotrophic bipinnaria larva metamorphoses after five days without going through a brachiolaria stage. The larvae swim close to the surface until settlement.

HABITAT, DEPTH AND AUSTRALIAN DISTRIBUTION

On silty sand, or coral rubble, 5–165 m. On the continental shelf from the Passage Islands, WA to Townsville, Queensland.

TYPE LOCALITY

South China Sea.

FURTHER DISTRIBUTION

The Red Sea, Sri Lanka, Indonesia and the Philippines, Hainan Island (southern China), Taiwan and Japan.

REFERENCES

Clark, H.L., 1946: 77.
Clark, A.M. and Rowe, 1971: 33.
Döderlein, 1917: 159; pl. 6 figs 5, 15–16; pl. 15 fig. 3.
Lemmens et al., 1995: 447–455 (food).
Mortensen, 1937: 43; pl. 10 fig. 3 (reproduction).

Superomarginal characteristics, Western Australian Museum (WAMZ6502).

REMARKS

Liao and Clark (1995) doubt that *Astropecten velitaris* from Australian waters are the same as those named *A. velitaris* from the South China Sea, where it is abundant, but R is always < 40 mm, while Australian specimens have R = 100+ mm.

ZEBRA SAND STAR

Astropecten zebra

Sladen, 1883

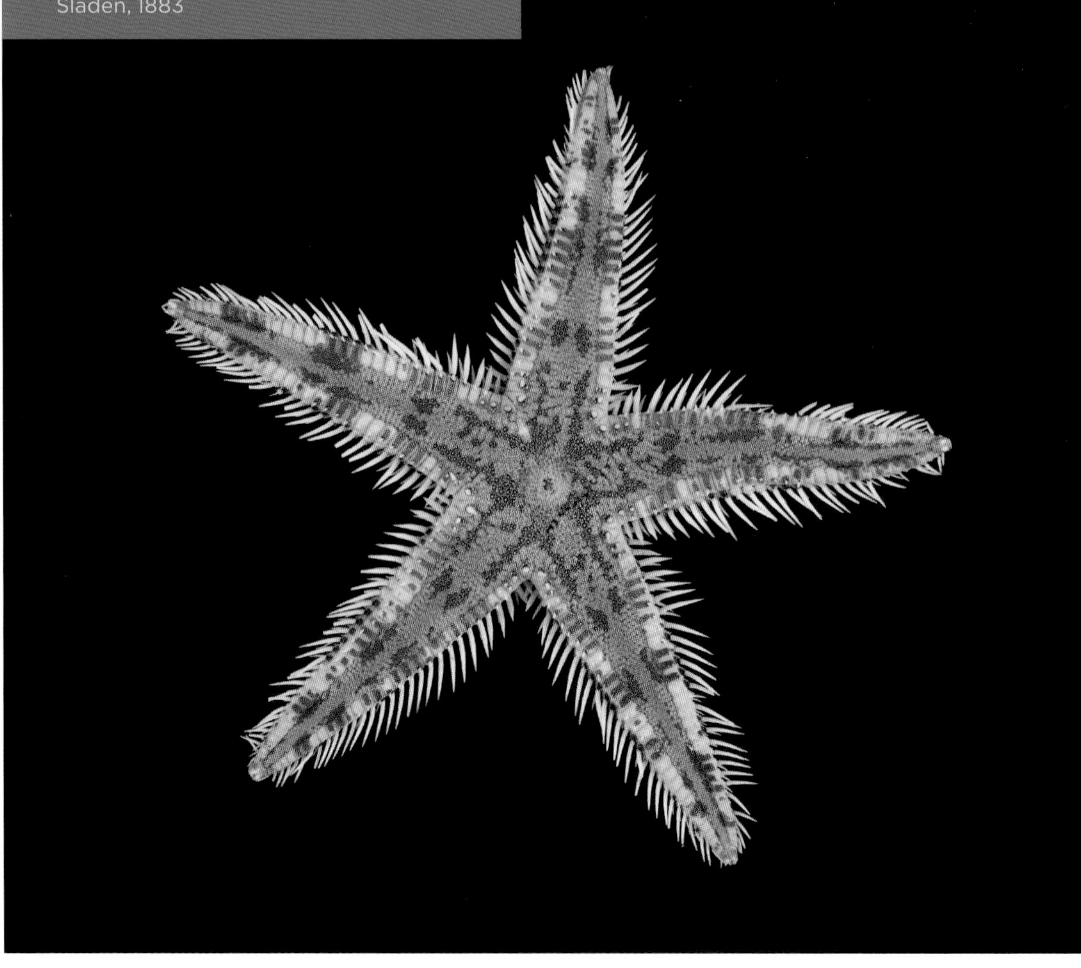

Western Australian Museum (WAMZ6498).

DESCRIPTION

Very similar to *Astropecten vappa* but **superomarginal plates have a short conical spine usually only on the first 3–5 plates**; arms are narrow, tapering from the base to a blunt tip; abactinal paxillae are large with 1–4 (or more) central spiniform granules surrounded by 8–12 (up to 15) peripheral spinelets; superomarginal plates are higher than broad, with rounded, papilliform granules on the upper surface, smaller more squamous granules on the vertical face and papilliform spinelets between adjacent plates; inferomarginal plates do not project beyond the superomarginals; inferomarginals bear a long, curved, slightly flattened marginal spine with a second spine half to two-thirds as long below it on the distal edge of the plate, followed by three shorter, flattened, pointed spines, the rest of the plate is covered by **round-ended, flat, scale-like spinelets**; there are three furrow spines, the middle curved spine is longest, two subambulacral flat, truncate spines, then three blunt-ended smaller spines behind and three smaller spines on the outer edge.

MAXIMUM SIZE

R/r = 125/19 mm (R = ~6.5r),
br at fifth superomarginal = 18 mm.

COLOUR

Beige to brown, with a dark V-shaped mark on the disc, the arms of the V towards the arms of the seastar; dark patches on the arms sometimes forming three indistinct bands. Actinal surface cream.

FOOD

Predominantly bivalve and gastropod molluscs, which are engulfed whole.

HABITAT, DEPTH AND AUSTRALIAN DISTRIBUTION

On sand or mud, 3–41 m. Houtman Abrolhos, WA to Great Barrier Reef, Queensland.

TYPE LOCALITY

Torres Strait, Queensland.

Superomarginal plate characteristics at the arm angle [top], Barrow Island, Western Australia (J. Keesing); inferomarginal plate characteristics [bottom], Western Australian Museum (WAMZ6498).

FURTHER DISTRIBUTION

Sri Lanka to Indonesia.

REFERENCES

Clark, H.L., 1946: 77.
John, 1948: 492–495.
Lemmens et al., 1995: 447–455 (food).
Shepherd, 1968: 735–737 (as *Astropecten vappa*).
Sladen, 1889: 212–214; pl. 36 figs 3–4; pl. 39 figs 7–9.

REMARKS

Shepherd (1968) included *Astropecten zebra* in the synonymy of *A. vappa* without any discussion, comparing *A. vappa* with *A. triseriatus* only. Clark and Rowe (1971) list both *A. vappa* and *A. zebra* as do Rowe and Gates (1995), giving the distribution of *A. zebra* in Australia as Houtman Abrolhos, WA to Yeppoon, Queensland and the tropical east Indo-West Pacific. *A. vappa* has the same distribution but extends to the south coast of Australia as well, and has a wider tropical east Indo-West Pacific distribution. Both species are included here.

GENUS *BOLLONASTER* McKnight, 1977

Resembling species of *Astropecten* in form but with three rows of actinal plates basally extending in a single series two-thirds the length of the arms. The marginal plates are massive, the superomarginals may have spines, and they have well developed marginal fascioles. The inferomarginals have an appressed comb of slender spines and the abactinal paxillae are large, flat-topped and crowded. The furrows have 5–7 divergent spines and numerous subambulacral spines. An unpaired actinal interradial plate may be present.

Type species: *Bollonaster pectinatus* (Sladen, 1883) (Australia)
Monotypic genus. Diagnosis as for the species (page 77).

GENUS *CRASPIDASTER* Sladen, 1889

Type species: *Craspidaster hesperus* (Müller and Troschel, 1840) (Indo-West Pacific)
Monotypic genus. Diagnosis as for the species (page 79).

Bollonaster pectinatus, Swan River estuary, Western Australia (S. Morrison).

COMB-LIKE SAND STAR
Bollonaster pectinatus
(Sladen, 1883)

Western Australian Museum (WAMZ6221).

DESCRIPTION

Bollonaster is distinguished from *Astropecten* by having **three rows of actinal plates in the arm angle**; the arms are fairly short, broad and pointed, the paxillar area takes up about two-thirds of the width of the arm; paxillae bear up to 15 central spinelets with a similar number of peripheral spinelets; the superomarginal plates appear as long as wide from above, covered with rounded slightly cylindrical granules (coarser than paxillar spinelets), each bears **a small conical spine (occasionally two)** near the inner margin of the plate in the arm angle, towards the outer edge after the third plate; the inferomarginal plates are covered with small flat, round-tipped close spaced scales and 1–2 flat spines on the distal edge; **there are up to six thin pointed marginal spines in an oblique row**, the second or third is longest; there are three furrow spines, the middle one slightly longer than the others, tapering, the other two are slightly flattened, blunt-ended; the subambulacral spines are in two groups of three radiating, flat, truncate spines.

MAXIMUM SIZE

R/r = 55/18 mm (R = 3–5r).

COLOUR

Pale pink with a reddish dorsal stripe on each arm, or pinkish brown with a darker disc and midline stripe.

FOOD

Carnivorous.

Furrow and actinal plate characteristics [top], Western Australian Museum (WAMZ6221); marginal plate characteristics [bottom], Cockle Creek, Tasmania (S. Grove).

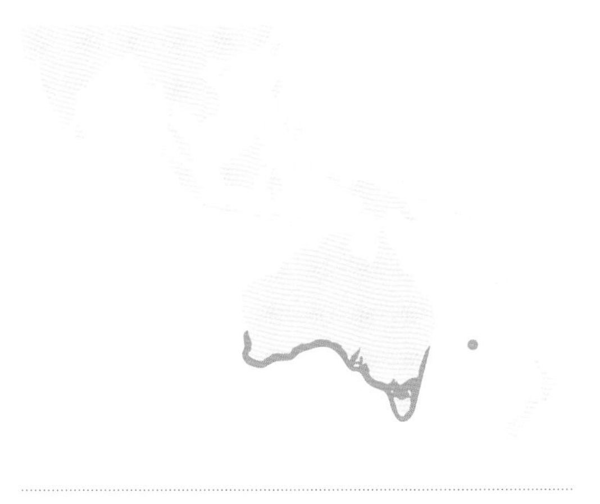

HABITAT, DEPTH AND AUSTRALIAN DISTRIBUTION

On fine sand or sandy mud, 2–280 m. Newcastle, NSW to City Beach, WA. Generally it lives deeper than 30 m but is found in shallow water in Tasmania. Southern Australian endemic, including Tasmania.

TYPE LOCALITY

Port Jackson, NSW.

FURTHER DISTRIBUTION

Australian endemic.

REFERENCES

Byrne et al., 2017: 239 (as *Bollonaster pectinatus).*

Clark, H.L., 1946: 74 (as *Astropecten schayeri, A. syntomus* and *A. pectinatus*).

Clark, H.E.S. and McKnight, 2000: 37 (tentatively as *Astromesites*).

Mah, 2019: as *Bollonaster pectinatus.*

McKnight, 1977: 113 (as *Bollonaster pectinatus*).

Rowe and Gates, 1995: 51 (as *Bollonaster pectinatus*).

Shepherd, 1968: 733 (synonymy *Astropecten schayeri* Döderlein, 1917 and *A. syntomus* H.L. Clark, 1928 with *A. pectinatus*).

Sladen, 1889: 202–203; pl. 33 figs 3–4; pl. 37 figs 4–6.

REMARKS

McKnight (1977) transferred *Astropecten pectinatus* to a new genus *Bollonaster* with type species *B. primigenius* (Mortensen, 1925), a New Zealand species; Clark and McKnight (2000) then transferred *B. primigenius* to *Astromesites* but did not determine the status of *B. pectinatus. Astromesites* as defined by Fisher (1913) lacks any spines on the superomarginals whereas *A. pectinatus* has small spines on all or most of the superomarginals. Rowe (personal communication) does not consider the presence of superomarginal spines invalidates the inclusion of *A. pectinatus* in *Astromesites.* Mah (2019) includes *A. pectinatus* in the genus *Bollonaster.*

EVENING SAND STAR
Craspidaster hesperus
(Müller and Troschel, 1840)

Western Australian Museum (WAMZ6729).

Singapore (R. Tan).

DESCRIPTION

Five arms; the **shape resembles a goniasterid more than an astropectinid, but with pointed tube feet**; it is flat and rigid, with a more pronounced disc than *Astropecten*; arms are moderately long and flat tapering to a narrow tip, bordered by large marginal plates with a paxillate abactinal surface; paxillae with up to 10–20 central granules and 16–20 peripheral spinelets, smallest have a single central granule; paxillae on the arms are in regular transverse series;

superomarginal plates are about twice as wide as long inter-radially, length and width about equal distally; the marginal plates are separated by well defined channels fringed by small webbed spinelets; **each inferomarginal has a single small flat truncate spinelet on its outer edge; the surface of the marginal plates is covered by large, well spaced hemispherical granules**; there are 5–6 tapering, radiating furrow spinelets and skin-covered small spinelets around the edges of the adambulacral plates and a short, sub-conical spinelet near its proximal edge; **there are about four large actinal plates next to the adambulacrals and a second row of 1–2 plates next to the large inferomarginals**; actinal plates have granules similar to the marginal plates; there is no anus and pedicellariae are absent.

Abactinal [top] and actinal plate characteristics [bottom], Singapore (R. Tan).

MAXIMUM SIZE

R/r = 55/15 mm (R = 3.6r).

COLOUR

Sand coloured with grey or blue tinges.

FOOD

Probably feeds on microorganisms found in the substrate.

HABITAT, DEPTH AND AUSTRALIAN DISTRIBUTION

On muddy sand, or mud, 12–196 m. In Australia it has only been recorded from two localities in the NT.

TYPE LOCALITY

Japan.

FURTHER DISTRIBUTION

Bay of Bengal, Indonesia to the Philippines, north to Vietnam, Hainan Island (southern China), Taiwan and Japan.

REFERENCES

Fisher, 1919: 60–61; pl. 9 fig. 3.

Liao and Clark, A.M., 1995: 81.

LUIDIIDAE

Luidiidae is a small family with only a single genus, *Luidia*, found in tropical and subtropical seas. Eight species are recorded from Australian waters, five of which are found in depths of less than 30 m.

Luidia australiae, Cape Peron, Western Australia (C. Bryce).

GENUS *LUIDIA* Forbes, 1839

Species of this genus generally have five to many arms. The tube feet ampullae are double, without suckers, having a knob-like tip or the ends rounded. The superambulacral plates are present. The abactinal surface is covered with paxillae and the superomarginal plates are not distinguished from the paxillae. The edge of the body is defined by large inferomarginal plates.

Type species: *Luidia ciliaris* (Philippi, 1837) (North Atlantic, Mediterranean Sea)

Five species are found in Australian waters in depths of less than 30 m.

KEY TO THE SPECIES OF *LUIDIA* IN SHALLOW AUSTRALIAN WATERS

1	Arms 5–6.	**2**
	Arms 7–10.	**3**
2	Arms five; no abactinal spines; adambulacral armament a furrow spine and five subambulacral spines proximally, reducing to three distally; large pedicellariae on oral plates, adambulacral pedicellariae present or absent; marginal spines short and stubby; R ≤ 65 mm; colour uniform, pale grey or light brown.	***L. hardwicki*** page 85
	Arms six; superficially similar to *L. maculata* but distinguished by having pedicellariae on oral plates; R ≤ 140 mm.	***L. hexactis*** page 87
3	Arms 7–8.	**4**
	Arms 9–10; some abactinal paxillae with short spines on dorsolateral rows; spines on some superomarginals, pedicellariae on adambulacral plates; R ≤ 195 mm; tropical waters.	***L. avicularia*** page 84
4	Arms seven; middle paxillae near arm ends are enlarged and conspicuous; R ≤ 275 mm; temperate waters.	***L. australiae*** page 83
	Arms usually eight (6–9); middle paxillae near arm ends not enlarged; R ≤ 350 mm; tropical waters.	***L. maculata*** page 88

Luidia australiae, Geographe Bay, Western Australia (S. Morrison).

SOUTHERN SAND STAR
Luidia australiae
Döderlein, 1920

DESCRIPTION
Seven arms tapering fairly evenly from the base to the end; in contrast to *Luidia maculata* **a number of middle paxillae near the arm ends are enlarged and conspicuous,** otherwise the two species are very similar apart from the number of arms, which is almost always seven in *L. australiae*, more variable but usually eight in Australian *L. maculata*; inferomarginal plates with a very large spine on the outer edge of the plate followed by a slightly smaller spine, and a third even smaller spine, then spinelets and trivalved pedicellariae; no large bivalved pedicellariae on the oral plates; up to five transverse and longitudinal rows of quadrangular lateral paxillae; adambulacral armature of two furrow spines and two subambulacral spines, and several large pedicellariae with 2–3 valves.

MAXIMUM SIZE
R/r = 275/35 mm (R = 7.8r).

COLOUR
Cream to light brown or grey, blotched with dark brown, dark green or black, sometimes in irregular bands on the arms.

FOOD
Small molluscs, ophiuroids and echinoids (especially heart urchins such as *Echinocardium cordatum*), together with mud and detritus. Swallows prey whole.

REPRODUCTION
Development is through a planktotrophic bipinnaria larva, without a brachiolaria stage.

HABITAT, DEPTH AND AUSTRALIAN DISTRIBUTION
On sand or mud where it buries itself by lateral arm movements, intertidal to 110 m. From Moreton Bay, Queensland and Lord Howe Island around southern Australia to the vicinity of Geraldton, WA. It is replaced at the Houtman Abrolhos and further north by *L. maculata*. Southern Australian endemic.

TYPE LOCALITY
Cockburn Sound, WA.

FURTHER DISTRIBUTION
Australian endemic.

REFERENCES
Clark, H.L., 1946: 71.

Döderlein, 1920: 266.

Paxillae characteristics at the arm base and tip, Western Australian Museum (WAMZ6028).

BIRD-LIKE SAND STAR
Luidia avicularia
Fisher, 1913

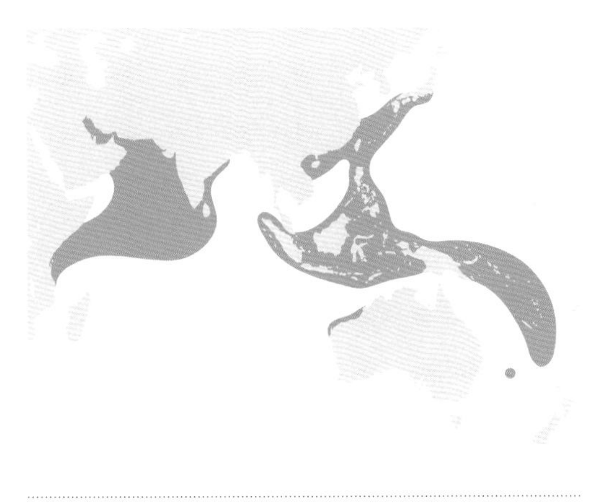

DESCRIPTION

Nine or ten arms, tapering, blunt-ended, **scattered paxillae with spines among the normal paxillae** of the three dorsolateral rows; a spine is present on some superomarginals; abactinal and superomarginal paxillae with 10–15 short, sharp central spinelets and ~30 peripheral spinelets on large specimens; small two-jawed pedicellariae on abactinal and superomarginal paxillae; inferomarginals encroach on abactinal surface and have 3–4 (up to six) marginal spines and sometimes a pedicellaria; one strongly curved furrow spine and three (up to five) subambulacral spines in a transverse row; below (dorsal to) the furrow spine is a characteristic pedicellaria with the upper jaw curved over the end of the slightly curved lower jaw (appearing beak-like); pedicellariae may replace some of the subambulacral spines; the oral plates generally have two large pedicellariae directed over the mouth area.

MAXIMUM SIZE

R/r = 195/18 mm (R = 10.5r),
br at base of arm = 13 mm,
br at widest part of arm = ~18 mm.

Western Australian Museum (WAMZ23103).

COLOUR

Uniform yellow-brown without patterning.

FOOD

Probably carnivorous.

REPRODUCTION

The gonads extend in a series of tufts along the arm and each gonoduct opens abactinally, near the base of the arm. Development is through a planktotrophic bipinnaria larva without a brachiolaria stage.

HABITAT, DEPTH AND AUSTRALIAN DISTRIBUTION

On soft substrates, 6–308 m. Off Carnarvon, Point Cloates and Port Hedland, WA. Elizabeth Reef and off Norfolk Island, Tasman Sea. Australian records are all from 100 m or more.

TYPE LOCALITY

Between Samar and the Masbate Islands, the Philippines.

FURTHER DISTRIBUTION

Indonesia, the Philippines, southern Japan and Taiwan.

REFERENCES

Fisher, 1919: 72; pl. 44 fig. 2; pl. 46 fig. 2a–c.
Liao and Clark, A.M., 1995: 69; pl. 3 fig. 3.

HARDWICK'S SAND STAR
Luidia hardwicki
(Gray, 1840)

Western Australian Museum (WAMZ8084).

DESCRIPTION

Five arms; large pedicellariae present on the oral plates and usually on the outer part of the adambulacral plates; two or three lateral series of paxillae form transverse rows with the superomarginals; a single enlarged marginal spine at the top of each inferomarginal plate, with smaller appressed spines on the ventral side of the plate; abactinal paxillae have 1–5 central spinelets and 10–12 peripheral spinelets; the superomarginal paxillae have up to ten central and ~20 peripheral spinelets.

MAXIMUM SIZE

R/r = 65/10 mm, (R = 5–8.5r).

COLOUR

Pale grey or light brown. Actinal surface cream to brown.

FOOD

Predominantly bivalve and gastropod molluscs but the diet includes other echinoderms, crustaceans and Foraminifera. They also scavenge dead animals and are sometimes cannibals. The prey is ingested whole.

REPRODUCTION

Development is through a planktotrophic bipinnaria larva without a brachiolaria stage.

HABITAT, DEPTH AND AUSTRALIAN DISTRIBUTION

On sand, mud and shelly sand, 8–420 m. From the continental shelf south of Albany to Shark Bay WA and around northern Australia to Queensland, and south to Jervis Bay, NSW.

TYPE LOCALITY

'Indian Ocean'.

FURTHER DISTRIBUTION

From the western Indian Ocean, Persian Gulf, east and west coasts of India, South-East Asia, south to Kermadec Islands, and north to Hainan Island and southern Japan.

REFERENCES

Chiu et al., 1990a: 907–933.

Clark, A.M., 1953a: 391–392; pl. 39 figs 2–3.

Clark, A.M., 1989: 242.

Clark, A.M. and Rowe, 1971: 44; pl. 4 fig. 2.

Fisher, 1919: 164–166; pl. 41 figs 5, 5a; pl. 45 figs 1–2 (as *Luidia prionota*).

Liao and Clark, A.M., 1995: 69–70; pl. 3 fig. 1.

Mah et al., 2009: 395.

Mah, 2019: as *Luidia prionota*.

Rowe and Gates, 1995: 74–75.

Sastry, 2007: 24–25.

Singapore (R. Tan).

REMARKS

Clark and Rowe (1971) note that the presence of pedicellariae on some adambulacral plates is the only feature separating *Luidia hardwicki* from *L. prionota* Fisher, 1913, but the two species are kept separate by Clark (1989), Rowe and Gates (1995) and Mah (2019). However, Liao and Clark (1995) regard *L. prionota* as a synonym of *L. hardwicki*, as followed here.

SIX-ARMED SAND STAR
Luidia hexactis
H.L. Clark, 1938

DESCRIPTION

Six arms; paxillae and colour pattern resemble *L. maculata*; arms long and slender; paxillae are in five longitudinal series of close set squares on each side of the arms, between these are five rather irregular series of smaller, more polygonal paxillae, which are covered with fine granules, none enlarged or spine-like; the inferomarginal plates are short and wide, the surface is covered by a longitudinal series of 2–6 spines (the longest is more than 2 mm long) and many sharp, small spines (1 mm or less) well spaced over the plate; **the adambulacral armature and pedicellariae distinguish it from *L. maculata***, which can be six armed; *L. hexactis* has two large furrow spines and two subambulacral spines; outer furrow spine straight, blunt-pointed, inner spine curved, sharply pointed; subambulacral spines taper to a slender point; behind them is a large pedicellaria and often a smaller pedicellaria; two actinal plates on each side each with a large pedicellaria; two small pedicellariae of four spinelets on oral plates deep in the mouth; the presence of pedicellariae on the oral plates distinguishes *L. hexactis* from *L. maculata*.

Furrow characteristics, Western Australian Museum (WAMZ6092).

MAXIMUM SIZE

R/r = 140/15 mm (R = ~9r, ~9br).

COLOUR

Beige blotched with large irregular areas of dark greenish grey. Actinal surface cream.

FOOD

Carnivorous.

REPRODUCTION

Development is through a planktotrophic bipinnaria larva without a brachiolaria stage.

HABITAT, DEPTH AND DISTRIBUTION

On soft substrates, 16–55 m. Cockatoo Island, WA to west of Edward River, Gulf of Carpentaria, Queensland and from Port Molle to Port Curtis, Queensland. Northern Australian endemic.

TYPE LOCALITY

Near Montgomery Reef, WA.

FURTHER DISTRIBUTION

Australian endemic.

REFERENCES

Clark, H.L., 1938: 73; pl. 17 fig. 1.

Clark, H.L., 1946: 71.

SPECKLED SAND STAR

Luidia maculata

Müller and Troschel, 1842

Montebello Islands, Western Australia (C. Bryce).

DESCRIPTION

Usually eight arms (6–9); abactinal paxillae squarish, forming 6–8 longitudinal rows on each side of the midline; paxillae (depending on size) with 6–20 short blunt spinelets surrounded by fine peripheral spinelets; the general surface smooth; 3–5 short stout inferomarginal spines all similar in size; a furrow spine and 2–3 large subambulacral spines on each plate; **no pedicellariae on the oral plates** but several three-bladed pedicellariae on the adambulacral plates and shorter pedicellariae on the inferomarginals.

MAXIMUM SIZE

R = 350 mm (R = 9–11r),
R usually < 250 mm.

COLOUR

Cream to buff grey, mottled or irregularly banded with dark grey to black.

FOOD

Observed feeding on other seastars, sand dollars and heart urchins (such as species of *Breynia* and *Echinocardium*), which it engulfs whole; also molluscs, crustaceans, worms, ophiuroids and holothurians (Birtles, personnal communication).

REPRODUCTION

In Japan reproduction takes place during July. The sexes are separate. Eggs are 174 µm in diameter and the embryos develop into planktotrophic bipinnaria larvae which commence metamorphosis 50 days after fertilisation without going through a brachiolaria stage; after 14 days metamorphosis is complete and resulting juveniles are about 700 µm in diameter with nine arms (the adult number in Japanese specimens).

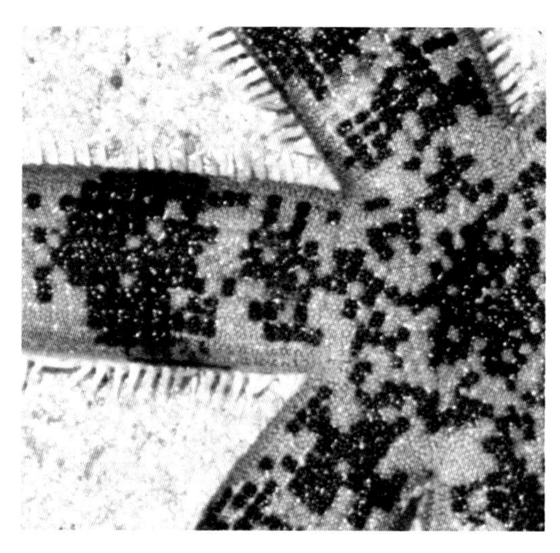

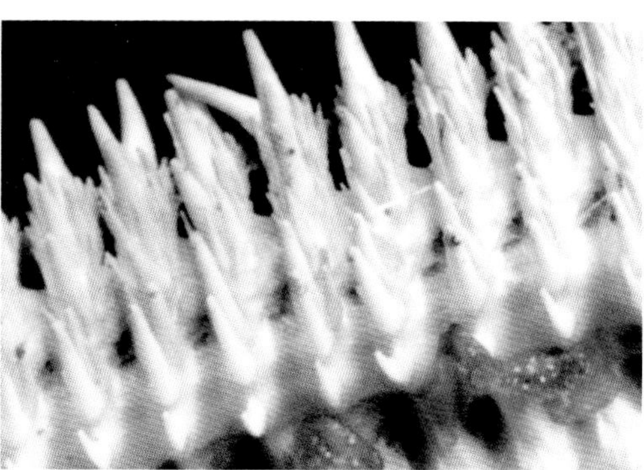

Paxillae characteristics [top], Montebello Islands, Western Australia (L. Marsh); furrow and inferomarginal characteristics [bottom], Western Australian Museum (WAMZ6071).

HABITAT, DEPTH AND AUSTRALIAN DISTRIBUTION

On sand or mud, intertidal to 150 m; moderately common. From the Houtman Abrolhos and Shark Bay, WA, around northern Australia to Port Molle, Queensland.

TYPE LOCALITY

Japan.

FURTHER DISTRIBUTION

East Africa to the Red Sea and Persian Gulf, through South-East Asia east to New Caledonia, south to North Island, New Zealand, and north to southern China, Japan and Taiwan.

REFERENCES

Clark, H.L., 1946: 71–72.
Clark, A.M. and Rowe, 1971: 43; pl. 4 fig. 3.
Döderlein, 1920: 262–266; pl. 18 figs 4, 13; pl. 19 fig. 16; pl. 20 figs 23–24.
Komatsu et al., 1994: 327–333; figs 1–2 (development).
Mah et al., 2009: 395.

VALVATIDA

Fromia monilis, Browse Island, Western Australia (M. Ekins).

Species of the Valvatida have tube feet, with sucking discs, in two rows, and a digestive system terminating with an anus. The body shape is varied, usually five armed. The discs are large, more or less arched (e.g. Oreasteridae) or small, with more or less cylindrical arms (e.g. Ophidiasteridae) or with flattened arms (Archasteridae). The abactinal ossicles are varied, including tabulate, pseudopaxillate, reticulate or imbricate. The marginals are usually well developed and the intermarginals are rare. Actinal plates are few to numerous and the body skin is smooth or covered by granules obscuring the plates in some species. Spines or spinelets are often present and pedicellariae of various forms occur, from fasciculate to bivalved.

The inclusion of Acanthasteridae in the Valvatida and the removal of the Metrodiridae to the Spinulosida (after Blake 1987) leaves a rather heterogeneous collection of 16 families.

KEY TO THE FAMILIES OF VALVATIDA IN SHALLOW AUSTRALIAN WATERS

1	Isolated, irregular, enlarged primary plates usually armed with massive single spines; the spinose marginal plates irregular, discontinuous series, offset vertically.	2
	Primary plates similar; if massive spines developed then regularly arranged; marginal plates regular, matching, at least distally, or concealed by a coarse continuous armament of tubercles or sometimes thick skin.	3
2	Arms normally five; madreporite single; spines covered in scales.	**Mithrodiidae** page 286
	Arms usually 9–18; multiple madreporites present; spines covered in finely granulated skin.	**Acanthasteridae** page 94

Mithrodia clavigera, Malapascua Island, the Philippines (M. Banks).

3	Disc relatively small; interradial area angular and arms narrow, tapering but rectangular in cross section; abactinal plates paxillae-like, the mid-radial rows noticeably enlarged.	**Archasteridae** page 102
	Body form often stellate to pentagonal; sometimes the disc small but then the arms cylindrical; mid-radial plates only enlarged if the arms are carinate.	**4**
4	Disc small; arms more or less cylindrical; interradial areas angular.	**9**
	Disc large; body form stellate to pentagonal, rarely cushion-like; interradial arcs curved.	**5**
5	Abactinal plates small, imbricating proximally or adradially; marginals diminutive, barely distinct, though defining a more or less sharp ventrolateral angle; armament fine.	**Asterinidae** page 108
	Abactinal plating slightly overlapping or sometimes forming an open reticulum, rarely concealed by coarse armament or thick skin; marginals relatively large, only occasionally forming a ventrolateral angle and then armed with coarse macroscopic spines.	**6**
6	Limits of plates obscured by continuous coating of coarse tubercles and granules; only the distalmost superomarginals conspicuously bare at the tips of the arms.	**Asterodiscididae** page 199
	Marginal and other plates more or less distinct, except rarely when the entire body is covered by thick skin without armament.	**7**
7	Skin more or less thick, obscuring the skeletal plates; body form more or less stellate; superomarginals or inferomarginals projecting to form a ventrolateral angle.	**10**
	Skin only rarely thick enough to obscure the plating and then superomarginals not markedly inset from inferomarginals, the two series together forming the edge of the body.	**8**

8	Internally, interbrachial septum membranous; discs of tube feet without spicules; papulae often single, rarely in large groups.	**Goniasteridae** page 220
	Interbrachial septum partially calcified; tube foot discs with spicules; papular areas extensive.	**Oreasteridae** page 358
9	Abactinal plates covered by smooth skin or granules; marginal plates usually prominent.	**Ophidiasteridae** page 292
	Abactinal plates form paxillae with slender glassy spinelets; marginal plates small, inconspicuous, an odd interradial marginal plate in each series.	**Chaetasteridae** page 218
10	Skin bears numerous to single spines or none; superomarginals form a prominent margin to the body.	**Asteropseidae** page 210

(Modified from Liao and Clark, 1995.)

Linckia laevigata, Wakatobi, Indonesia (J. Fromont).

ACANTHASTERIDAE

Acanthasteridae is a small family of only one genus confined to the Indo-Pacific from east Africa to western Central America. There are currently two recognised species, both of which found in Australian waters in depths of less than 30 m.

Acanthaster planci, Montebello Islands, Western Australia [top] (J. Keesing); feeding on a colony of *Acropora* coral, Great Barrier Reef, Queensland [bottom] (J. Fromont).

GENUS *ACANTHASTER* Gerrais, 1841

Acanthasterids have large discs, relatively short and numerous arms, numerous madreporites, and upright two-jawed pedicellariae. The skeleton is reticulate and has prominent spines.

Type species: *Acanthaster planci* (Linnaeus, 1758) (Indo-Pacific)
Two species are recognised worldwide, both of which are found in shallow Australian waters; *A. brevispinus* (page 96) and *A. planci* (page 98).

Acanthaster planci, Ko Tao, Thailand (M. Jackson).

SHORT-SPINED CROWN-OF-THORNS SEASTAR

Acanthaster brevispinus

Fisher, 1917

Western Australian Museum (WAMZ99826).

DESCRIPTION

Like *Acanthaster planci* this species is multi-armed, usually with 14–16 short arms arising from a very large disc; **abactinal surface of disc bears numerous skin-covered spinelets, 2 mm or less in length**, commonly ending in a slightly expanded three-edged pointed tip; the **spines of the arms are more widely spaced**, 5–10 mm in length; the spines are not articulated to an elevated column as in *A. planci*; the abactinal surface has a spaced, fine granulation extending part way up the spines; furrow spines 2–3, shorter than the length of the plate (longer in *A. planci*); **abactinal pedicellariae are thick, round-tipped** so that the pedicellariae may be nearly as broad as high (longer and more slender in *A. planci*).

MAXIMUM SIZE

R/r = 90/51 mm (R = 1.76r).

COLOUR

Uniform dark red.

FOOD

Scallops, which are trapped under the arched disc. Disc remains arched while feeding; also everts the stomach on benthic algal biofilms and on corals (*Cycloseris* and *Leptoseris*) (Birtles, personnal communication). Yuasa et al. (2017) reported the species feeding on soft corals and in particular *Dendronephthya* species.

Great Barrier Reef, Queensland [top] (J. Lucas); abactinal granules [bottom], Muséum National d'Histoire Naturelle (MNHNIE201474).

REPRODUCTION

Development is through planktotrophic bipinaria and then brachiolaria larvae, which metamorphose after about four weeks.

HABITAT, DEPTH AND AUSTRALIAN DISTRIBUTION

On sand or rubble, 18–50 m. Inner continental shelf off Ningaloo Reef, north-western Australia and inter-reefal areas off north Queensland, from off Townsville to the Capricorn Group, and Elizabeth Reef, Tasman Sea.

TYPE LOCALITY

Sulu Archipelago, the Philippines.

FURTHER DISTRIBUTION

Philippines, Japan and Seychelles (western Indian Ocean).

REFERENCES

Fisher, 1919: 442–444; pl. 117, pl. 118 fig. 1; pl. 131 figs 6, 6a–d.

Lucas and Jones, 1976: 409–412 (food).

Yuasa et al., 2017: 1009–1010 (food).

CROWN-OF-THORNS SEASTAR
Acanthaster planci
(Linnaeus, 1758)

Western Australian Museum (WAMZ99861).

DESCRIPTION
Adults are large, 30–40 cm diameter but up to 70 cm; the disc is large and 15–18 (8–21) fairly short arms radiate from it; the disc and arms are covered with a soft skin that also covers the **stout spines which are up to 30 mm long and articulated on a pedicel 5–20 mm** high so that the total length is up to 50 mm; the spines have a razor sharp triangular tip and venom glands are contained in the skin covering them; marginal plates are not obvious; furrow spines are in a fan of four (sometimes five), two large flattened central spines and 2–3 very small ones on either side; there are two subambulacral spines, one behind the other.

MAXIMUM SIZE
R/r = 350/200 mm (R = 1.75r).

COLOUR
Usually blue-grey with a reddish tint where the respiratory papulae are expanded. The spines are red to orange.

FOOD
Coral, particularly plate and staghorn *Acropora*, and *Montipora*. The stomach is everted through the mouth and spread over the coral, digesting an area as large as its

body disc in 4–6 hours. Although *Pocillopora* coral is a preferred food item colonies are vigourously defended by *Trapezia* crabs.

REPRODUCTION

Spawning takes place in Australia from November to January. The sexes are separate and a female can release up to 60 million small eggs (100–200 µm). Development is through planktotrophic bipinnaria larvae, which swim near the surface, changing into a brachiolaria before settling after 2–3 weeks. After metamorphosis, the juveniles feed on encrusting coralline algae on coral rubble for 4–6 months until they are ~10 mm diameter and change to a diet of coral polyps. Sexual maturity is reached after 1.5–2 years at a diameter of about 200 mm.

HABITAT, DEPTH AND AUSTRALIAN DISTRIBUTION

On coral reefs, 0–65 m (generally less than 30 m); they favour lagoons and windward reef slopes below the surf zone and are less often found on reef flats, often hidden by day. In Australia *A. planci* is found from Ningaloo Reef, WA around northern Australia and the Great Barrier Reef, Queensland south to the Solitary Islands and Lord Howe Island, Tasman Sea.

TYPE LOCALITY

Goa, India.

Abactinal spine characteristics, Western Australian Museum (WAMZ99861).

FURTHER DISTRIBUTION

Found throughout the tropical Indo-Pacific from east Africa to the Red Sea, South-East Asia, north to Taiwan and Japan, south to the Kermadec Islands, and east to Hawaii and Central America (Panama). Four molecular clades have been distinguished within the species range, one occurring in the Red Sea, two in the Indian Ocean and one in the Pacific.

REFERENCES

Blake, 1979: 313 (genus).
Fabricius et al., 2010: 14–16.
Haszprunar and Spies, 2014: 272–279.
Haszprunar et al., 2017: 1–10.
MacNeil et al., 2017: 7.
Marsh and Slack-Smith, 2010: 190–193.
Moran, 1986: 379–480.
Moran, 1990: 95–96 (summary of data).
Potts, 1981: 55–86.
Pratchett et al., 2000: 36 (*Trapezia*).
Pratchett, 1999: 171 (disease).
Sweatman, 2008: R598–R599.
Vogler et al., 2008: 696–698.
Vogler et al., 2012: 1–8.
Wilkinson, 1990: 93–172.
Wilkinson and MacIntyre, 1992: 51–122.
Yamaguchi, 1974: 139–146.
Yamaguchi, 1977b: 283–296.

REMARKS

The crown-of-thorns seastar has become notorious for its destruction of coral over wide areas of the tropical Indo-Pacific and Australian coral reefs when populations reach excessive numbers. Reviews by Potts (1981) and Moran (1986, 1990) provide early overviews of the phenomenon while Wilkinson (1990) and Wilkinson and MacIntyre (1992) cover later research. More recent research addresses monitoring strategies and management plans (Sweatman, 2008), and potential causes for the outbreaks (Fabricius et al., 2010).

Using molecular techniques, Vogler et al. (2008) demonstrated that *Acanthaster planci* is a species complex composed of

Puncture wounds from the venomous spines cause intense pain and multiple stings can cause recurrent vomiting. Embedded spines must be surgically removed to prevent later complications. Dampier Archipelago, Western Australia (C. Bryce).

Cocos (Keeling) Islands [top] (C. Bryce); the 'electric blue' variety of the northern Indian Ocean [bottom], Sulawesi, Indonesia (C. Carbillet).

four evolutionary lineages with restricted ranges, with northern and southern Indian Ocean sister groups, and Pacific and Red Sea clades more weakly related. Vogler et al. (2012) reported on detailed studies on the genetic structure and paleohistory of the Indian Ocean sister species. Haszprunar and Spies (2014) reviewed existing species names as potential names for the four clades, and highlighted the need to locate type material, and fresh collect from type localities. They determined the northern Indian Ocean 'electric blue' colour variety to conform to *A. planci*, and the southern Indian Ocean species to *A. mauritiensis*, although the latter is still considered to be a synonym of the former species in WoRMS (Mah, 2019). Haszprunar et al. (2017) identified that more work was required on the Red Sea species, still to be named, and the Pacific Ocean species, proposing *A. solaris* (Schreber, 1793) as a potential species name for the latter. The type locality for *A. solaris* was originally thought to be the Strait of Magellan, Chile but this is considered an error (Haszprunar and Spies 2014; Mah, 2019). This clade includes all specimens examined from tropical Australia (WA, NT and Queensland). Haszprunar et al. (2017) recommended DNA analysis of specimens under study is essential until such time as unequivacol Linnean names are available for the Pacific and Red Sea species, and at this stage there has not been a formal redescription or diagnosis of *A. solaris*. However, some authors have adopted the name *Acanthaster* cf. *solaris*, for example in population studies on the Great Barrier Reef (MacNeil et al., 2017). WoRMS (Mah, 2019) has not adopted the resurrected species name *A. solaris* for the Pacific species, and this position is followed here.

ARCHASTERIDAE

Archasteridae is a small family with a single genus in the Indo-Pacific region. Two species are found in Australian waters.

Storm-wracked *Archaster angulatus*, Cottesloe, Western Australia (L. Marsh).

GENUS *ARCHASTER* Müller and Troschel, 1840

Archasterids differ from astropectinids in having tube feet with sucking discs and blunt marginal spines. The abactinal plates are similar to paxillae and arranged in oblique transverse rows on each side of the carinal series of larger plates. The superomarginal plates are almost completely vertical in alignment.

Type species: *Archaster typicus* Müller and Troschel, 1840 (Indo-Pacific)

Three species occur worldwide, two of which are found in shallow Australian waters; *A. angulatus* (page 103) and *A. typicus* (page 106).

ANGULAR SAND STAR

Archaster angulatus

Müller and Troschel, 1842

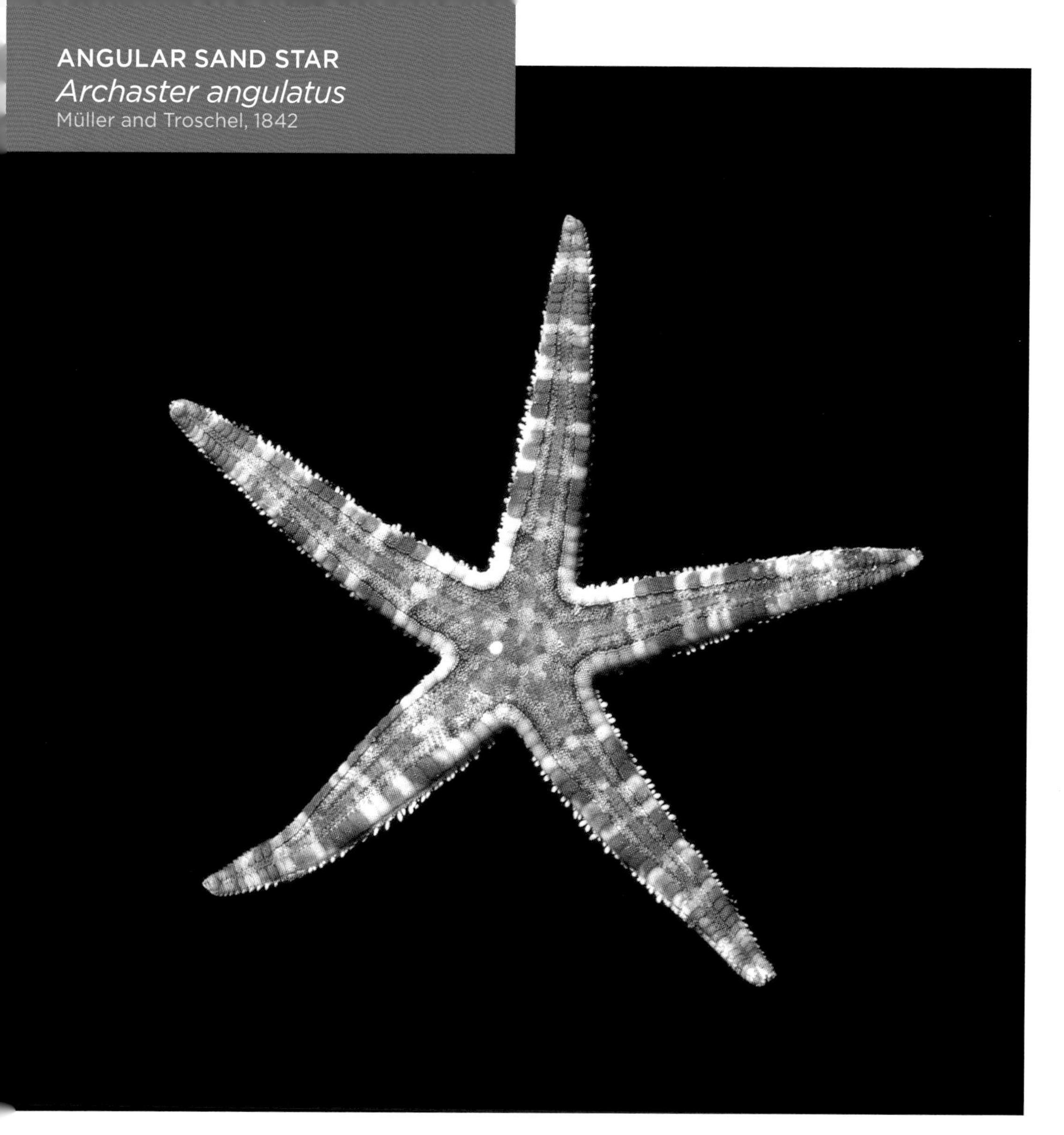

Dampier Archipelago, Western Australia (B. Wilson).

DESCRIPTION

Small disc, five long tapering arms, smooth in outline, tube feet with sucking discs; mid-radial (carinal) row of paxillae-like plates conspicuously enlarged and regular, the paxillae-like plates covered by cylindrical, round-tipped spinelets, so closely packed as to appear granuliform superficially, lateral paxillae-like plates in distinct angled rows of six; superomarginals covered by papilliform spinelets; **inferomarginals armed with 2–5 short spatulate spines or enlarged squamiform spinelets at the outer end of the plate, appressed or hardly projecting**; usually two actinal plates in each series; mostly three furrow spines per plate backed by two rows of subambulacral spines and sometimes an embedded pedicellaria.

MAXIMUM SIZE

R/r = 150/20 mm (R = 7.5r).

Pseudo-copulation, Cockburn Sound, Western Australia [top] (P. Baker) and Geographe Bay, Western Australia [bottom] (S. Morrison).

COLOUR

Light beige or yellow-brown, variegated or indistinctly banded with pink, blue-grey, orange, brown or purple. Actinal surface cream.

FOOD

Fine organic particles in the sediment.

REPRODUCTION

Pseudo-copulation occurs as in *Archaster typicus* and has been observed in Geographe Bay, WA during November and December (Coleman, personal communication) and a number of south-west sites from October to January (Lawrence et al., 2010; Keesing et al., 2011); juveniles were seen in January (Coleman, personal communication). Partly or fully spawned individuals were found in January at Port Beach and Whitford Rock near Perth, WA (Lawrence et al., 2010).

HABITAT, DEPTH AND AUSTRALIAN DISTRIBUTION

Buried in sand among seagrass, in moderately sheltered waters and in sand among corals in back reef areas of coral reef, intertidal to 50 m; common. Found on all coasts of WA westwards from Doubtful Islands Bay on the south coast, common in Geographe Bay, Cockburn Sound and off beaches near Perth, at the Houtman Abrolhos and north-western coast to Broome. Also found in the NT and Queensland south to Hayman Island (Whitsunday Group).

Inferomarginal characteristics, Western Australian Museum (WAMZ17513).

TYPE LOCALITY

'Mauritius' (may be erroneous according to Sukarno and Jangoux (1977)). Probably Western Australia.

FURTHER DISTRIBUTION

It has a disjunct distribution, Mozambique, the Philippines and the Ryukyu Islands.

REFERENCES

Clark, H.L., 1938: 75–77; pl. 17 fig. 2 (as *Archaster laevis*).

Jangoux, 1972: 163–172; figs 1–9.

Keesing et al., 2011: 1163–1173 (food, reproduction).

Lawrence et al., 2011: 1–9 (reproduction).

Sukarno and Jangoux, 1977: 830–834; figs 1, 2b, 3b; pl. 5 fig. 5; pl. 6 figs 1–6.

TYPICAL SAND STAR
Archaster typicus
Müller and Troschel, 1840

Solomon Islands (B. Wilson).

DESCRIPTION
Superficially similar to *Astropecten* but tube feet have suckers; arms shorter than those of *A. angulatus*; abactinal paxillae-like plates in the form of a rounded cone, radial paxillae-like plates larger than lateral ones, paxilla-like spines elongate, club-shaped; superomarginals have a uniform cover of papilliform spinelets; each **inferomarginal plate has a single prominent flattened marginal spine**, not sharply pointed; inferomarginals covered in flat, round-ended spinelets; three furrow spines, the middle is longest, curved; two elongate, flattened, round-ended subambulacral spines, often with an upright two-jawed pedicellaria between them.

MAXIMUM SIZE
R/r = 95/15 mm (R = 6.3r).

COLOUR

Variable, depending on the substrate. Those from mud are generally grey with darker arm bands, those from sand are cream with irregular brown arm bands.

FOOD

Fine organic particles in the sediment. Has been observed everting stomach over substrate.

REPRODUCTION

Spawning has been reported in February, May, September and October in Indonesia. During the reproductive season males mount the females and are induced to release sperm by the spawning female, fertilisation is external. Eggs are small (190 μm), developing into a planktotrophic bipinnaria larva, which swims near the bottom, changing to a brachiolaria after six days and attaching to the bottom. Metamorphosis to juveniles takes place after 14–24 days.

Motupore Island, Papua New Guinea [top] (R. Steene); inferomarginal plate characteristics [bottom], Western Australian Museum (WAMZ17551).

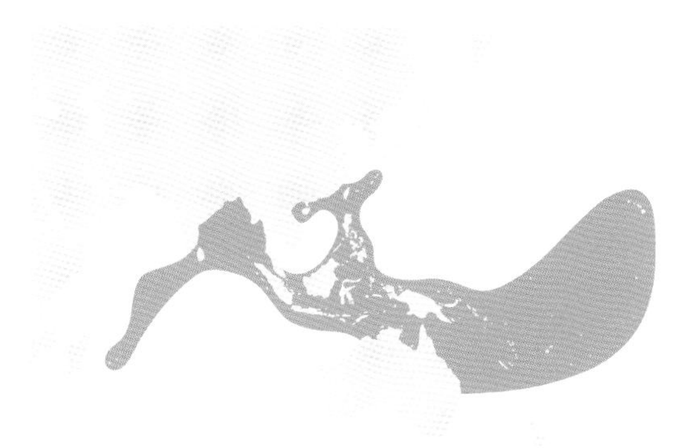

HABITAT, DEPTH AND AUSTRALIAN DISTRIBUTION

On sand or mud, 0–3 m. From Ashmore Reef, Timor Sea to Angourie, NSW.

TYPE LOCALITY

Sulawesi, Indonesia.

FURTHER DISTRIBUTION

Common through South-East Asia, from Singapore and Indonesia to the Andaman Islands north to Hainan Island (southern China) and Taiwan and the Ryukyu Islands, east to Caroline Islands, New Caledonia, Fiji and Samoa.

REFERENCES

Chen and Run, 1991: 257–260 (reproduction).

Mukai et al., 1986: 371–372.

Run et al., 1988: 247–253 (reproduction).

Sukarno and Jangoux, 1977: 822–829; figs 2a, 3a; pl. 4 figs 1, 5.

Yamaguchi, 1977b: 283–296 (reproduction).

ASTERINIDAE

Asterinids have five or more arms, and some species are fissiparous. The body can be pentagonal to stellate to having finger-like arms (*Nepanthia* and *Pseudonepanthia*). Minute marginal plates are present and in some species the inferomarginals form a prominent margin to the body. The abactinal skeleton is composed of imbricated plates bearing grouped or single spinelets that are granuliform to acicular with single or multiple points. Fasciculate pedicellariae are sometimes present and actinal interradial plates are in regular transverse or longitudinal series.

There is great diversity in reproductive methods within several genera of this family, from broadcast spawning to intragonodal brooding. Genetic and reproductive studies have revealed a number of cryptic species in the genera *Cryptasterina* and *Meridiastra* resulting in many taxonomic changes. Byrne (2006) has reviewed life history, diversity and evolution in the Asterinidae while O'Loughlin and Waters (2004) made a systematic revision of the genera of Asterinidae based on molecular and morphological studies. The genus *Ailsastra* O'Loughlin and Rowe was added in 2005.

The family is cosmopolitan with 24 genera of which 13 genera and 42 species are known from Australia in depths of less than 30 m. H.L. Clark (1946) recognised nine genera and 38 species of Asterinidae many of which have been synonymised or assigned to different genera, while new species and genera have been described.

Pseudonepanthia troughtoni, King George Sound, Western Australia (C. Bryce).

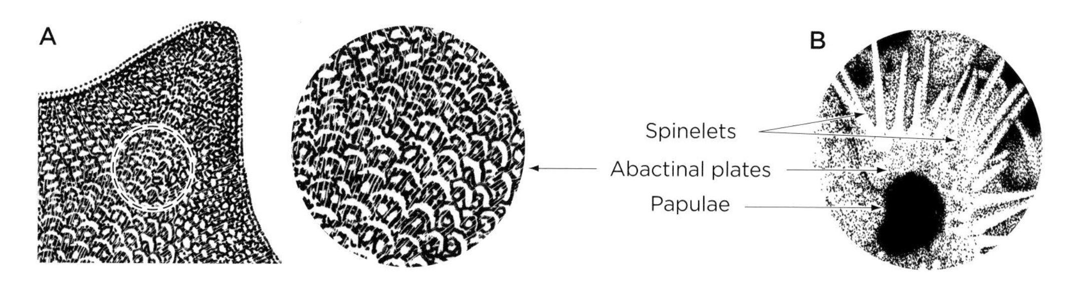

Figure 10: Crescent-shaped, imbricating abactinal plates, found in species of Asterinidae. A) The cresentic abactinal plate formation of *Meridiastra occidens*. B) Abactinal plate spinelets (e.g. *Indianastra sarasini*).

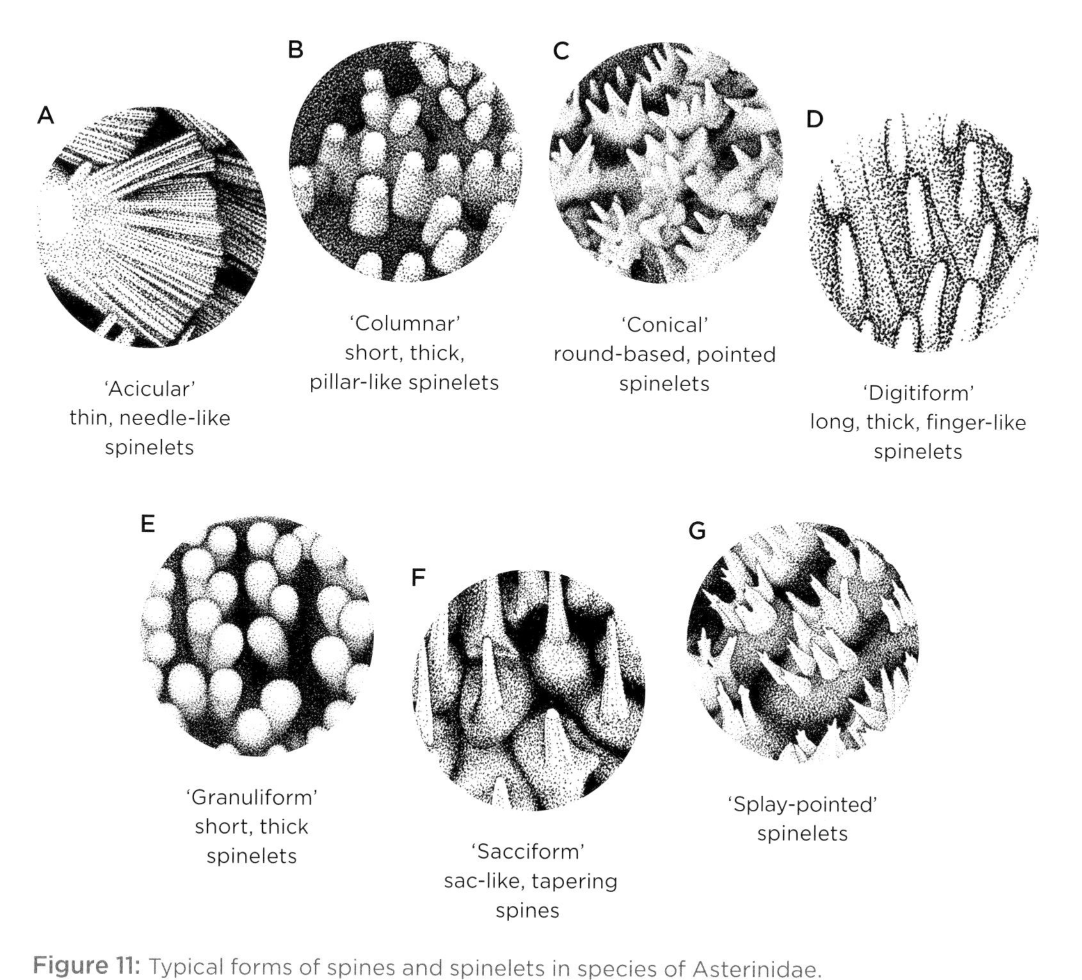

Figure 11: Typical forms of spines and spinelets in species of Asterinidae.
A) Acicular (e.g. *Paranepanthia grandis*). B) Columnar (e.g. *Meridastra occidens*).
C) Conical (e.g. *Aquilonastra cassini*). D) Digitiform (e.g. *Parvulastra dyscrita*).
E) Granuliform (e.g. *Meridiastra gunnii*). F) Sacciform (e.g. *Disasterina longispina*).
G) Splay-pointed (e.g. *Aquilonastra corallicola*).

KEY TO THE GENERA OF ASTERINIDAE IN SHALLOW AUSTRALIAN WATERS

1	Arms narrow at base, subcylindrical or digitiform.	2
	Arms broad at base or not discrete, not subcylindrical or digitiform.	3
2	Arms 4–7 more or less flat actinally with distinct to slight marginal edge; abactinal plates irregular in arrangement, secondary plates present, pedicellariae and crystal bodies present; inferomarginal plates project slightly; superambulacral, transactinal and superactinal plates present.	***Nepanthia*** page 164
	Arms 4–10 (five in shallow water species), not flat actinally, without a marginal edge; abactinal plates fairly regular in arrangement, secondary plates absent; pedicellariae and crystal bodies absent; inferomarginal plates do not project; superambulacral plates present, transactinal plates rare, superactinal plates absent.	***Pseudonepanthia*** page 189
3	Body flat, interradial area very thin; abactinal spinelets sacciform; interradial area supported by many long, thin interior articulating projections from abactinal and actinal plates.	***Anseropoda*** page 115
	Body slightly to very convex, interradial area not very thin.	4
4	Lacking superambulacral, transactinal and superactinal plates.	5
	One or both superambulacral and superactinal plates present.	6
5	Arms 5–8, generally not discrete, not fissiparous, margin usually straight or slightly incurved (deeply incurved in one species); abactinal spinelets granuliform.	***Meridiastra*** page 148
	Arms 6–7, discrete, narrow basally, fissiparous; spinelets stout, glassy, conical or splay-tipped.	***Ailsastra*** page 113

6	Arms five, discrete; abactinal and actinal plates with dense subpaxilliform tufts of glassy acicular subsacciform spinelets and spines; superambulacral plates absent.	***Paranepanthia*** page 171
	Abactinal spinelets and actinal spines not in dense subpaxilliform tufts; superambulacral plates present.	**7**
7	Abactinal spinelets glassy, acicular or sacciform, or splay-tipped sacciform or long thin conical.	**8**
	Abactinal spinelets opaque, granuliform or digitiform, or short thin to thick columnar.	**11**
8	Arms 5–9; abactinal plates with numerous firmly attached glassy spinelets; superomarginal spinelets same as those on abactinal plates.	***Aquilonastra*** page 117
	Abactinal plates with very fine glassy spinelets, numerous to none, weakly attached; superomarginal spinelets few to none.	**9**

Meridiastra oriens, Lord Howe Island, Tasman Sea (N. Coleman).

9	Arms five; abactinal plates loosely contiguous leaving non-plated spaces; distal abactinal plates in series, perpendicular or zigzag to margin; superomarginal plates small, not in regular series; inferomarginal plates with stout sacciform spinelets.	***Disasterina*** page 139
	Abactinal plates imbricate, always contiguous; distal abactinal plates not in perpendicular series to margin; superomarginal plates in distinct regular series; inferomarginal spinelets not stout sacciform.	**10**
10	Arms five, body low; abactinal plates small, thin, deeply notched for papulae, up to six series along each side of arms; inferomarginal spinelets acicular, in dense integument covered tufts.	***Indianastra*** page 144
	Arms five, body high; abactinal plates large, thick, shallow notches or crescentiform, up to three series along side of arms; inferomarginal spinelets discrete, not in dense integument covered tufts.	***Tegulaster*** page 193
11	Arms five, form subpentagonal to medium armed stellate; large papular spaces; numerous papulae and secondary plates per space; spinelets granuliform or digitiform; up to three spines on mid-interradial actinal plates.	***Patiriella*** page 181
	Form pentagonal to subpentagonal; small papular spaces, few papulae and secondary plates per space.	**12**
12	Arms five; superomarginal plates typically in prominent series, longitudinally subrectangular; inferomarginal plates project narrowly to define the margin; abactinal papular spaces predominantly with one papula; mid-arm and distal actinal interradial plates commonly with one spine.	***Cryptasterina*** page 134
	Arms five; superomarginal and inferomarginal plates typically subequal; inferomarginal plates project prominently to define margin; abactinal papular spaces commonly with more than one papula; mid-arm and distal actinal interradial plates commonly with two spines.	***Parvulastra*** page 173

(Modified from O'Loughlin and Waters, 2004.)

GENUS *AILSASTRA* O'Loughlin and Rowe, 2005

Species of this genus are small, stellate seastars with R < 16 mm and predominantly five arms, although one species is fissiparous with 5–7 arms. The arms are convex abactinally and flat actinally, and the inferomarginals project to form an acute ventrolateral margin. The carinal plates, which are present for most of the arm, are doubly papulate and imbricated by adjacent plates laterally so that the carinals appear sunken. Other abactinal plates closely imbricate and are not notched, and there are no secondary plates. Papulae on the arms are one per papular space in a longitudinal series. Spinelets are glassy, thick or thin, conical, sacciform or subsacciform, pointed or splay-tipped and occur in clusters or single or double series across the proximal plates. Plates have prominent crystal bodies. The oral plates are distinctively small and constricted distally with up to six oral spines. Superambulacral and superactinal plates are absent from the type species.

Type species: *Ailsastra paulayi* O'Loughlin and Rowe, 2005 (Indonesia)

Six species are recognised, all from the Indian Ocean and adjacent areas, but only one species, *A. heteractis* (page 113), has been found in Australia.

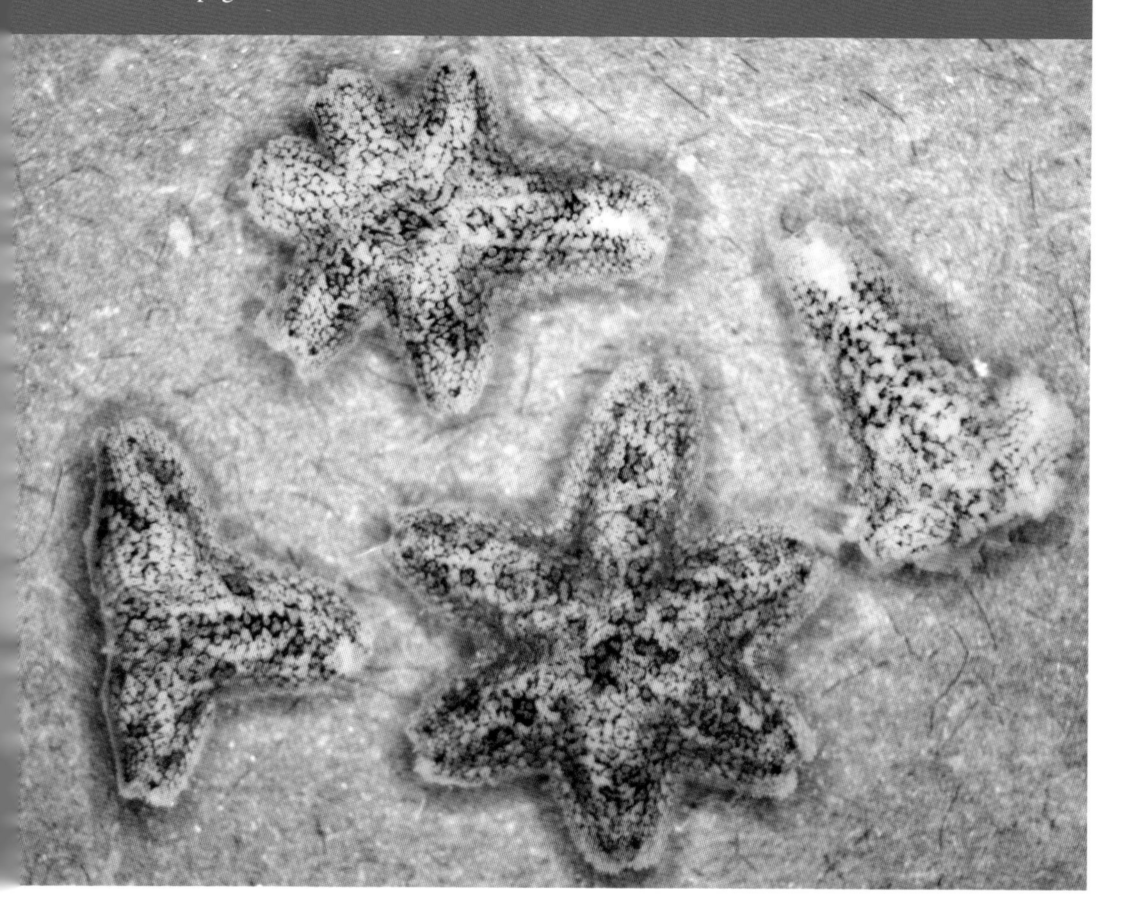

Ailsastra heteractis, Heron Island, Queensland (M. Byrne).

HETERACTIS SEASTAR

Ailsastra heteractis

(H.L. Clark, 1938)

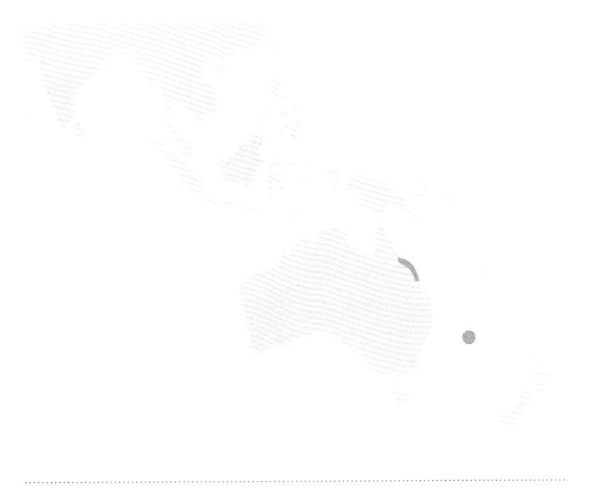

DESCRIPTION

Six or seven discrete high arms with narrow bases; blunt distally; multiple inconspicuous madreporites; fissiparous; integument evident; arm lengths often unequal; carinal series of plates imbricated laterally by adjacent plates, carinals doubly papulate except basally, other arm plates slightly indented for single large papula; single longitudinal series of papulae on each upper side of arms, four longitudinal series across mid-arm; papular spaces large, disc not delineated; spinelets glassy, stout conical to subsacciform, splay-tipped, up to eight per plate in single or double series across plates or in apical tufts, up to 150 μm long; superomarginal plates with up to four spinelets, inferomarginals with up to 10 longer splay-tipped webbed spinelets; single large proximal actinal pore in each interradius; interradial actinal plates with 1–3 subsacciform elongate pointed spines; 3–4 proximal furrow spines; 3–4 subambulacral spines; 2–4 adradial actinal spines; 4–6 oral spines; up to three suboral spines on small oral plates; pedicellariae absent.

MAXIMUM SIZE

R/r = 7.6/4.3 mm (R = 1.75r).

COLOUR

Salmon pink or light orange. Actinal surface whitish.

REPRODUCTION

Paired interradial gonopores on the abactinal surface. Reproduces asexually by fissipary.

HABITAT, DEPTH AND AUSTRALIAN DISTRIBUTION

On the undersurface of rocks on sand, intertidal to 20 m; rare. Broadhurst Reef (18°56'S) and Heron Island (23°24'S), Great Barrier Reef, Queensland; Lord Howe Island and Middleton Reef, Tasman Sea. Australian endemic.

TYPE LOCALITY

Neds Beach, Lord Howe Island, Tasman Sea.

FURTHER DISTRIBUTION

Australian endemic.

REFERENCES

Clark, H.L., 1938: 152; pl. 22 fig. 5 (as *Asterina heteractis*).

Clark, H.L., 1946: 133 (as *Asterina heteractis*).

O'Loughlin and Rowe, 2005: 188; figs 1, 2e, 3e, 5d–f.

O'Loughlin and Waters, 2004: 11, 13, 15 (as *Asterina heteractis*).

GENUS *ANSEROPODA* Nardo, 1834

Seastars of this genus have a body that is very thin with the margin variably curved. They have 5–18 arms that are short or not discrete, and broadly rounded or pointed. The arms have narrow raised radial areas with single papulae that are scattered or in a single longitudinal series, sometimes ringed by secondary plates. The abactinal plates are thin, in longitudinal and oblique series, and are not notched. Plates have subpaxilliform glassy, sacciform spinelets that are few or in a tuft. Actinal plates are in longitudinal and oblique series and actinal spines are few to numerous per plate, glassy and sacciform. Superambulacral plates are lacking, superactinal plates may be present or absent. Abactinal and actinal interradial plates meet internally by long, thin articulating projections that are extensive in the interradial area. Adambulacral plates have a fan of furrow spines that are usually webbed, and there are several subambulacral spines in a straight or curved series.

Type species: *Anseropoda placenta* (Pennant, 1777) (north-eastern Atlantic)

There are 16 species worldwide but only one, *A. rosacea* (page 116), is found in shallow water in Australia and a second one, *A. macropora*, in deeper continental shelf waters.

Anseropoda rosacea, Geraldton, Western Australia (Department of Fisheries).

ROSEATE SEASTAR
Anseropoda rosacea
(Lamarck, 1816)

DESCRIPTION
This large, very flat asterinid is easily recognised from a photograph because of its many arms (usually 16) and extreme thinness.

MAXIMUM SIZE
R = 130 mm (R = ~1.14r).

COLOUR
Grey with black speckling or mottled orange-pink. Actinal surface is a creamy white with rust colour along the ambulacral groove. Dried specimens are cream, speckled with orange-red.

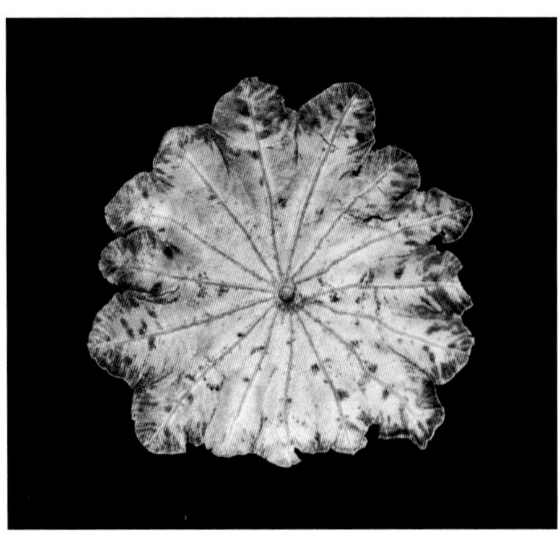

Western Australian Museum (WAMZ8130).

FOOD
Small echinoderms, crustaceans and molluscs. Also a scavenger and sometimes taken in rock lobster pots where it is attracted to the bait.

HABITAT, DEPTH AND AUSTRALIAN DISTRIBUTION
Found just below the surface of the sand, 0–145 m. From the south coast of WA and Dongara on the mid-west coast, around northern Australia to southern Queensland and off the east coast of Tasmania.

TYPE LOCALITY
Not recorded.

FURTHER DISTRIBUTION
Bay of Bengal, Indonesia, Hainan Island (southern China) and south-western Japan.

REFERENCES
Clark, A.M. and Rowe, 1971: 65.

O'Loughlin and Waters, 2004: 4–5 (key), 7, 10; figs 7b, 8b.

REMARKS
A five armed species of *Anseropoda, A. macropora,* has been collected at ~100 m depth on the outer shelf, off north-western Australia.

GENUS *AQUILONASTRA* O'Loughlin, 2004

Species of this genus have a stellate form and five discrete arms (5–8 in fissiparous species) that are broad at the base and tapering and rounded distally. The interradial margin is deeply incurved and flat actinally but is high and convex abactinally. The abactinal plates are in a longitudinal series and are not perpendicular to the margin. The papulate areas are extensive, with the papulae predominantly single and large, and in longitudinal series along the sides of the arms. The abactinal plates have crystal bodies, and the abactinal spinelets and actinal spines are predominantly fine, glassy, conical or sacciform, or splay-tipped sacciform. They occur in numerous (10–40 per plate) bands or tufts. The actinal plates are in longitudinal, sometimes oblique series. The superambulacral plates are present for the whole arm, a part of the arm, or absent in pedomorphic (adults with juvenile characteristics) species. Superactinal plates are present.

Type species: *Aquilonastra cepheus* (Müller and Troschel, 1842) (Indo-Pacific)

Other *Aquilonastra* species occurring in Australia are: *A. anomala* (H.L. Clark, 1921); *A. byrneae* O'Loughlin and Rowe, 2006; *A. corallicola* (Marsh, 1977); *A. coronata* (von Martens, 1866); *A. rosea* (H.L. Clark, 1938); and *A. scobinata* (Livingstone, 1933). Two new species from the Kimberley, north-western Australia were described by O'Loughlin and Bribiesca-Contreras (2015). A further 22 species are found predominantly in the Red Sea to northern Indo-Pacific (O'Loughlin and Rowe, 2006).

The genus name is from the Latin *aquilonalis* ('northern') and *astrum* ('star').

Aquilonastra byrnea, One Tree Island, Queensland (M. Byrne).

Aquilonastra cepheus, Scott Reef, Western Australia (C. Bryce).

KEY TO THE SPECIES OF *AQUILONASTRA* IN SHALLOW AUSTRALIAN WATERS

1	Arms five, equal or subequal; madreporite single.	2
	Arms more than five (up to nine); shape asymmetrical; always more than one madreporite.	6
2	Abactinal plates paxilliform, spinelets in dense clusters.	3
	Abactinal plates not paxilliform, spinelets not in dense clusters.	4
3	Spinelets pencil-like, clusters often crescentic; gonopores actinal, clearly visible; R < 18 mm; south-eastern Australia.	***A. scobinata*** page 132
	Spinelets long, splay-tipped, in dense round clusters; gonopores abactinal, sometimes obscured; R < 17 mm; south-western Australia.	***A. rosea*** page 131
4	Arms rounded distally; spinelets short, conical to digitiform, splay-tipped; R < 15 mm; north-eastern Australia.	***A. byrneae*** page 123
	Arms tapered, elongate.	5

5	Spinelets of one form, rugose, long, thin tapering to a fine point; $R < 30$ mm; Sri Lanka, Indonesia to northern Australia.	***A. cepheus*** page 125
	Spinelets of two forms, long thick digitiform or pointed spinelets surrounded by short thin, pointed spinelets on the scattered paxilliform plates; central area of disc delimited by a ring or pentagon of elevated plates; $R < 37$ mm; Japan to northern Australia.	***A. coronata*** page 129
6	$R \leq 7$ mm	**7**
	$R > 7$ mm	**8**
7	Not fissiparous, six arms; pedicellariae absent; abactinal spinelets splay-pointed; $R \leq 6$ mm; north-western Australia.	***A. alisonae*** page 120
	Fissiparous, 5–6 arms; pedicellariae absent; abactinal spinelets mostly conical; $R \leq 7$ mm; north-western Australia.	***A. cassini*** page 124
8	Fissiparous, 2–8 arms; abactinal spinelets of one form, long, commonly splay-tipped; $R < 12$ mm; Indo-Pacific, Christmas Island to Hawaii.	***A. anomala*** page 121
	Fissiparous, 6–7 arms; plates of basal part of arms with two forms of spinelets; on the apex of each plate are 1–3 thick digitiform blunt spinelets with thin, conical pointed spinelets peripherally; $R < 16$ mm; West Pacific, Indonesia, Cocos (Keeling) Islands, Capricorn Group, Great Barrier Reef.	***A. corallicola*** page 127

Holotypes of *Aquilonastra alisonae* [left] (WAMZ26200) and *A. cassini* [right] (WAMZ26198), Western Australian Museum.

ALISON'S SEASTAR

Aquilonastra alisonae

O'Loughlin and Bribiesca-Contreras, 2015

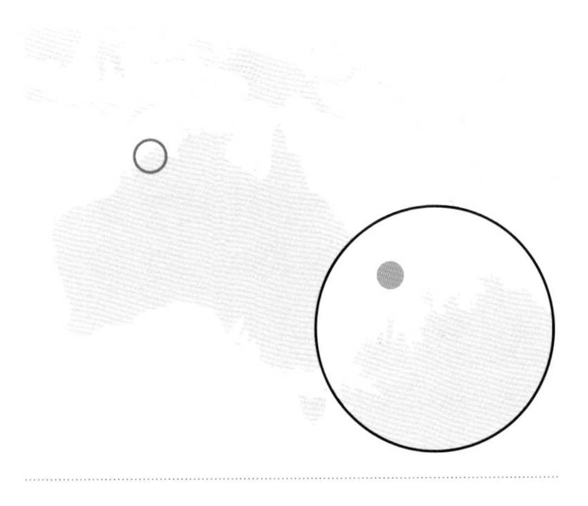

DESCRIPTION

A small asterinid seastar, non-fissiparous, disc not discretely demarcated; single large, conspicuous madreporite (only detected on holotype) above junction of bases of two arms; no abactinal or actinal gonopores detected; stellate seastar with six slightly unequal short arms, arms wide basally, merging at bases, tapering to a rounded end distally; lacking doubly-papulate carinal plates; interradial junction of arms sub-acute, arms low convex abactinally, flat actinally, margin acute; glassy convexities on cleared abactinal and actinal plates; up to about 10 predominantly abactinal splay-pointed spinelets per abactinal plate, commonly in two transverse series across proximal edge and middle of plate; pedicellariae absent; spinelets not clustered into groups on plates; superomarginal and inferomarginal plates subequal; inferomarginal plates not projecting noticeably; internal superambulacral and superactinal plates present; abactinal disc plates imbricate irregularly with those of arms; upper arm plates irregular in form, not in regular series, no carinal series of plates, most upper arm plates widely concave proximally; single papula per papular space; rare secondary plates; four prominent longitudinal series of papulae across arms, short lower series of smaller papulae along arms, up to 11 papulae per series along upper arm, irregular; superomarginal plates with up to eight splay-pointed spinelets per plate in two series; actinal surface with six oral, long thin, slightly cylindrical to spatulate spines per plate; five digitiform furrow spines; four digitiform to splay-pointed subambulacral spines; seven conical to splay-pointed actinal spines on central plates; up to nine, predominantly splay-pointed inferomarginal spines; actinal interradial plates in slightly irregular longitudinal and oblique series.

MAXIMUM SIZE

R/r of holotype = 6.2/3 mm (R = 2r).

COLOUR

Not recorded.

HABITAT, DEPTH AND AUSTRALIAN DISTRIBUTION

Rock substrate, fore reef, 0–6 m. Long Reef, Kimberley, WA. North-western Australian endemic.

TYPE LOCALITY

Long Reef, Kimberley, WA.

FURTHER DISTRIBUTION

Australian endemic.

REFERENCE

O'Loughlin and Bribiesca-Contreras, 2015: 27–40; figs 1–4.

ANOMALUS SEASTAR

Aquilonastra anomala

(H.L. Clark, 1921)

Lord Howe Island, Tasman Sea (N. Coleman).

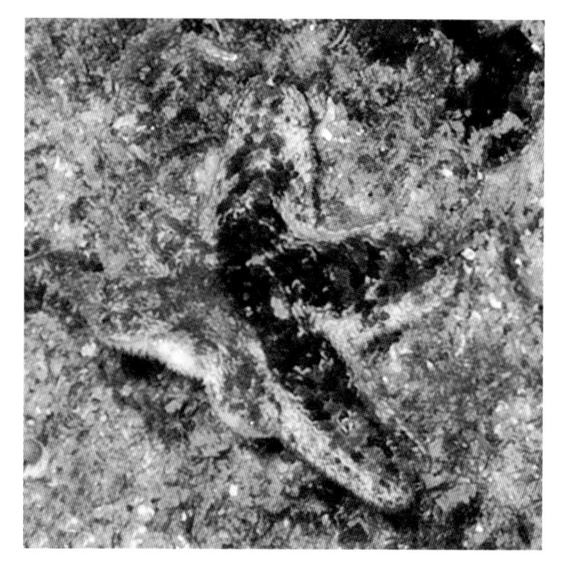

Singapore (R. Tan).

DESCRIPTION

A small fissiparous *Aquilonastra* species with 2–8 stellate arms, (usually 7–8) of unequal length, often four or five short and three long arms; pedicellariae with differentiated conical valves, larger than spinelets, sometimes present interradially; abactinal surface without carinal series of plates, upper arms with two irregularly arranged longitudinal series of singly papulate plates that are domed and angled over papulae more than notched; rarely two papulae per plate, some secondary plates intergrade with primary plates; spinelets long, thin, conical-pointed to predominantly splay-tipped sacciform; up to 20 spinelets projecting over the surface of each proximal abactinal plate, predominantly in a transverse double band, rare clustering of spinelets; distal interradial plates with up to about six splay-tipped spinelets; superomarginal and inferomarginal plates subequal; superomarginals with up to eight spinelets, inferomarginals with up to 16 slightly larger spinelets; all marginal spinelets are long, splay-tipped sacciform; interradial actinal spines glassy, thin, conical, pointed, spinous, 3–4 (up to seven) in transverse series or tufts; six furrow spines, five subambulacral spines; seven oral spines, five suboral spines; up to four inconspicuous madreporites; the relatively long commonly splay-tipped abactinal spinelets are distinctive.

MAXIMUM SIZE

R/r = 12.5/~7 mm (R = 1.5–2r).

COLOUR

Variable, often green or red-brown variegated with white and rusty red or orange, sometimes all red.

FOOD

Compound ascidians and algal substrate, biofilm, possibly sponges.

REPRODUCTION

Reproduces asexually by fissiparity. Gonopores are abactinal inferring pelagic development but sexual reproduction is unknown.

REMARKS

The validity of *Aquilonastra anomala* as distinct from *A. burtoni* and *A. cepheus* has often been questioned but O'Loughlin and Waters (2004) have shown conclusively that *A. anomala* is a valid species.

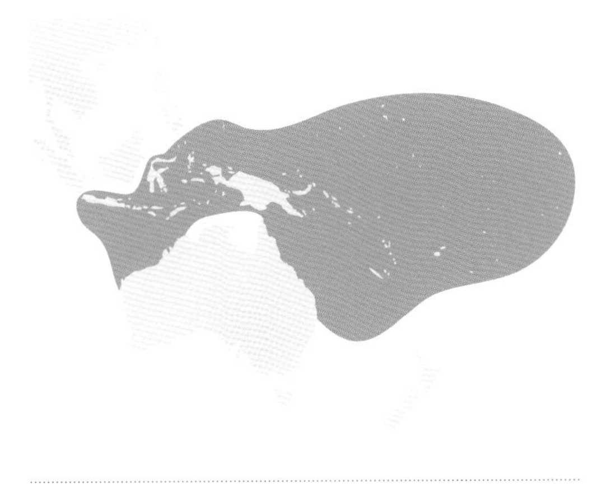

HABITAT, DEPTH AND AUSTRALIAN DISTRIBUTION

Found on the underside of rock or coral slabs and in crevices of rock or coral, sometimes among algae, usually on coral reefs, intertidal to 56 m on the outer reef slope. In WA, Scott Reef, Cape Leveque and Long Reef, Kimberley, Ashmore Reef, Timor Sea, also known from the NT and Great Barrier Reef, Queensland to 24°S, Christmas Island and Lord Howe Island.

TYPE LOCALITY

Murray Island, Torres Strait, Queensland.

FURTHER DISTRIBUTION

Christmas Island (Indian Ocean), Palau, Caroline Islands and Fiji, Marshall Islands to Hawaii.

REFERENCES

Clark, H.L., 1921: 95–96; pl. 7 fig. 8; pl. 23 fig. 5; pl. 26 figs 2–3.

O'Loughlin and Rowe, 2006: 260–261; figs 1, 2a, 7a.

O'Loughlin and Waters, 2004: 4–5 (key), 11, 13–15; fig. 1.

Yamaguchi, 1975: 20–21 (food).

BYRNE'S SEASTAR
Aquilonastra byrneae
O'Loughlin and Rowe, 2006

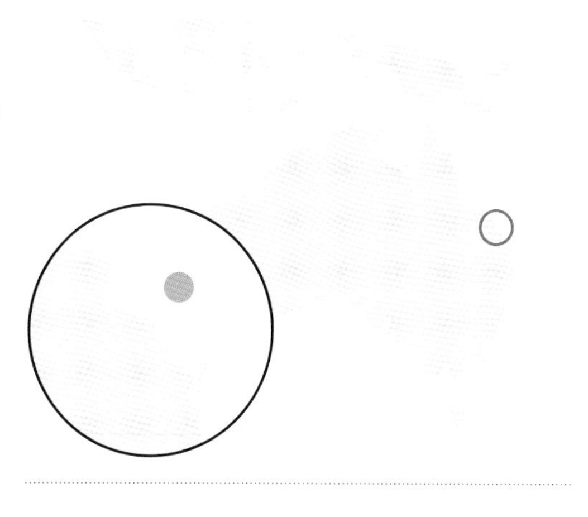

DESCRIPTION
A non-fissiparous five armed species of *Aquilonastra* with arms narrow basally, broadly to narrowly rounded distally; gonopores actinal, interradial pairs close, near margin; proximal carinal plates doubly papulate and with four tufts of spinelets with 3–5 spinelets per tuft, numerous proximal secondary plates, up to three per space; single large papula per space, rarely two; spinelets small, short from conical to digitiform in shape, not splay-tipped; up to 16 spinelets across the projecting edge of proximal abactinal plates, 1–2 on bare distal mid-plate, up to 12 on proximal surface of distal interradial plates; superomarginal plates smaller than inferomarginals; actinal interradial plates with five (up to eight) thick bluntly conical spines, seven furrow spines, six subambulacral, seven oral and six suboral spines.

MAXIMUM SIZE
R/r = 15/8 mm (R = 1.9r).

COLOUR
Predominantly mottled green to greenish brown. Disc sometimes cream, surrounded by a dark brown ring. Actinal surface cream with green patches.

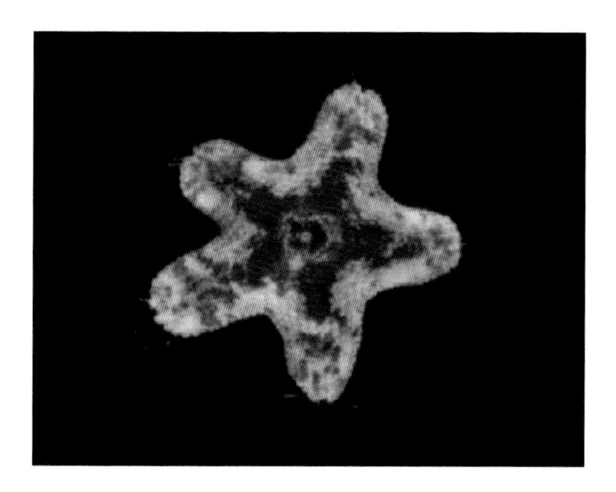

One Tree Island, Queensland (M. Byrne).

REPRODUCTION
Gonopores open on the actinal surface. Protandric hermaphrodite with eggs ~400 µm in diameter, negatively buoyant with benthic lecithotrophic development. Late stage oocytes and mature testes are found in June; winter spawning is nearly complete by October.

HABITAT, DEPTH AND AUSTRALIAN DISTRIBUTION
On the underside of coral slabs in the rubble zone — algal pavement interface in the low intertidal on coral reefs. Only known from One Tree and Tryon islands in the Capricorn Group, Great Barrier Reef, Queensland.

TYPE LOCALITY
One Tree Island, Queensland.

FURTHER DISTRIBUTION
Possibly Guam.

REFERENCES
Byrne, 2006: 245, 248, 251; table 2.

O'Loughlin and Rowe, 2006: 259–260, 270; figs 1, 2e, 5b, 7c.

CASSINI ISLAND SEASTAR
Aquilonastra cassini
O'Loughlin and Bribiesca-Contreras, 2015

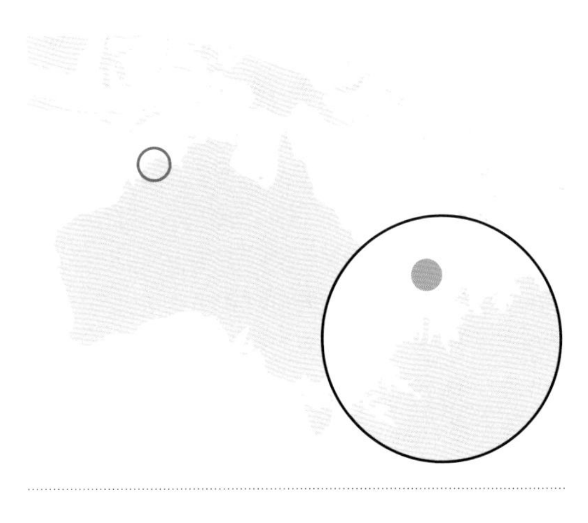

DESCRIPTION

A small asterinid seastar, asymmetrical, 5–6 sub-equal arms, sub-digitiform, narrow and rounded distally, slightly widened basally, arms merging at bases; interradial junction of arms sub-acute, arms low convex abactinally, slightly convex actinally, margin acute; disc not discretely demarcated; three small, inconspicuous, abactinal madreporites (on holotype); fissiparous; no abactinal or actinal gonopores detected; pedicellariae not detected; inferomarginal plates significantly larger than superomarginal plates; inferomarginal plates project noticeably at margin; internal superambulacral and superactinal plates present; abactinal disc plates imbricate irregularly with those of upper arms; upper arm plates proximal to disc irregular; lacking secondary plates; regular carinal series of plates along some upper arms only, up to 11 carinal plates per series, each carinal plate with paired single papular spaces; plates on sides of arms with single papular space, single series of papulae adcarinally on arms, up to 15 per series; short lower series of smaller papulae along arms, four prominent longitudinal series of papulae across arms; abactinal spinelets predominantly conical; disc with 3–6 spinelets per plate, each carinal plate with cluster of 3–5 spinelets on crown of plate; adcarinal plates with up to seven spinelets across angled plate, proximal and distal interradial abactinal plates with predominantly four (3–6) spinelets, conical to splay-pointed; superomarginal plates with 4–5 splay-pointed spinelets per plate; actinal surface with 3–4 oral spines; 0–1 sub-oral, digitiform, slightly spatulate spine, with minute distal spinelets; 3–4 proximal furrow spines; 2–3 subambulacral spines; four predominantly conical actinal spines with pointed distal ends; up to about 11 predominantly splay-pointed inferomarginal spines; actinal plates in slightly longitudinal and oblique series.

MAXIMUM SIZE

R/r of holotype = 7/2.5 mm (R = 2.8r).

COLOUR

Not recorded.

HABITAT, DEPTH AND AUSTRALIAN DISTRIBUTION

Rock substrate, 1.8–3 m. Cassini Island, Kimberley, WA. North-western Australian endemic.

TYPE LOCALITY

Cassini Island, WA.

FURTHER DISTRIBUTION

Australian endemic.

REFERENCE

O'Loughlin and Bribiesca-Contreras, 2015: 27–40; figs 5–8.

SKIRTED SEASTAR

Aquilonastra cepheus

(Müller and Troschel, 1842)

Scott Reef, Western Australia (C. Bryce).

DESCRIPTION

A stellate asterinid with convex disc and usually five broadly based, tapering blunt-ended arms; the proximal carinal plates are doubly papulate with some large secondary plates; abactinal plates have a crescentic ridge bearing slender, sometimes radiating, glassy, rugose long, thin, not splay-tipped spinelets, up to 32 in double series on large radial plates, up to 24 on proximal abactinal plates, up to 15 on mid-interradial plates; actinal plates each with a cluster of 4–8 short, thick, conical spinelets; 6–8 furrow spines and

Heron Island, Queensland (J. Keesing).

4–8 subambulacral spines in a fan on each adambulacral plate; 6–9 oral spines; 2–8 suboral spines; papulae single, on disc and upper part of arms; madreporite single; non-fissiparous; gonopores abactinal.

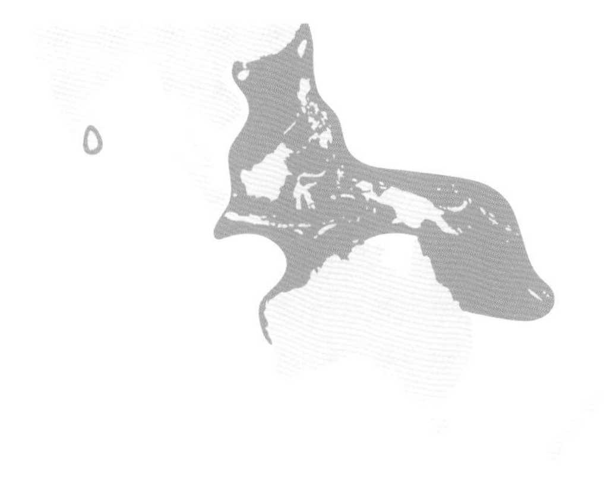

MAXIMUM SIZE

R/r = 30/12 mm (R = 1.8–2.5r).

COLOUR

Variable, often shades of grey, green or brown blotched with darker shades, sometimes with patches of dull red, occasionally all red.

FOOD

Encrusting sponges and compound ascidians.

REPRODUCTION

Gonopores are abactinal, indicating that it is a broadcast spawner.

HABITAT, DEPTH AND AUSTRALIAN DISTRIBUTION

Usually on coral reefs, under rock or coral slabs, or on sand among coral, 0–70 m. From the Houtman Abrolhos, WA to Darwin, NT and on the offshore reefs off WA and Ashmore Reef, Timor Sea. Rare on the North West Shelf and Kimberley coast. In Queensland on the Great Barrier Reef south to Lady Elliott Island. Common on the Great Barrier Reef.

TYPE LOCALITY

Jakarta, Indonesia.

FURTHER DISTRIBUTION

Sri Lanka, Indonesia, the Philippines and South China Sea to Papua New Guinea.

REFERENCES

Clark, A.M. and Rowe, 1971: 68.

O'Loughlin and Rowe, 2006: 270–271; figs 1, 2f–h, 5c–d.

Sloan, 1980: 73.

Abactinal plate characteristics, Western Australian Museum (WAMZ6886).

Juvenile, Hibernia Reef, Timor Sea (C. Bryce).

CORAL-DWELLING SEASTAR
Aquilonastra corallicola
(Marsh, 1977)

Cocos (Keeling) Islands (N. Coleman).

DESCRIPTION

A fissiparous species of *Aquilonastra* usually with 6–7 arms (up to eight); arms elongate, wide basally, tapered, narrowly rounded distally; some abactinal pedicellariae with 2–3 differentiated thick valves; abactinal gonopores; plates of the abactinal surface of arms irregular in arrangement, a few carinal plates are double papulate; 1–3 papulae per papular space, papulae in two longitudinal series along each side of arm; 0–2 secondary plates per papular space proximally; upper arm plates paxilliform with two forms of spinelets, 1–3 thick digitiform, blunt, apical spinelets per plate, up to 12 thin conical pointed spinelets peripherally on each plate, spinelets not splay-tipped; superomarginal plates each with up to six thin pointed spinelets, inferomarginals with 1–4 thick blunt apical spinelets, up to eight smaller, thin conical pointed spinelets; actinal interradial spines five per plate, rugose, tall digitiform and short, thin, pointed; five furrow and five subambulacral spines, six oral and six suboral spines; the number of spines is to some extent size dependent; up to five madreporites.

MAXIMUM SIZE

R/r = 16/7 mm (R = 2.4r).

COLOUR

Mottled red and yellow.

FOOD

The type specimens were observed to be feeding on *Acropora echinata* at 10 m.

REPRODUCTION

Reproduces asexually by fission, sexual reproduction is not known but gonopores are abactinal, inferring pelagic development. Mature ovaries were found in a specimen with R = 6 mm in February in Palau.

HABITAT, DEPTH AND AUSTRALIAN DISTRIBUTION

On coral reefs, among living branching corals and under dead coral rubble, 0–32 m. In Australia it is only known from One Tree Island, Great Barrier Reef, Queensland and from the Cocos (Keeling) Islands.

TYPE LOCALITY

Palau, Caroline Islands.

FURTHER DISTRIBUTION

Guam and Saipan, Mariana Islands, Indonesia, Fiji, Singapore.

REFERENCES

Marsh, 1977: 271–275; tables 1–2; figs 8–9 (as *Asterina corallicola*).

Oguro, 1983: 224–225; figs 7–9, 14 (as *Asterina corallicola*).

O'Loughlin and Rowe, 2006: 258–259, 273; figs 1, 4b, 5g–h, 8c.

O'Loughlin and Waters, 2004: 11, 13–15.

Abactinal plate characteristics, paratype, Western Australian Museum (WAMZ1704).

REMARKS

This species is distinguished from *Aquilonastra anomala* by the narrower more elongate arms, two forms of spinelets on paxilliform upper arm plates, and its mottled red and yellow colour.

CROWNED SEASTAR
Aquilonastra coronata
(von Martens, 1866)

Kimberley, Western Australia (B. Wilson).

DESCRIPTION

A five-rayed non-fissiparous species of *Aquilonastra* with broad based tapering arms, narrowly rounded distally; numerous abactinal pedicellariae commonly present with 2–4 curved, pointed valves; abactinal surface is uneven with scattered high paxilliform plates; abactinal spinelets of two forms on paxilliform plates, up to six long, thick digitiform spinelets or pointed spinelets on apex of each plate with about 10 short, thin pointed spinelets (not splay-tipped) around margin of plate; the central area of the disc is clearly differentiated with an elevated ring or pentagon of five radial and five interradial plates; doubly papulate carinal plates lacking; upper arm with irregular zigzag series of small primary and secondary plates lacking papulae; 0–4 secondary plates per papular space on upper arms intergrade with primary plates, 1–4 papulae per papular space; superomarginals smaller than inferomarginals with up to 10 thin, rugose, pointed (not splay-tipped) spinelets per plate; inferomarginals with up to four long, thick digitiform or pointed spinelets on apex of each plate, with about 12 thin spinelets around margin of plate; actinal interradial plates usually carry five (up to seven) conical, pointed, often thick and thin spinelets on each plate; seven furrow spines, seven subambulacral spines, oral plates bear eight peripheral and eight suboral spines; madreporite single.

MAXIMUM SIZE

R/r = 37/14.8 mm (R = 2–2.5r).

COLOUR

Variable, often mottled light and dark olive green, with blotches of dark red; beige, mottled with brown and red; orange or completely rust red.

REPRODUCTION

Eggs are large (420 µm) and development is through non-feeding pelagic brachiolaria larvae.

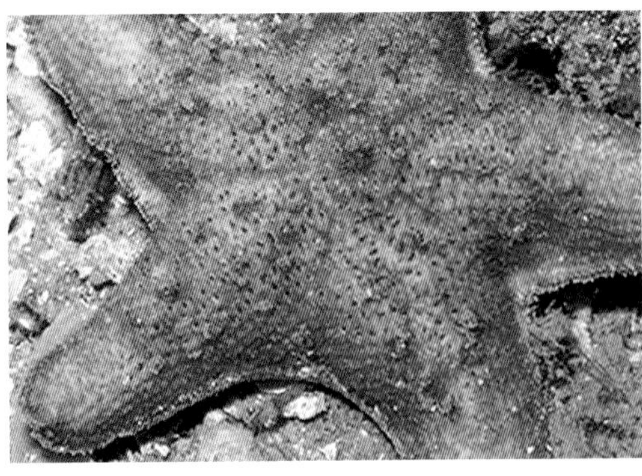

Bali, Indonesia [top] (N. Coleman); Singapore [bottom] (R. Tan).

HABITAT, DEPTH AND AUSTRALIAN DISTRIBUTION

Found on the underside of boulders on mud or silty sand, 0–10 m. From the Dampierland Peninsula, north of Broome, WA, to Darwin, NT and Thursday Island, Torres Strait.

TYPE LOCALITY

Batjan, Molucca Islands and Larentuka, Flores, Indonesia.

FURTHER DISTRIBUTION

Singapore, Indonesia and the Philippines and north to Taiwan and southern Japan, east to Ponape in the Caroline Islands and west to Sri Lanka. Common in suitable habitats.

REFERENCES

Clark, H.L., 1938: 145–148; pl. 12 fig. 1.
Clark, A.M. and Rowe, 1971: 68; pl. 9 fig. 6.
Fisher, 1919: 411–416; pl. 115 figs 1–3; pl. 116 figs 1–2; pl. 131 figs 4, 4a, 5, 5a.
O'Loughlin and Rowe, 2006: 273–274; figs 1, 2l, 8d.

REMARKS

Aquilonastra coronata is the muddy shore counterpart of *A. cepheus*, which is found on coral reefs.

ROSY SEASTAR
Aquilonastra rosea
(H.L. Clark, 1938)

DESCRIPTION

A small secretive species, non-fissiparous, stellate with five arms, wide basally, tapering to a narrowly rounded tip; carinal plates are doubly papulate proximally, abactinal plates not in two 'fields'; abactinal plates have a high raised rounded column or ridge with a concave papular notch; longitudinal series of plates and papulae along sides of arms, predominantly a single large papula per space, few secondary plates; disc delineated by five wide radial plates and five short interradial plates; spinelets sacciform, long, splay-tipped, spinelets in paxilliform dense radiating tufts of more than 30 per plate; superomarginal and inferomarginal plates with subequal tufts of spinelets; actinal interradial plates have about 20 spines in tufts, spines are sacciform, splay-tipped; there are seven furrow spines backed by 12 subambulacral spines (in tufts); oral plates bear nine oral and 12 suboral spines.

MAXIMUM SIZE

R/r = 17/9 mm (R = 1.4–2.6r).

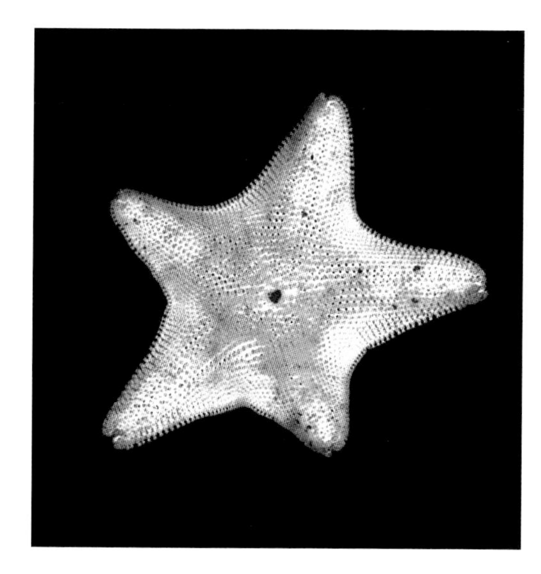

Ludlow, Western Australia (N. Coleman).

COLOUR

Cream, mottled with rose pink or orange-red, sometimes all pink, usually there is a circle of raised cream plates around the centre of the disc. Actinal surface is cream, mottled with pink.

HABITAT, DEPTH AND AUSTRALIAN DISTRIBUTION

Cryptic, clinging to the underside of rocks in shallow water, in the weed mat on rock platforms or in crevices of weed-covered dead coral, 0–110 m. From Cape Naturaliste to the Houtman Abrolhos, WA. South-western Australian endemic.

TYPE LOCALITY

Rottnest Island, WA.

FURTHER DISTRIBUTION

Australian endemic.

REFERENCES

Clark, H.L., 1938: 161–162; pl. 22 fig. 8 (as *Paranepanthia rosea*).

Clark, H.L., 1946: 137 (as *Paranepanthia rosea*).

O'Loughlin and Rowe, 2006: 281; figs 1, 3f, 6h.

O'Loughlin and Waters, 2004: 4–5 (key) (transfer to new genus *Aquilonastra*).

RASP FILE SEASTAR
Aquilonastra scobinata
(Livingstone, 1933)

Cape Nelson, Victoria (L. Altoff).

DESCRIPTION

Five pointed arms, wide basally; body stellate non-fissiparous, moderately elevated; the carinal plates form a zigzag series proximally; the abactinal plates are crescentric, imbricated, form three longitudinal rows mid-radially, and these plates are larger and less crescentic than those of the disc; papulae are generally single, plates are not notched; in the lower interradial region the plates are nearly circular, only slightly imbricated; on the crescentic surface of the abactinal plates are 10–20 delicate, minute splay-tipped sacciform spinelets; the superomarginal plates each have a tuft of up to 10 splay-tipped spinelets, the inferomarginals have about 20 spinelets and project beyond the superomarginals; actinal interradial plates each carry up to seven slender spinelets; there is a pair of uncalcified areas distal to the oral plates in each interradius; the furrow spines are in curved combs of 4–6 fairly long slender spinelets; the subambulacral spinelets are in oblique combs of 4–6; the oral plates each have six slender peripheral spinelets and 3–6 suboral spinelets.

MAXIMUM SIZE

R/r = 18/8 mm (R = 2.25r).

COLOUR

Pale cream with dark flecks to light or dark brown.

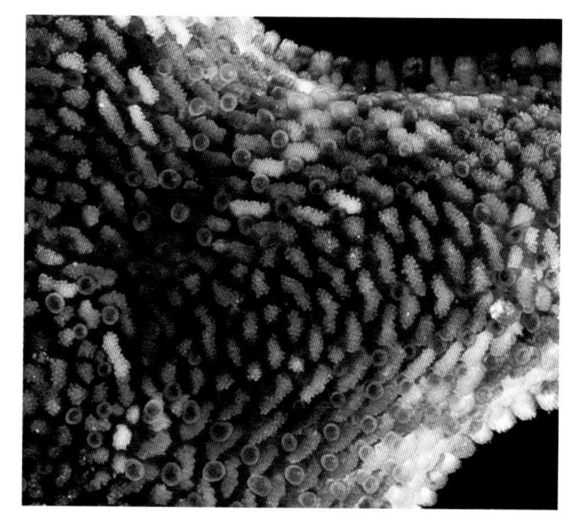

Carinal plate characteristis [top], Cape Nelson, Victoria; Port Fairy Abalone Farm, Victoria [bottom] (L. Altoff);

REPRODUCTION

Hermaphrodites with a preponderance of female gonads. Gonopores are on the actinal surface. Eggs are ~300 μm in diameter, brooded below the adult, developing lecithotrophically.

HABITAT, DEPTH AND AUSTRALIAN DISTRIBUTION

Under rocks and in crevices in the lower intertidal zone to shallow subtidal. Circum-Tasmania and Victoria to Port MacDonnell, SA. Southern Australian endemic.

TYPE LOCALITY

Tasmania.

FURTHER DISTRIBUTION

Australian endemic.

REFERENCES

Dartnall, 1970a: 1–4; figs 1–3.

Livingstone, 1933: 1–2; pl. 5 figs 9–12, 15.

O'Loughlin and Rowe, 2006: 283; figs 1, 3i, 6k.

O'Loughlin and Waters 2004: 4–5 (key), 13 (transfer to new genus *Aquilonastra*).

GENUS *CRYPTASTERINA* Dartnall, Byrne, Collins and Hart, 2003

Seastars of this genus have five arms with a form that is either pentagonal, or have short arms that are rounded distally with an R ≤ 20 mm. The abactinal surface is low convex. Abactinal plates are in longitudinal series, many are U-shaped more than crescentric, and they are deeply notched for papulae. A longitudinal series of predominantly single papulae occurs along the sides of the arms with extensive non-papulate interradial areas. There are few (1–3) papulae per papular space. Abactinal spinelets are granuliform and spaced over the plates. Pedicellariae and crystal bodies are lacking. The superomarginal series are prominent and longitudinally subrectangular. The inferomarginals tend to project narrowly. There is commonly a non-plated actinal area distal to the oral plates and the actinal interradial plates are in an oblique series. Actinal spines are short, conical, usually one per plate, rarely 2–3 distally. There are few, very small, superambulacral plates that never occur as a series. Superactinal plates are sometimes present in multiple plate struts. There is an adradial row of spineless plates next to the adambulacrals.

Type species: *Cryptasterina pentagona* (Müller and Troschel, 1842) (Indo-Pacific)
Three species occur worldwide, two of which are found in shallow Australian waters; *C. hystera* (page 135) and *C. pentagona* (page 137).

(Dartnall et al., 2003; and O'Loughlin and Waters, 2004.)

Cryptasterina pentagona, Mission Beach, Queensland (P. Marsh).

HYSTERA SEASTAR

Cryptasterina hystera

Dartnall and Byrne, 2003

Yeppoon, Queensland (M. Byrne).

DESCRIPTION

Form is pentagonal to sub-pentagonal; five arms, body thick, flat actinally, flattened dome abactinally, acute angle at margin; abactinal surface with papulate areas more extensive than non-papulate areas, few secondary plates; abactinal plates closely imbricate, rarely with more than one papular pore in each papular space; projecting proximal edges of plates mostly crescentric, lobed; metapaxillar ridge low, not prominent, disc not distinct; proximal interradial areas papulate, similar to radial areas; distal interradial areas without papulae; **abactinal spinelets granular**, about 150 µm long, **broader at base with constricted waist and capitate spiny tip**; carinal and disc plates carry 7–12 spinelets in an irregular double row; distal interradial plates with a cluster of

5–7 spinelets; superomarginal plates elongate in the radial axis carrying a double row of up to eight spinelets; inferomarginal plates with a projecting flange of about seven webbed spinelets form the disc margin; actinal plates and spines in regular series from furrow to margin; interradial bare area proximal to the oral plates carries 2–5 loosely imbricate plates often without spines, other actinal plates with a tapered conical spinelet up to 650 µm long, near the edge of the disc a few plates may have two spines; adambulacral plates with furrow spines usually in webbed pairs occasionally three, subambulacral spines one (sometimes two), 500 µm long proximally, shorter distally; five oral spines on each oral plate and one suboral spine; pedicellariae absent; not fissiparous; viviparous, intragonadal brooder.

MAXIMUM SIZE

R/r = 12/8 mm (R = 1.7–1.22r).

COLOUR

Dark olive green, paler on the actinal surface.

FOOD

Probably feeds on the benthic biofilm.

REPRODUCTION

Cryptasterina hystera has ovotestes with a mosaic of oogenic and spermatogenic areas. Reproduction takes place during September to November. Eggs are large, gold or orange colour (440 µm diameter) and larvae develop within the gonad but are capable of independent development through a buoyant lecithotrophic brachiolaria, which takes 16 days to develop to the juvenile stage. In aquaria juveniles brooded within the gonad emerged from the abactinal gonopore with two pairs of tube feet, a mouth and functional digestive system.

HABITAT, DEPTH AND AUSTRALIAN DISTRIBUTION

Under rocks in the mid intertidal in mangrove habitats and under small intertidal rocks on mud and muddy sand. Only known from the Queensland coast between Yeppoon and Bargara (24°49'S), near Bundaberg. Southern Queensland endemic.

TYPE LOCALITY

Statue Bay, Queensland.

FURTHER DISTRIBUTION

Australian endemic.

REFERENCES

Byrne et al., 2003: 285–294 (reproduction).

Dartnall et al., 2003: 1–14.

O'Loughlan and Waters, 2004: 18–19; figs 1, 11a.

PENTAGONAL SEASTAR
Cryptasterina pentagona
(Müller and Troschel, 1842)

Lindeman Island, Queensland (G. Edgar).

DESCRIPTION

Arms five; form pentagonal to slightly stellate; body thick, flat orally, flattened dome abactinally, acute angle at margin; abactinal surface with papulate areas more extensive than non-papulate areas; few secondary plates; abactinal plates closely imbricate, rarely with more than one papular pore per space; projecting proximal edges of plates mostly crescentic, lobed; metapaxillar ridge low, not prominent; disc not distinct; distal inter-radial areas non-papulate; abactinal spinelets granular, about 150 µm long, broader at base with constricted waist and capitate spiny tip; carinal and disc plates carry 7–12 spinelets in an irregular double row; distal interradial plates with a cluster of 5–7 spinelets; superomarginal plates elongate in the radial axis with a double row of up to eight spinelets; inferomarginals with a projecting flange of about seven webbed spinelets forming the disc margin; actinal plates and spines in regular series from furrow to margin; proximal interradial plates with 2–5 loosely imbricate plates, often without spines; other actinal plates with a tapered conical spinelet up to 650 µm long; adambulacral plates with furrow spines usually in webbed pairs, occasiaonlly three per plate; subambulacral spines usually single, 700 µm long proximally, 300 µm long distally; oral plates with five oral spines and one suboral spine; no pedicellariae; small superambulacral plates and a few superactinal plates present.

MAXIMUM SIZE

R/r = 22/13 mm (R = 1.7r).

COLOUR

Mottled dull brown and green red or black, often darker in the centre, with a five armed star pattern. Actinal surface dull brownish green. Tube feet pale straw with off-white suckers.

FOOD

Algae and detritus, biofilm.

REPRODUCTION

Gonopores are on the abactinal surface. Sexes are separate and spawn in October-November. The yolky eggs are large (413 µm diameter) and development is through non-feeding planktonic brachiolaria larvae. Development to the settled juvenile stage takes six days at 30°C or nine days at 23°C; the mouth opens three weeks after settlement.

HABITAT, DEPTH AND AUSTRALIAN DISTRIBUTION

On or under rocks in the high intertidal. Found on the east coast of Queensland from Mission Beach to Airlie Beach. Dartnall et al. (2013) reported that the Mission Beach population was lost following Cyclone Larry (2006) but adults were found there in April 2012 following Category 5 Cyclone Yasi in February 2011. The seastars were in shallow tide pools on a rock/sand platform with large boulders at Bingil Bay north of Mission Beach.

TYPE LOCALITY

The type locality for *Cryptasterina pseudoexigua* (now *C. pentagona*) is Airlie Beach, central Queensland. The type locality of *C. pentagona* (Müller and Troschel, 1842) is Java, Indonesia.

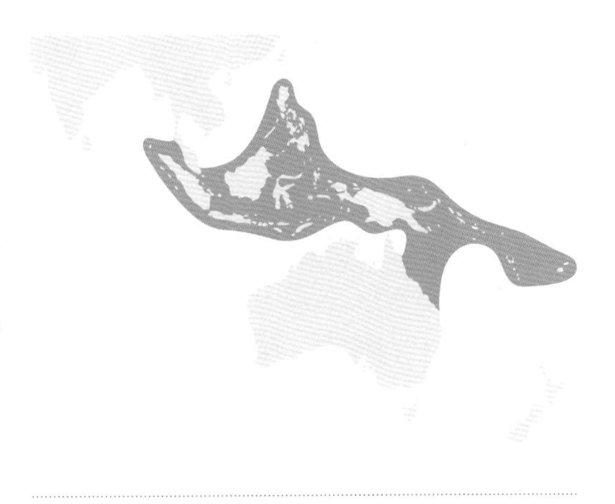

FURTHER DISTRIBUTION

Indonesia, the Philippines, Papua New Guinea, Vanuatu, Solomon Islands, Palau, Malaysia. Further distribution uncertain as many records may refer to *Cryptasterina pacifica.*

REFERENCES

Byrne and Barker, 1991: 332–345 (reproduction).

Byrne and Cerra, 1996: 17–26 (reproduction).

Dartnall, 1971: 43–46; pls 3a–b, 4a–b; fig. 1 (as *Patiriella obscura* and *P. pseudoexigua*).

Dartnall et al., 2003: 7–10.

Hart et al. 2003: 1109–1116 (development, as *Patiriella pseudoexigna*).

O'Loughlin and Waters, 2004: 4–5 (key), 18–19; table 2; fig. 2f.

Rowe and Gates, 1995: 41 (as *Patiriella pseudoexigna*).

REMARKS

Cryptasterina pentagona and *C. hystera* are morphologically identical apart from colour and size but have different reproductive methods and are genetically distinct.

GENUS *DISASTERINA* Perrier, 1875

Seastars of this genus have five arms and a thin body covered by a thick integument. Their form is medium to long armed and stellate, with wide based arms tapering strongly to the tips. The seastars are flat actinally, of low height and convex. The abactinal plates are thin, predominantly irregular in shape, size and arrangement. They are loosely imbricate, may or may not be contiguous, thereby leaving non-plated spaces. Distally the plates are in a few irregular longitudinal series, sometimes weakly notched and crescentiform. The interradial area is thin distally and there are small non-papulate plates in perpendicular or zigzag series extending to the margin of the body. Papulae are few and irregular on the arms. Abactinal spinelets are few, rare or completely lacking; they can be glassy, sacciform, short or long. Superomarginal plates are small and not in distinct series. Inferomarginal plates project widely and are loosely contiguous, with a distal fringe of a few large sacciform spinelets. The actinal interradial plates are in oblique series with the proximal actinal interradial areas commonly not plated. The interradial actinal plates have 1–2 long sacciform spines. Superambulacral plates irregularly present mid-arm and distally and not in a complete series. Superactinal plates are present as single plate struts.

Type species: *Disasterina abnormalis* Perrier, 1875 (Indo-West Pacific)

Six species are now included in *Disasterina*, including the type species and one endemic Australian species, *D. longispina* (H.L. Clark, 1938). Species previously assigned to *Disasterina*, *D. leptalacantha* and *D. praesignis* (with junior synonym *D. spinulifera*), were removed to *Tegulaster* by O'Loughlin and Waters, 2004. Two species are found in shallow Australian waters; *D. abnormalis* (page 140) and *D. longispina* (page 142).

(O'Loughlin and Waters, 2004.)

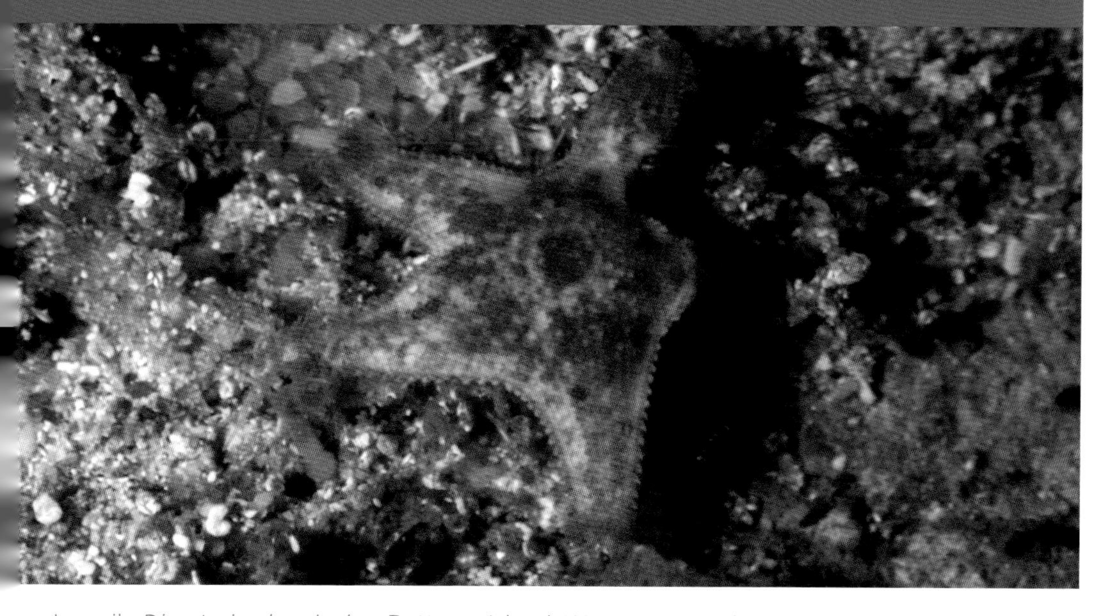

Juvenile *Disasterina longispina*, Rottnest Island, Western Australia (C. Bryce).

ABNORMALIS SEASTAR
Disasterina abnormalis
Perrier, 1875

New Caledonia (P. Laboute).

Milne Bay, Papua New Guinea (R. Steene).

DESCRIPTION

The disc and five arms are slightly elevated, the shape stellate; **the body is covered by a thick skin that obscures the plates, which are visible after drying; the abactinal plates of the disc and top of the arms are small and irregular in shape, often with skin showing between them; plates commonly bare or with a few sacciform spinelets around the disc**; the interradial area is marked by larger plates; the single papulae are scattered over the disc and upper part of the arms; the inferomarginals each have a pair of sacculate, pointed spines (young animals have up to four); the actinal plates are in oblique series and each has a single elongate sacculate spine; there is an uncalcified area, sometimes with a floating plate distal to the oral plates; there are 3–4 webbed furrow spines per plate and a single subambulacral spine, like those of the actinal plates; oral plates carry up to six webbed peripheral spines and a single, long, sacculate suboral spine; superambulacral plates are irregularly present mid-arm and distally; superactinal plates present as single struts.

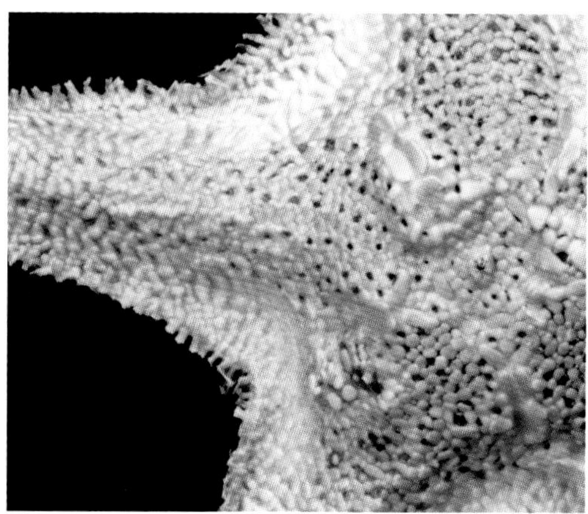

Abactinal plate characteristics, New Caledonia [top] (P. Laboute), Western Australian Museum (WAMZ6754) [bottom].

MAXIMUM SIZE

R/r = 30/12 mm (R = 2–2.5r).

COLOUR

Mottled olive green and white. Marginal spinelets and arm tips violet, or mottled pink and white.

HABITAT, DEPTH AND AUSTRALIAN DISTRIBUTION

On the underside of rocks on coral reefs, 0–2 m. Scott Reef and Rowley Shoals, WA and the Great Barrier Reef, south to the Capricorn Group, Queensland.

TYPE LOCALITY

New Caledonia.

FURTHER DISTRIBUTION

Indonesia, Papua New Guinea, New Caledonia.

REFERENCES

Clark, H.L., 1946: 138–139.

Livingstone, 1933: 7–8; pl. 4 figs 6–7; pl. 5 fig. 13.

O'Loughlin and Waters, 2004: 4–5 (key), 19–20; figs 2g, 10e, 12a.

LONG-SPINED SEASTAR

Disasterina longispina

(H.L. Clark, 1938)

Hopetoun, Western Australia (C. Bryce).

DESCRIPTION

Five arms, stellate, arms taper evenly to a rounded tip; the body is covered by fleshy skin obscuring the plates; the abactinal plates are small, more or less irregular in shape and arrangement, irregularly overlapping; papulae on disc and arms except for the mid-radial line, extending about half way to the margin interradially; **abactinal plates near the madreporite, and sometimes others on the central disc bear short, blunt or pointed spinelets, 1–4 per plate; interradially there may be few or many, mostly longer spinelets, 1–7 per plate; the distal plates of the arms usually have long, slender spines**; crystal bodies present on abactinal plates; the superomarginal plates are round, bare; the inferomarginals form the margin of the body, each with 4–6 elongate spinelets in a more or less vertical row; the actinal plates each bear a single, long sacculate spine and there is sometimes an uncalcified area of membrane near the oral plates; there are 4–5 furrow spines and two, sometimes three long subambulacral spines; the oral plates bear 5–7 webbed marginal spines and 1–3 suboral spines; superambulacral plates irregularly present mid-arm and distally; superactinal plates present as single plate struts.

MAXIMUM SIZE

R/r = 36/12.4 mm (R = 2–2.9r).

COLOUR

Mottled with cream, pink and magenta in varying proportions with scattered spots of rusty red. Actinal surface pale mottled pink.

HABITAT, DEPTH AND AUSTRALIAN DISTRIBUTION

Under boulders or coral slabs on rock platforms, intertidal to 3 m; uncommon. From east of Hopetoun on the south coast to the Houtman Abrolhos, WA. South-western Australian endemic.

TYPE LOCALITY

Rottnest Island, WA.

FURTHER DISTRIBUTION

Australian endemic.

REMARKS

Although the type locality is Rottnest Island, *Disasterina longispina* has not subsequently been found there. Most specimens are from Cape Naturaliste and the Pelsaert Group of the Houtman Abrolhos.

REFERENCES

Clark, H.L., 1938: 158; pl. 21 figs 1–2 (as *Manasterina*).

Clark, H.L., 1946: 139–140 (as *Manasterina*).

O'Loughlin and Waters, 2004: 4–5 (key), 19–20; figs 5f, 10f, 12b (*Manasterina* transferred to *Disasterina*).

Rowe and Gates, 1995: 36 (as *Manasterina*).

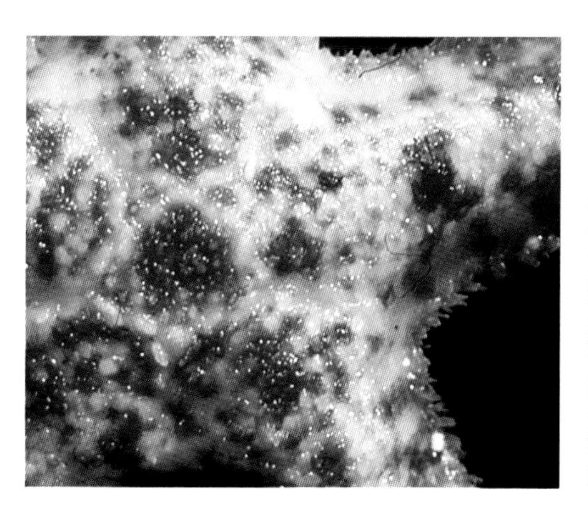

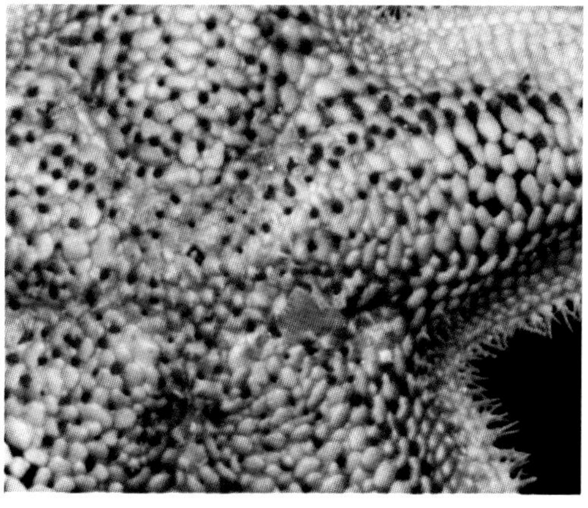

Abactinal plate characteristics, Houtman Abrolhos, Western Australia [left] (C. Bryce), Western Australian Museum [right] (WAMZ6757).

GENUS *INDIANASTRA* O'Loughlin 2004

Seastars of this genus have five arms that are petaloid to subpentagonal. The body is thin, of small size and covered with an integument. The abactinal and actinal interradial plates are in longitudinal series, the abactinal plates are deeply notched for short, crescentiform papulae. Along the sides of the arms there are numerous, regular longitudinal series of plates with papulae. The abactinal spinelets are inconspicuous but fine, glassy and acicular to subsacciform. There can be few to numerous per plate, in tufts or scattered over the plate, but not on ridges or domes, and they are fragile and readily lost. Pedicellariae are sometimes present over the papulae. The superomarignal plates are bare or with a few spinelets, the inferomarginals have distal, sub-paxilliform dense tufts of acicular spinelets, covered by integument. Actinal interradial plates are in longitudinal series, the plates have small clusters of webbed, short sacciform spines. The seastars lack a superambulacral series of plates and have rare single plates distally. The superactinal plates are single plate struts.

Type species: *Indianastra sarasini* (de Loriol, 1897) (Indo-West Pacific)
Asterina lutea H.L. Clark, 1938 and *A. nuda* H.L. Clark, 1921 are junior synonyms of the type species. *Asterina inopinata* (Livingstone, 1933) (and junior synonym *A. perplexa* H.L. Clark, 1938) are now included in *Indianastra.* Two species are found in shallow Australian waters; *I. inopinata* (page 145) and *I. sarasini* (page 146).

The name *Indianastra* is derived from the region of occurrence of the genus (Indo-West Pacific) and the Latin *astrum* ('star').

(O'Loughlin and Waters, 2004.)

Indianastra sarasini, Western Australian Museum (WAMZ6831).

INOPINATA SEASTAR
Indianastra inopinata
(Livingstone, 1933)

DESCRIPTION

Five arms, **body fairly thin and depressed, nearly pentagonal; abactinal plates are crescentic, imbricated, bare except for a few minute glassy spinelets** bordering their crescentic upper edge on the disc and radial areas, and occasionally a central tuft of 1–4 similar spinelets on interradial plates; papular pores are confined to the disc and radial areas with 1–2 rows each side of the median radial plates, no pores in the interradial areas; the margin is formed by the inferomarginal plates, each of which has a circular group of small crowded spinelets, which form a prominent marginal flange; the actinal interradial plates have a curved or straight row of 2–4 stout, sharply pointed spinelets, webbed for half their length; furrow spines are long, slender, in a crescentic fan of 5–6, webbed; there are 2–3 webbed subambulacral spines placed at an angle to the furrow; there are eight webbed oral spines and three webbed suboral spines; superambulacral plates lacking; margin supported by single superactinal plates and projections from abactinal plates.

Holotype, Australian Museum (AMJ3077).

MAXIMUM SIZE

R = 20 mm (R = 1.2r).

COLOUR

White, except for a small dark spot on each of the arm tips, or mottled blue and white.

REPRODUCTION

Gonoducts open on the abactinal surface.

HABITAT, DEPTH AND AUSTRALIAN DISTRIBUTION

Under rocks, 0–25 m. Found in south-eastern Australia from Mooloolaba, Queensland to Victoria and Lord Howe Island. South-eastern Australian endemic.

TYPE LOCALITY

Long Reef, Collaroy, NSW.

FURTHER DISTRIBUTION

Australian endemic.

REFERENCES

Clark, H.L., 1938: 155; pl. 22 fig. 4 (as *Asterina perplexa*).

Clark, H.L., 1946: 131 (as *Asterina perplexa*).

Livingstone, 1933: 3–5; pl. 5 figs 1–8; pl. 14.

O'Loughlin and Waters 2004: 4–5 (key), 20–22; fig. 13a.

SARASIN'S SEASTAR
Indianastra sarasini
(de Loriol, 1897)

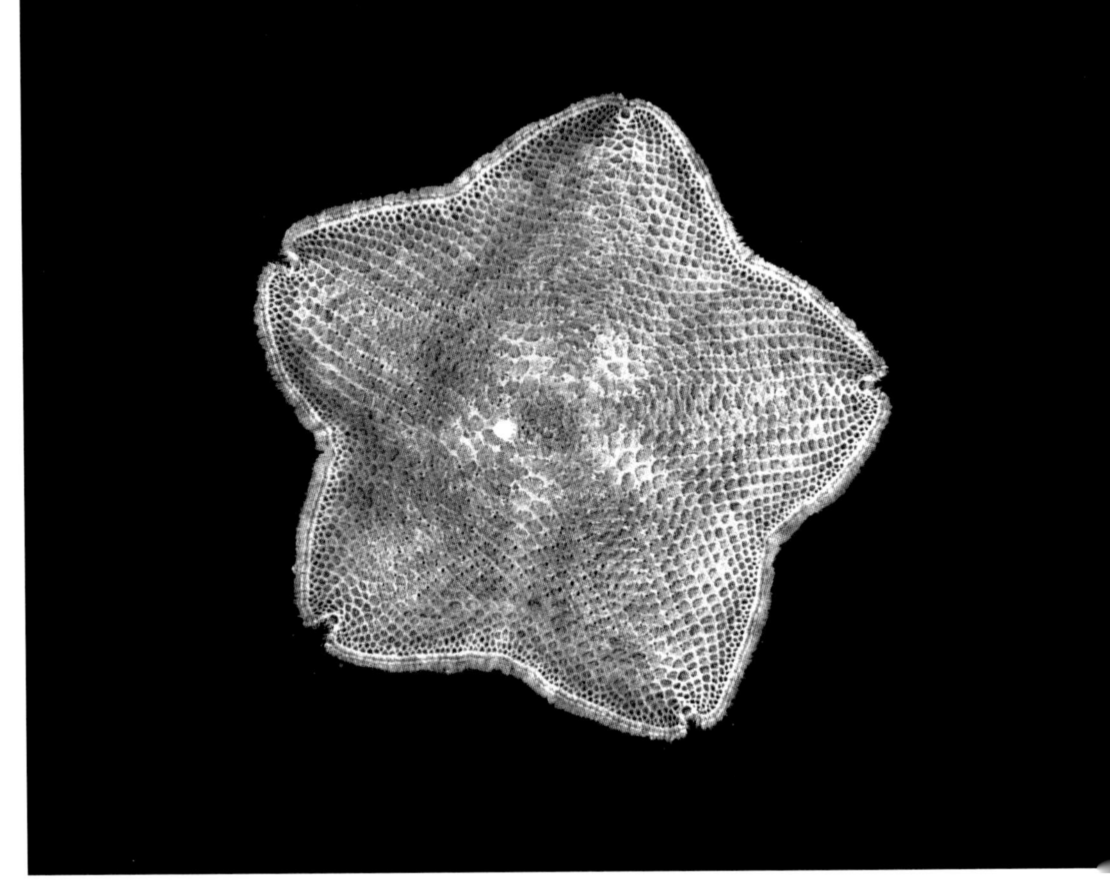

Darwin, Northern Territory (N. Coleman).

DESCRIPTION

Arms five, broad and rounded with the interradial area slightly concave and notched in the interradial line; disc nearly flat to moderately elevated, integument covered; abactinal plates imbricated with minute hyaline spinelets in a line curved around the distal side of each papula; simple pedicellariae with short valves over papulae; papulae in a petaloid pattern on disc and arms with a distinct interradial band without papulae; **crystal bodies on plates**; inferomarginal plates with tufts of numerous very slender spinelets; actinal plates in longitudinal series with a group of slender, short, sharp, sacciform spinelets, the largest in a comb of 5–7; furrow spines in a fan of 7–8 (between 5–9) very slender webbed spines, subambulacrals in a fan of six (3–9) similar spines; 8–11 oral spines, 4–9 suboral spines in a webbed fan; superambulacral plates not in a series, rarely single plates distally; margin supported by a series of single superactinal plates and projections from abactinal plates.

MAXIMUM SIZE

R/r = 26/20 mm (R = 1.15–1.4r).

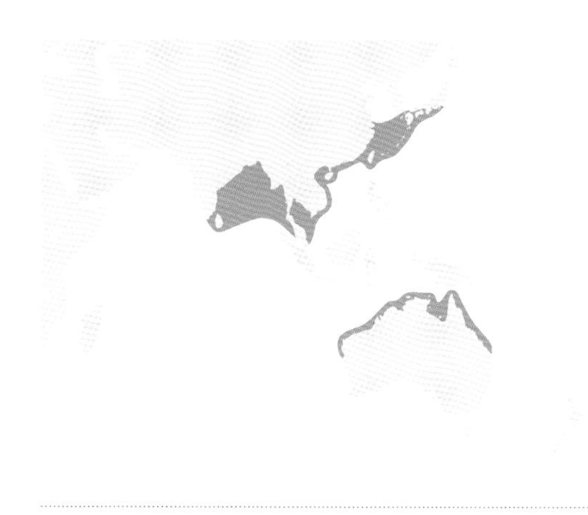

COLOUR
Yellow to bright orange-vermilion or greyish orange.

HABITAT, DEPTH AND AUSTRALIAN DISTRIBUTION
On the under side of rocks or coral, 0–42 m. Coastal northern Australia from Dorre Island, Shark Bay, WA, NT to the Great Barrier Reef and Queensland coast south to Byron Bay, NSW.

TYPE LOCALITY
Sri Lanka.

FURTHER DISTRIBUTION
From Sri Lanka and the Bay of Bengal to Hong Kong and Japan; not yet recorded from Indonesia or the Philippines.

REFERENCES
Clark, H.L., 1938: 153–155; pl. 12 fig. 2 (as *Asterina lutea*).

Clark and Rowe, 1971: 67–68; pl. 9 figs 7–8 (as *Asterina sarasini*).

O'Loughlin and Waters, 2004: 4–5 (key), 20–22; figs 2h, 6b, 13b–f (new genus *Indianastra*).

Rowe and Gates, 1995: 35 (as *Asterina sarasini*).

REMARKS
H.L. Clark (1938) described this species from Broome as *A. lutea* but Clark and Rowe (1971) suggested that it is probably synonymous with the geographically separated species *A. sarasini*, the course followed here. The two species were formally synonymised by Rowe (in Rowe and Gates, 1995) who added *A. nuda* H.L. Clark, 1921 and *A. orthodon* Fisher, 1922 to the synonymy; this was confirmed by O'Loughlin and Waters (2004) and transferred to their new genus *Indianastra.*

Indianastra sarasini and *I. inopinata* are closely related but *I. sarasini* is a widespread tropical species while *I. inopinata* is endemic to south-eastern Australia, the two species overlap only in south-eastern Queensland and northern New South Wales. Alive they differ in colour; *I. sarasini* is a uniform yellow to orange-red while *I. inopinata* is blue and white. Morphological differences are slight but *I. inopinata* has only 2–3 subambulacral spines while *I. sarasini* has six (3–9). *I. sarasini* has crystal bodies on abactinal plates, there are none in *I. inopinata.*

GENUS *MERIDIASTRA* O'Loughlin, 2002

Seastars of this genus predominantly have 5–8 arms. The interradial margin is straight or incurved so the arms are not discrete. The abactinal surface can be low to high, convex and flat actinally. The abactinal plates are in longitudinal series along the arms, the carinal series are present for at least part of the arm length. The papulae are small and not in longitudinal series. The abactinal spinelets are granuliform. There are no pedicellariae, and the abactinal plates have crystal bodies. The actinal plates are in longitudinal series with 1–3 actinal spines per plate in the mid-interradius. The spines are digitiform, tapering or conical. The superambulacral and superactinal plates are lacking, and the interradial margin is supported by interior projections from the abactinal and actinal plates.

Type species: *Meridiastra atyphoida* (H.L. Clark, 1916) (Australia)

Species of *Meridiastra* occur in the waters of southern Australia, New Zealand, the central and eastern Pacific, Mexico and Panama.

(O'Loughlin, 2002; O'Loughlin and Waters, 2004; and O'Loughlin et al., 2003.)

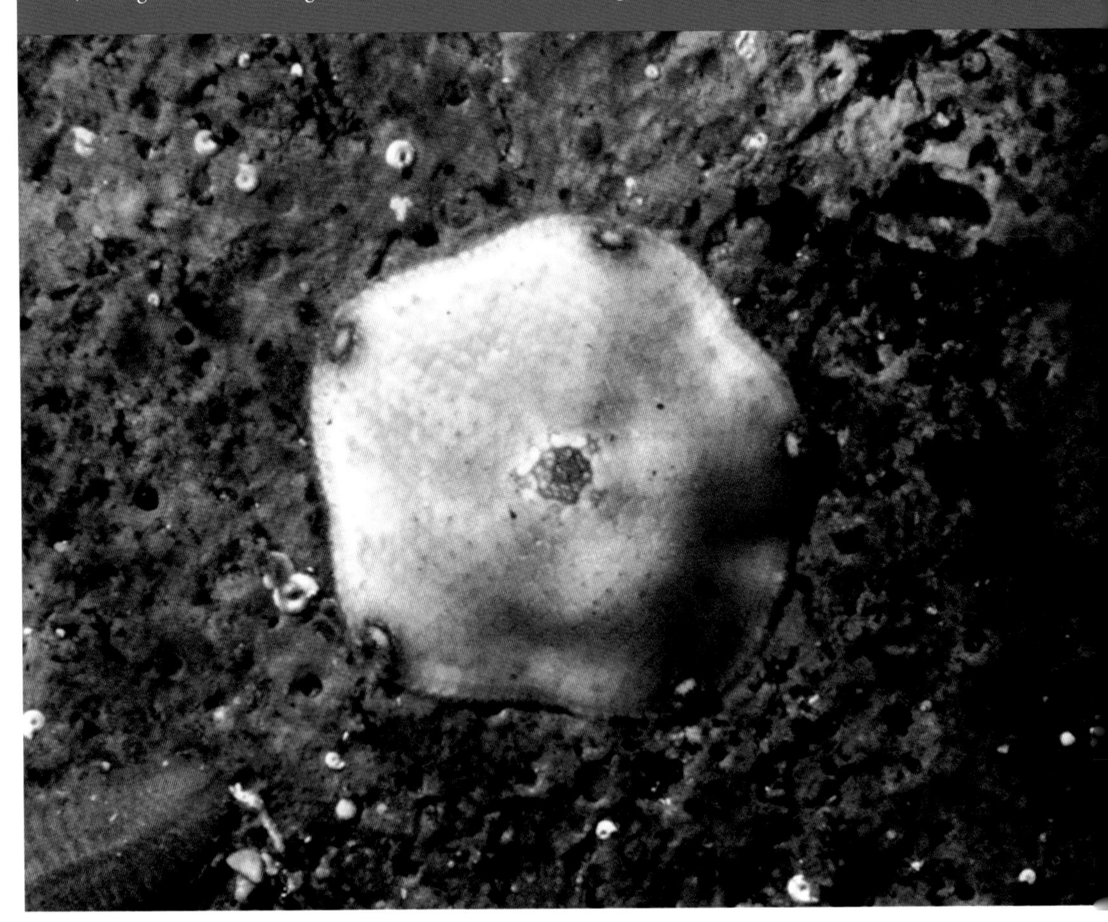

Meridiastra atyphoida, Gulf St Vincent, South Australia (G. Edgar).

KEY TO THE SPECIES OF *MERIDIASTRA* IN SHALLOW AUSTRALIAN WATERS

1	Pentagonal form; single madreporite; not fissiparous; abactinal plates in regular series.	2
	Arms 6–8.	3
2	Gonopores abactinal; abactinal spinelets predominantly in single series on proximal plate edge; subambulacral spines predominantly single; single suboral spine commonly present; live colour variable cream, yellow, orange, mauve, red, brown.	***M. atyphoida*** page 151
	Gonopores actinal; abactinal spinelets commonly in irregular double series on proximal plate edge; subambulacral spines predominantly paired; suboral spines rarely present; live colour white with small black spots abactinally.	***M. nigranota*** page 159
3	Arms 6–8, form irregular (post fissipary); more than one madreporite; fissiparous; only distal abactinal plates in regular series.	***M. fissura*** page 154
	Form regular, not fissiparous.	4
4	Arms eight.	***M. calcar*** page 152
	Arms six.	5
5	Subambulacral spines predominantly one per plate, up to twice the length of adradial actinal spines; actinal interradial spines very short, commonly bulbous; abactinal surface uneven; abactinal spinelets prominently spinous, commonly low to subcapitate; abactinal and actinal colour of adults consistently uniform crimson, purple to brownish; tube feet orange.	***M. gunnii*** page 155
	Subambulacral spines predominantly 2–3 per plate, not up to twice the length of adradial actinal spines; actinal interradial spines not very short or bulbous; abactinal surface even; abactinal spinelets moderately to minutely spinous, commonly columnar; abactinal colour variable, not uniform crimson, purple to brownish nor with orange tube feet.	6

6	At least a few suboral spines commonly present; adradial actinal spines up to about two-thirds the length of subambulacral spines; actinal interradial spines short and fine; carinal plates normally doubly papulate for less than two-thirds arm length, commonly less than half arm length; abactinal spinelets relatively small and fine, predominantly narrowing distally, minutely spinous; abactinal colour variable, frequenly overall maroon red (not reported with grey or blue, or with black disc); actinal colour off-white, commonly with prominent flecking.	***M. medius*** page 158
	Suboral spines rarely present; adradial actinal spines commonly more than two-thirds length of subambulacral spines; actinal interradial spines digitate; carinal plates normally doubly papulate for more than half arm length, commonly more than two-thirds arm length; abactinal spinelets relatively coarse, predominantly columnar, slightly narrowing to slightly widening distally, moderately spinous; abactinal colour variable, not overall maroon red.	**7**
7	Form variable, commonly distinctive short arms with interradial margin deeply indented; subambulacral spines commonly projecting fairly prominently; abactinal spinelets coarse, columnar, moderately spinous, commonly widened distally; abactinal colour variable, grey, brown or blue, sometimes all orange.	***M. occidens*** page 160
	Form variable, commonly subhexagonal with interradial margin slightly incurved; subambulacral spines not projecting significantly; abactinal spinelets fairly coarse, columnar, moderately spinous, commonly narrowing distally; abactinal colour commonly pale, with white, pink, mauve, orange or bright red flecks, disc commonly black, but not with blue or grey (although the latter reported in NSW).	***M. oriens*** page 162

(Modified from O'Loughlin, 2002; and O'Loughlin et al., 2003.)

MODEST SEASTAR
Meridiastra atyphoida
(H.L. Clark, 1916)

DESCRIPTION
A small, nearly pentagonal asterinid with a thin, slightly domed body, flat orally; abactinal surface with five rows of nearly bare (apart from their covering of skin), overlapping rhombic, oval or oblong plates radially and five chevrons of rhombic plates in each interradius, secondary plates rare; papulae single on disc and inner two-thirds of each radial area; **there are up to 10 minute conical spinelets on the proximal edges of the plates** (up to 20 scattered over the plates in juveniles) but these are easily lost; the inferomarginal plates project beyond the superomarginals and have a fringe of minute spinelets; **the actinal plates, except those next to the adambulacrals, each have a single short, blunt spine,** rarely two, in mid-interradius; 1–3 very small spines on distal plates; there are three (rarely four) webbed furrow spines and one large subambulacral spine on each adambulacral plate; 4–6 oral spines, one suboral spine usually present on each oral plate; no pedicellariae.

MAXIMUM SIZE
R/r = 15/11.5 mm (R = 1.3r).

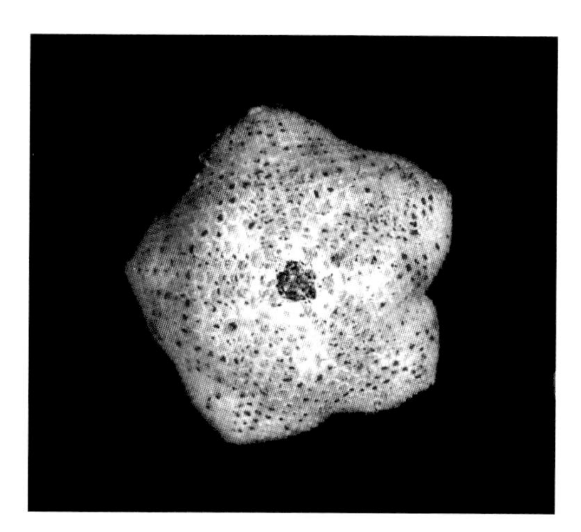

Bruny Island, Tasmania (N. Coleman).

COLOUR
Variable, often pale cream, yellow, pink, red-brown, red-orange, mauve or light brown, sometimes spotted or mottled; disc and arm ends dark grey-green to black. Actinal surface white.

REPRODUCTION
Paired interradial gonopores open on the abactinal surface.

HABITAT, DEPTH AND AUSTRALIAN DISTRIBUTION
Found under rocks and in crevices in intertidal rock pools and among algae, 0–59 m. From south-eastern Tasmania and East Gippsland, Victoria and SA, to Esperance, WA. Southern Australian endemic.

TYPE LOCALITY
Off Cape Jervis, SA.

FURTHER DISTRIBUTION
Australian endemic.

REFERENCES

Clark, H.L., 1916: 57; pl. 17 figs 1–2.
Clark, A.M., 1966: 324–325.
Dartnall, 1970a: 1–4; figs 1–3.
Marsh and Pawson, 1993: 281.
O'Loughlin, 2002: 280–283; fig. 2.
O'Loughlin and Waters, 2004: 4–5 (key), 22–23; figs 2i, 10g.

EIGHT-ARMED SEASTAR
Meridiastra calcar
(Lamarck, 1816)

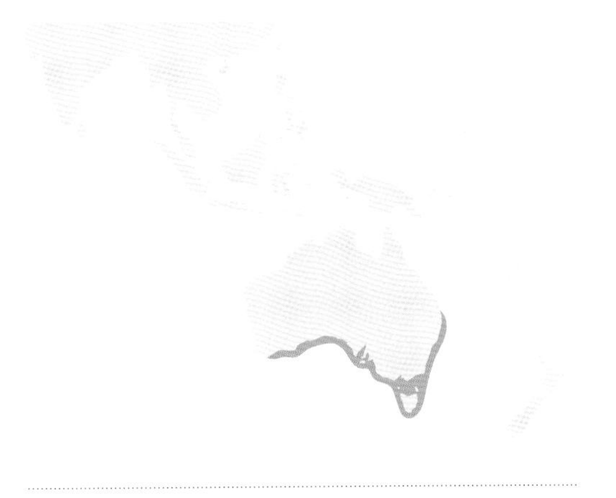

DESCRIPTION
This species is clearly distinguished from other *Meridiastra* **by its eight narrow arms and humped disc** (occasionally there are 7–11 arms); **abactinal spinelets granuliform**; abactinal plates with crystal bodies on the plates; **actinal plates each bear a single spinelet**; furrow spines are paired; pedicellariae absent; superambulacral and superactinal plates lacking.

MAXIMUM SIZE
R/r = 50/28 mm (R = 1.8r).

COLOUR
Highly variable, often mottled shades of blue, green, orange or brown or a mixture of colours. Actinal surface pale.

FOOD
Omnivorous, small gastropods and bivalves, detritus, algae, including drift algae. Scavenges dead animals.

REPRODUCTION
In NSW spawning takes place intermittently between August and December. The sexes are separate, gonopores are abactinal and the eggs are large and yolky (415 µm). The larva is a non-feeding planktonic brachiolaria that lives on its yolk reserves, settling after a few days and attaching with its brachiolar arms. Metamorphosis takes place after 12–16 days.

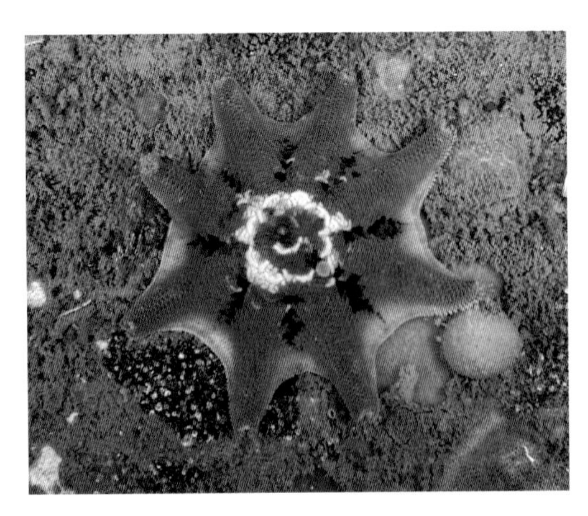

Mornington, Victoria (N. Coleman).

HABITAT, DEPTH AND AUSTRALIAN DISTRIBUTION
Intertidal rock platforms among algae and rock pools on moderately exposed coasts, 0–3 m. Found from Currumbin (28°08'S), Queensland around southern Australia, including Tasmania, to Albany on the south coast of WA. Southern Australian endemic.

TYPE LOCALITY
'Australian Seas'. Probably King George Sound, WA.

FURTHER DISTRIBUTION
Australian endemic.

REFERENCES

Byrne, 1991: 499–502; fig. 1a (reproduction).
Byrne, 1992: 297–316 (as *Patiriella calcar*).
Byrne and Anderson, 1994: 564–576 (hybridisation with *Meridiastra oriens*).
Clark, H.L., 1946: 134–135 (as *Patiriella calcar*).
Dartnall, 1971: 46 (as *Patiriella calcar*).
O'Loughlin and Waters, 2004: 4–5 (key), 22–23.
Shepherd, 1968: 746 (food, as *Patiriella calcar*).

Mornington, Victoria [top] (N. Coleman); Cockle Creek, Tasmania [bottom] (G. Edgar).

CLEFT SEASTAR
Meridiastra fissura
O'Loughlin, 2002

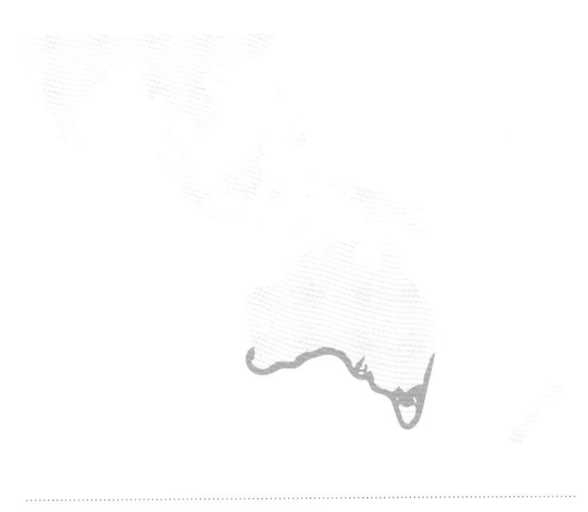

DESCRIPTION

Small, body thin, slightly domed abactinally, flat actinally; **arms 6–8, fissiparous**, irregular (post fissipary); **more than one madreporite; up to five anal openings**; abactinal plates overlapping except proximally, in regular series distally only; papular spaces small, papulae usually single, papulate areas similar in extent to non-papulate areas; secondary plates apically and distally, never separating carinals; five longitudinal series of radial plates mid-arm; carinal series in mid-arm only; proximal plates often slightly notched for single papula; **plates with up to 30 small rugose often webbed glassy spinelets**; inferomarginals extend beyond superomarginals and have a marginal fringe of seven fine distal spinelets; **actinal interradial plates with single often reduced spines**; three (rarely four) webbed furrow spines; two webbed subambulacral spines per plate; 4–6 oral spines, no suboral spines; no pedicellariae.

MAXIMUM SIZE

R/r = 9/6.9 mm (R = 1.3r).

Fremantle, Western Australia (N. Coleman).

COLOUR

Fawn to cream or white distally with a few orange, red to dark reddish brown spots; madreporites yellow. Actinal surface white.

REPRODUCTION

There are one to two gonopores in each abactinal interradial area.

HABITAT, DEPTH AND AUSTRALIAN DISTRIBUTION

On exposed rock platforms and rocky sublitoral areas, 0–30 m. From Sydney, NSW around southern Australia and southern Tasmania to Garden Island, near Fremantle, WA. Southern Australian endemic.

TYPE LOCALITY

Flinders, Victoria.

FURTHER DISTRIBUTION

Australian endemic.

REFERENCES

O'Loughlin, 2002: 283–285; fig. 3a–f.

O'Loughlin and Waters, 2004: 4–5 (key), 22–23.

GUNN'S SEASTAR

Meridiastra gunnii

(Gray, 1840)

Western Australian Museum (WAMZ9436).

DESCRIPTION

Form variable; **predominantly six arms** (rarely 5–9); arms rounded to pointed with interradial margin incurved to hexagonal; **body thick, domed abactinally, flat actinally**; margin fairly thin; no pedicellariae; gonopores abactinal; on abactinal surface papulate areas more extensive than non-papulate areas; secondary plates abundant, irregular in size and shape; proximal radial and interradial plates imbricate; proximal papular spaces large with 16–20 papulae and 16–20 secondary plates outside disc; **abactinal plates thick, raised, crescentic in papulate areas, carinally with a double notch**; carinal series more or less regular to near end of arms; distal interradial non-papulate plates closely imbricate, domed; disc variably distinct bordered by crescentic radial plates and smaller interradial plates; **abactinal plates granular, covered by crystal bodies**, lacking spine bearing ridge; **abactinal**

Geographe Bay, Western Australia (S. Morrison).

spinelets granuliform, longer distally; lacking internal superambulacral plates, but with distal abactinal and actinal interradial plates with internal tapered projections. Projecting inferomarginal plates form margin and bear up to 11 spinelets per plate; actinal plates in regular series, curving acutely from furrow to margin; actinal plates carry 1–2 thick short spines, up to four distally; adradial plates usually with one short, thick spine up to half the length of subambulacral spines; **2–3 furrow spines, fairly thick; one, rarely two, thick subambulacral spines**; 4–6 oral spines; suboral spines rare.

MAXIMUM SIZE

R/r = 56/35 mm (R = 1.1–1.6r).

Marmion Lagoon, Western Australia (G. Edgar).

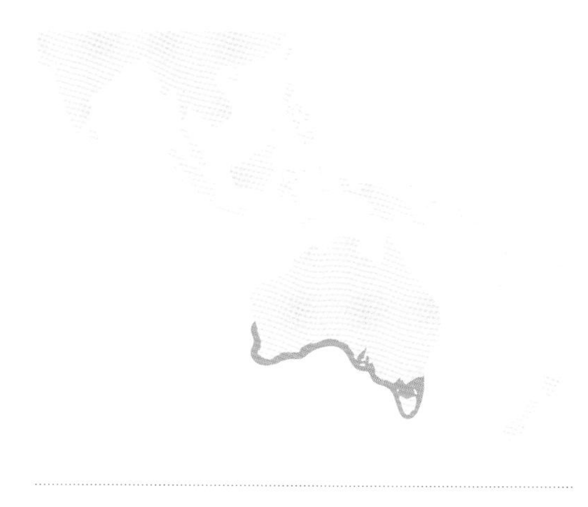

COLOUR

Dark crimson, purple to chocolate brown on abactinal and actinal surfaces. Tube feet orange.

FOOD

Red algae (such as *Laurencia*), bryozoans, compound ascidians and sponges. Also small molluscs, crustaceans and detritus. Scavenges dead animals.

REPRODUCTION

Spawning has been observed in January. Sexes are separate and gonopores open on the abactinal surface. Large eggs develop into lecithotrophic brachiolaria with a brief planktonic phase before becoming demersal. The larvae move slowly over the substrate, attaching after 10 days (from fertilisation), and metamorphose by 12–13 days. The newly metamorphosed juveniles have five pairs of tube feet and circular, slightly five lobed outline; the sixth arm starts to develop one day after metamorphosis (Marsh, unpublished).

HABITAT, DEPTH AND AUSTRALIAN DISTRIBUTION

On algal covered rock among seagrass, sometimes on the underside of boulders, in sheltered waters, 0–29 m. From eastern Victoria to Dongara and the Houtman Abrolhos, WA, including Tasmania. Southern Australian endemic.

TYPE LOCALITY

Sandy Bay, Derwent River estuary, Tasmania.

FURTHER DISTRIBUTION

Australian endemic.

REFERENCES

Clark, H.L., 1938: 166; pl. 22 figs 2–3 (as *Patiriella brevispina*).

Clark, H.L., 1946: 135 (as *Patiriella brevispina*).

Dartnall, 1970b: 74–75 (image of lectotype of *Patiriella gunnii*).

Dartnall, 1971: 47 (as *Patiriella brevispina*).

O'Loughlin et al., 2003: 49

O'Loughlin and Waters, 2004: 4–5 (key), 22–23; figs 5a, 10i.

Shepherd, 1968: 747–748 (food, as *Patiriella brevispina*).

REMARKS

The type series of *Asterina gunnii* Gray 1840 from the Derwent River estuary, Tasmania consisted of four specimens. Dartnall (1970b) selected one of these as a lectotype, which was later found by O'Loughlin et al. (2003) to represent what had been known as *Patiriella brevispina*; thus *brevispina* has become a junior synonym of *gunnii*, which has been included in the new genus *Meridiastra*. Species previously regarded as *Patiriella gunnii* have been shown to represent three additional species (O'Loughlin et al. 2003): *M. medius, M. occidens* and *M. oriens.*

MEDIUS SEASTAR
Meridiastra medius
(O'Loughlin, Waters and Roy, 2003)

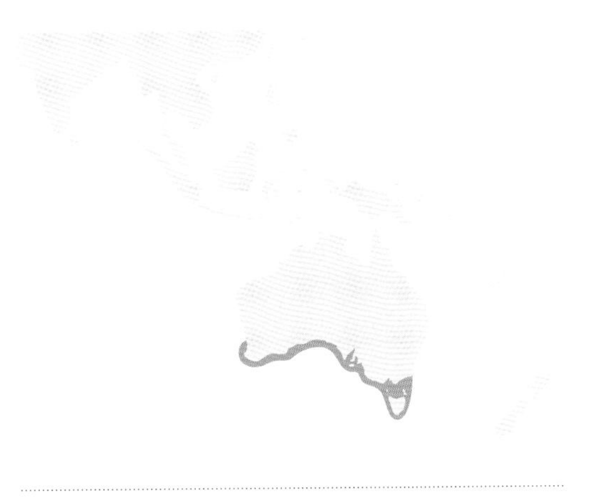

DESCRIPTION
Predominantly six armed; form variable, from six short, pointed arms with incurved interradial margin to hexagonal; body has a flattened dome abactinally, flat actinally; ***Meridiastra medius* differs from other six armed *Meridiastra* species in rarely having doubly papulate plates carinally for more than half the arm length; abactinal spinelets small and fine**; suboral spines commonly present; adradial row of actinal interradial spines up to about two-thirds the length of subambulacral spines, two per plate, minutely spinous distally; actinal interradial spines short and thin; furrow spines slender, webbed, tapering, 3–4 (up to five) per plate, minutely spinous distally; 5–7 oral spines, 1–12 suboral spines, variable between geographic locations; 2–3 subambulacral spines, thick, tapering, minutely spinous distally.

MAXIMUM SIZE
R = 38 mm.

COLOUR
Highly variable, often maroon or reddish brown, pale brown, orange or pink sometimes with flecks of other colours, sometimes mottled; never grey, blue or black. Actinal surface off-white with prominent maroon flecking.

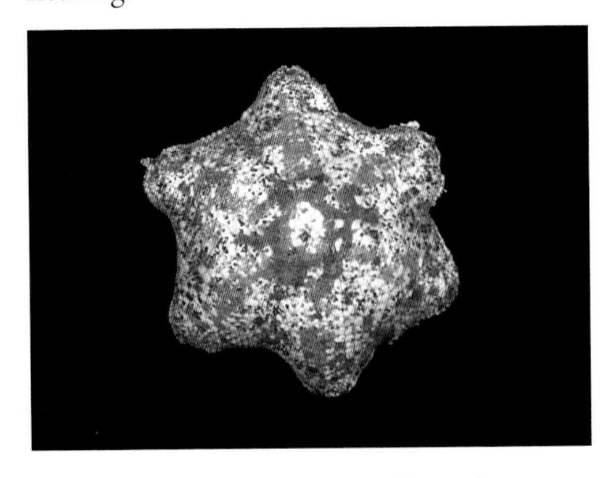

Bruny Island, Tasmania (N. Coleman).

FOOD
Algae, biofilm, byrozoans and ascidians.

REPRODUCTION
Gonopores on the abactinal surface.

HABITAT, DEPTH AND AUSTRALIAN DISTRIBUTION
Under rocks, 0–34 m. From west of Wilsons Promontory, Victoria, Bass Strait, Port Arthur and Bruny Island, Tasmania to Fremantle, WA. Southern Australian endemic.

TYPE LOCALITY
Griffitts Island, Victoria.

FURTHER DISTRIBUTION
Australian endemic.

REFERENCES
O'Loughlin and Waters, 2004: 4–5 (key), 22–23.
O'Loughlin et al., 2003: 184–188; figs 1, 4a–f.
Shepherd 2014: 235.

REMARKS
This species was previously included in *Patiriella gunnii*. The specific name is from the Latin *medius* meaning 'between the two', referring to its distribution across southern Australia, between the species *Meridiastra oriens* and *M. occidens*.

FRECKLED SEASTAR
Meridiastra nigranota
O'Loughlin, 2002

DESCRIPTION

Small, pentagonal or with slightly incurved interradial margins; **body thin, slightly domed abactinally, flat actinally; not fissiparous**; abactinal plates closely overlapping in regular series; papular spaces small, papulae single, papulate areas less extensive than non-papulate areas; secondary plates rare, apical and marginal only, never separating carinal plates; each radius has five longitudinal series of plates in mid-arm; distinct carinal series of about 12 plates to three-quarters of arm length; distally 2–12 zigzag radial plates; projecting interradial plates fan-shaped; **abactinal plates with more than 20 fine pointed rugose glassy spinelets with spiny tips** clustered on proximal edge in irregular single and double transverse series; on interradial plates the spinelets are clustered on the projecting centre of the plate; inferomarginal plates extend beyond superomarginals; both series of marginal plates have about six fine, webbed spinelets fringing the margin; **actinal interradial plates with one spine**, often two in mid-interradius, 1–3 distally; actinal series next to adambulacrals with none to very few commonly reduced spines; three webbed furrow spines, 1–2 tall subambulacral spines; 4–6 oral spines, suboral spines rarely present; no pedicellariae.

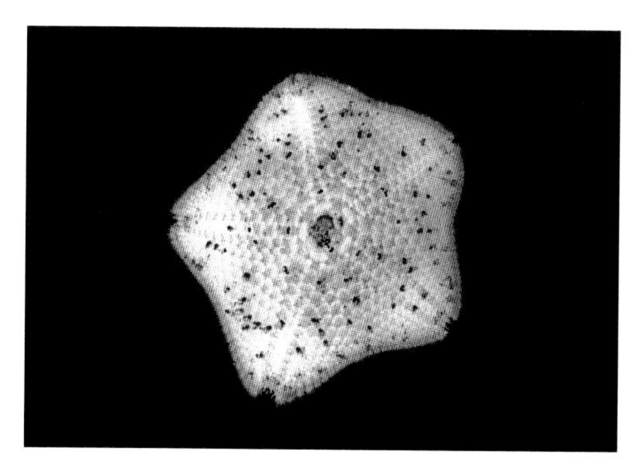

Port Phillip, Victoria (N. Coleman).

MAXIMUM SIZE

R = 13 mm (R = 1.3r).

COLOUR

White to fawn or pale grey with irregular black to dark reddish or greenish black spots and faint red flecking around ends of arms; disc dark greenish black and madreporite white. Actinal surface white.

REPRODUCTION

Paired interradial gonopores open on the actinal surface.

HABITAT, DEPTH AND AUSTRALIAN DISTRIBUTION

On ocean platforms, rocky sublittoral, 0–30 m. Tasmania south to the Tasman Peninsula, from Wilsons Promontory, Victoria around southern Australia to Rottnest Island, WA. Southern Australian endemic.

TYPE LOCALITY

Flinders, Victoria.

FURTHER DISTRIBUTION

Australian endemic.

REFERENCES

Dartnall, 1970a: 19, fig. 1 (as *Asterina atyphoida*).
Marsh and Pawson, 1993: 281 (part, Rottnest Island specimens as *Asterina atyphoida*).
O'Loughlin, 2002: 287–289; fig. 5.
O'Loughlin and Waters, 2004: 4–5 (key), 22–23.

WESTERN SEASTAR
Meridiastra occidens
(O'Loughlin, Waters and Roy, 2003)

Esperance, Western Australia (C. Bryce).

Common colour variations, Fremantle, Western Australia (L. Marsh).

DESCRIPTION

A flattened predominantly six armed asterinid (rarely 4–9 arms); form variable, from six pointed arms with interradial margin deeply incurved to rarely nearly hexagonal; body flat actinally; flattened dome abactinally; no pedicellariae; **the abactinal surface is paved with slightly crescentic imbricating plates, those of the carinal series are usually doubly notched for papulae, the remainder are crescentic**; small secondary plates are present between the larger plates; the abactinal plates bear groups of small, blunt thorny spinelets ~28 per plate near centre of disc and 6–8 per plate near margin; the inferomarginal plates form the edge of the body, each bears up to 11 spinelets; the **actinal plates carry 1–2 spines proximally, 2–3 near the margin**; adradial with one (rarely two) thick digitate spines about four-fifths the length of subambulacral spines; adambulacral plates bear three furrow spines (2–4 near mouth, one distally) and two subambulacral spines; oral plates bear five (between 4–6) oral spines, suboral spines rare.

MAXIMUM SIZE

R/r = 38/23 mm (R = ~1.7r).

COLOUR

Similar to *Meridiastra calcar*, blue, green, grey, orange, sometimes brown, often mottled, or a mixture of colours. Actinal surface off-white and tube feet pale straw.

FOOD

In WA it is omnivorous and scavenges dead animals, but it preferentially feeds on algae, particularly *Ulva*, possibly extracting small organisms.

REPRODUCTION

Sexes are separate and spawning takes place between March and May. Gonopores are abactinal. Eggs are 360–400 µm in diameter, developing into a planktonic non-feeding brachiolaria larva that attaches to the bottom and metamorphoses after 12–14 days.

HABITAT, DEPTH AND AUSTRALIAN DISTRIBUTION

Found on the underside of boulders in moderately sheltered bays; common under ledges in shallow pools and among algae on moderately exposed intertidal rock platforms, 0–14 m. From Port Fairy, Victoria around southern and south-western Australia to Kalbarri (27°45'S) on the mid-west coast of WA. Southern Australian endemic.

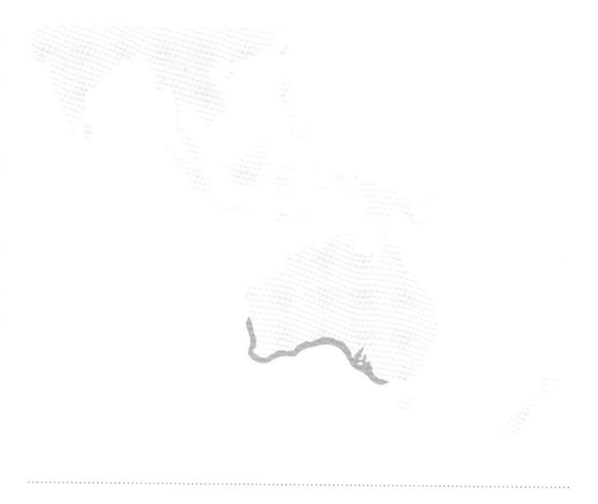

TYPE LOCALITY

Cottesloe near Fremantle, WA.

FURTHER DISTRIBUTION

Australian endemic.

REFERENCES

Dartnall, 1970b: 74–75; pl. 1 (redescription of type series of *Patiriella gunnii* and image of lectotype).

Dartnall, 1971: 46 (as *Patiriella gunnii*).

Grice and Lethbridge, 1988: 399–407 (reproduction, as *Patiriella gunnii*).

O'Loughlin and Waters, 2004: 4–5 (key), 22–23; fig. 10h.

O'Loughlin et al., 2003: 183 (key), 188–191; figs 5a–f, 7d.

REMARKS

The specific name is from the Latin *occidens* meaning 'west', in reference to the predominantly western distribution of this species in southern Australia. *Meridiastra occidens* was included in *Patiriella gunnii* prior to the revision by O'Loughlin et al. (2003) and the generic revision by O'Loughlin and Waters (2004).

EASTERN SEASTAR

Meridiastra oriens

(O'Loughlin, Waters and Roy, 2003)

Batemans Bay, New South Wales (N. Coleman).

DESCRIPTION

Predominantly six arms (4–8); form variable, with six short, pointed to rounded arms with interradial margin incurved to sub-hexagonal; **body a flattened dome abactinally**, flat actinally; no pedicellariae; carinal plates commonly doubly papulate for about two-thirds of the arm length; **abactinal spinelets commonly columnar and moderately spinous** distally creating a fairly coarsely spinous surface; furrow spines 2–4 per plate commonly slightly longer than subambulacral spines; actinal spines continuous in declining height with subambulacral spines; **actinal interradial spines digitate, usually one per plate**, distally 3–4; two thick, often unequal subambulacral spines; adradial actinal plates with 1–2 thick spines proximally; oral plates with five (4–6) spines, suboral spines normally absent.

MAXIMUM SIZE

R = 39 mm.

COLOUR

Highly variable, commonly pale or light coloured, often with a dark coloured disc, some fairly uniform white, pink, mauve, orange or bright red sometimes with dark red, brown or white flecks or mottling with these colours plus green or yellow; no blue in colour pattern. Actinal surface off-white with rare colour flecks.

Batemans Bay, New South Wales (N. Coleman).

FOOD

Small gastropods and encrusting ascidians.

REPRODUCTION

Gonopores are abactinal. Broadcast spawning takes place between August and December. Mature ova are 360 μm in diameter and development takes place through a planktonic brachiolaria that metamorphoses 12–16 days after fertilisation.

HABITAT, DEPTH AND AUSTRALIAN DISTRIBUTION

Under subtidal boulders, 0–30 m. From Rockhampton, Queensland to Nuyts Archipelago, SA; Tasmania and Lord Howe Island; one record from Cottesloe, WA. The distribution overlaps that of three other species of six armed *Meridiastra* in SA. South-eastern Australian endemic.

TYPE LOCALITY

Black Reef, Recherche Bay, Tasmania.

FURTHER DISTRIBUTION

Australian endemic.

REFERENCES

Byrne, 1991: 499–508.
Byrne, 1992: 297–316 (as *Patiriella gunnii*).
Byrne and Anderson, 1994: 564–576 (hybridisation with *Meridiastra calcar*).
Dartnall, 1971: 46 (as *Patiriella gunnii*).
O'Loughlin and Waters, 2004: 4–5 (key), 22–23; fig. 4c.
O'Loughlin et al., 2003: 183 (key), 191–193; figs 6a–f, 7e.

REMARKS

The species name is from the Latin *oriens* meaning 'east', in reference to the predominantly easterly distribution in southern Australia. *Meridiastra oriens* was included in *Patiriella gunnii* prior to the revision by O'Loughlin et al. (2003) and the generic revision by O'Loughlin and Waters (2004).

GENUS *NEPANTHIA* Gray, 1840

Species of this genus have 4–7 subcylindrical arms more or less flat actinally, with a slight or distinct marginal edge. The plates on the abactinal surface of the arms are irregular in arrangement, and secondary plates are present. Abactinal and actinal interradial plates have dense clusters of thick or thin glassy spinelets, which are commonly on spinelet-bearing elevations. Pedicellariae are present and there are crystal bodies on the plates. **Inferomarginal plates project slightly** giving an angular edge to the arms. There are six or more furrow spines per plate, the actinal spines are predominantly thin and glassy, sometimes sacciform. Superambulacral, transactinal and superactinal plates are present, embedded in the interior lining in most species.

Type species: *Nepanthia maculata* Gray, 1840 (Philippines, Indonesia, Australia)
Other Australian species now included in *Nepanthia* are *N. belcheri, N. crassa* and *N. fisheri* (the latter from deeper than 30 m). Four Australian species included in *Nepanthia* by Rowe and Marsh (1982) were reassigned to the genus *Pseudonepanthia* A.H. Clark, 1916 by O'Loughlin and Waters (2004): *P. briareus, P. gracilis, P. nigrobrunnea* and *P. troughtoni.*

Morphological characters that distinguish *Nepanthia* from *Pseudonepanthia* are: arms flat actinally, with a marginal edge; presence of secondary plates, pedicellariae and crystal bodies; six or more furrow spines per plate, and presence of transactinal and superactinal plates. The genus is confined to the central Indo-West Pacific and Australia, north to Vietnam and Myanmar.

Nepanthia belcheri, Singapore (R. Tan).

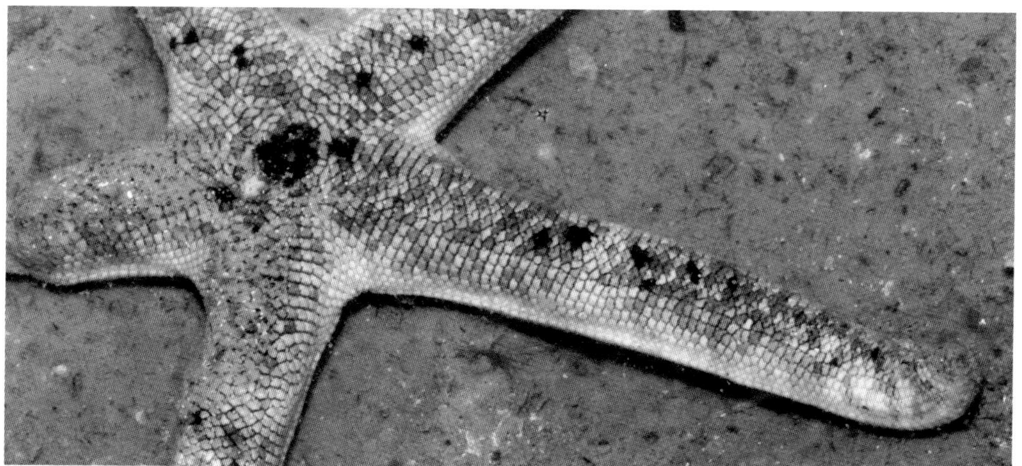

Nepanthia belcheri [top] and *N. maculata* [bottom], Singapore (R. Tan).

KEY TO THE SPECIES OF *NEPANTHIA* IN SHALLOW AUSTRALIAN WATERS

1	Spinelets thorny with 5–8 points; western Australia.	***N. crassa*** page 168
	Spinelets not as above.	**2**
2	Arms often slightly tapering, unequal in length, fissiparous.	***N. belcheri*** page 166
	Arms subcylindrical, not tapering, not fissiparous.	***N. maculata*** page 170

BELCHER'S SEASTAR
Nepanthia belcheri
(Perrier, 1875)

Western Australian Museum (WAMZ9003).

DESCRIPTION

A fissiparous species with 4–7 (usually 5–6) subcylindrical to slightly tapering moderately stout arms; papulae single, among small secondary plates in the space provided by the notched primary plates; **abactinal plates have a raised crescentic ridge carrying spinelets composed of 3–4 pillars joined by about six bridges, each pillar ends in a sharp point**; some spinelets are modified to form pedicellariae, which close over the papulae on the lateral areas of the arms; crystal bodies are present on the lower parts of the plates; **furrow spines are in a fan of usually 8–9, with a subambulacral fan of 7–12 spines** and 3–7 additional small spines; inferomarginal plates project slightly at the margin; actinal plates are closely covered by thorny, tapered radiating spinelets.

MAXIMUM SIZE

R/r = 65/14 mm (R = 2–5.3r).

COLOUR

Very variable, often grey-green or fawn background colour mottled with red, green, brown or black.

FOOD

Sponges, hydroids, soft corals and detritus.

REPRODUCTION

Both sexual and asexual reproduction takes place in *Nepanthia belcheri*. In Queensland fission appears to predominate with 45% of a population showing evidence of recent fission in early winter (April to June); hermaphroditism was the normal condition. Sexual reproduction takes place during early summer (October to November). Eggs are 450 µm in diameter suggesting pelagic lecithotrophic development. The lifespan was extrapolated from a growth curve and was estimated at four or more years (Kenny, 1969).

Abactinal plate characteristics [top], Western Australian Museum (WAMZ9003); abactinal spinelet characteristics [bottom], Western Australian Museum (WAMZ6974).

HABITAT, DEPTH AND AUSTRALIAN DISTRIBUTION

Clings to the underside of boulders on muddy sand and rock substrate, 0–50 m. In Australia it is found from off Ningaloo Reef, WA, around northern and eastern Australia south to Port Jackson, NSW, and east to Lord Howe Island.

FURTHER DISTRIBUTION

Mergui Archipelago (Myanmar), Singapore, Vietnam, the Philippines, eastern Indonesia and north to Taiwan.

REFERENCES

Clark, H.L., 1938: 174, 176; pl. 20 figs 1–2 (as *Nepanthia magnipinna*); pl. 20 figs 4–5 (as *Nepanthia variabilis*).

Döderlein, 1926: 20; pl. 4 figs 2, 2a (as *Nepanthia polyplax*).

Kenny, 1969: 51–55 (asexual reproduction).

O'Loughlin and Waters, 2004: 4–5 (key), 23–25; figs 4f, 6c.

Ottesen and Lucas, 1982: 223–233, (asexual and sexual reproduction).

Rowe and Marsh, 1982: 99–103; figs 1, 3d–e, 5a–b, 6a–b.

CRASSA SEASTAR

Nepanthia crassa

(Gray, 1847)

Geographe Bay, Western Australia (S. Morrison).

DESCRIPTION

Small disc with five stout, subcylindrical to slightly tapering, blunt-ended arms, flat actinally with the margin defined by inferomarginal plates; primary **abactinal plates subcrescentic, imbricating when young, tumid to hemispherical, surrounded by numerous secondary plates when large**; crystal bodies on plates; papulae single; **primary plates covered by thorny spinelets with 5–8 points**, secondary plates have tufts of spinelets; furrow spines in a graduated fan of nine (7–10) spines with a fan of 7–10 subambulacral spines and 5–7 spinelets; actinal plates are closely covered with radiating spinelets; fasciculate pedicellariae on plates of the lateral field; not fissiparous.

MAXIMUM SIZE

R/r = 72/17 mm (R = 2.8–5.1r).

COLOUR

Variable, often uniform blue, blue-green or orange, more often mottled browns or pink to red and brown.

REPRODUCTION

On 7 January 2011, four specimens of were collected in Cockburn Sound, WA. One female spawned very large opaque grey eggs 633–848 µm (mean 780 µm) in diameter, inferring lecithotrophic development (Lane, personal communication).

HABITAT, DEPTH AND AUSTRALIAN DISTRIBUTION

On algal covered rock platforms, jetty piles, sand and seagrass or muddy sand in sheltered waters, 0–38 m. From Cape Naturaliste to Point Cloates (22°40'S), WA. It is separated by North West Cape from the range of *Nepanthia maculata*. Western Australian endemic on the central west coast.

TYPE LOCALITY

'Western Australia'.

FURTHER DISTRIBUTION

Australian endemic.

REFERENCES

Clark, H.L., 1946: 143 (as *Parasterina crassa*).
O'Loughlin and Waters, 2004: 4–5 (key), 23–25.
Rowe and Marsh, 1982: 97–99; figs 1, 3a–c, 5g, 5j, 6d–e (transfer of species back to genus *Nepanthia*, as designated by Fisher 1941).

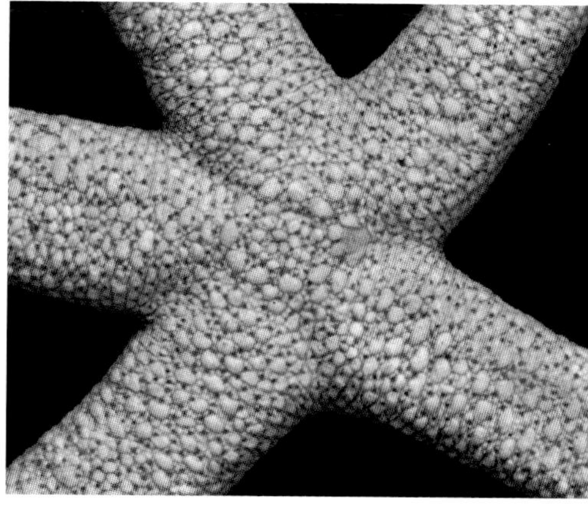

Abactinal plate characteristics, Geographe Bay, Western Australia [left] (C. Bryce), Western Australian Museum [right] (WAMZ9208).

SPOTTED SEASTAR
Nepanthia maculata
Gray, 1840

DESCRIPTION

The disc is small with five regular subcylindrical arms; abactinal plates crescentic to rhomboidal, notched for single papulae; secondary plates few or absent; **plates covered with slender spinelets with 1–3 acute points**; fasciculate pedicellariae may be present between the plates; crystal bodies sometimes present on plate margins; furrow spines in a graduated fan of 7–8 spines, a fan of 9–12 subambulacral spines, and 9–12 additional spinelets on the adambulacral plates.

Western Australian Museum (WAMZ9010), abactinal plate characteristics [bottom].

MAXIMUM SIZE

R/r = 94/15 mm (R = 5.4–8.3r).

COLOUR

Usually cream or beige with either dark blue, violet, green or brown spots in the disc centre and on the arms.

HABITAT, DEPTH AND AUSTRALIAN DISTRIBUTION

On muddy to coarse sand with rubble or coral, in sheltered waters, 0–106 m. Exmouth Gulf, north-west coast, WA, and inner shelf to Weipa, Gulf of Carpentaria, Queensland.

TYPE LOCALITY

Migupou, the Philippines.

FURTHER DISTRIBUTION

Indonesia and the Philippines.

REFERENCES

Clark, H.L., 1938: 175, pl. 20 fig. 3 (as *Nepanthia tenuis*).

O'Loughlin and Waters, 2004: 4–5 (key), 23–45; fig. 2j.

Rowe and Marsh, 1982: 106–107; figs 1, 2f, 5e, 6h–i.

GENUS *PARANEPANTHIA* Fisher, 1917

Seastars of this genus have five arms of medium length that are wide basally and pointed or rounded distally. They are flat actinally with low to elevated arms. The **abactinal and actinal interradial plates have subpaxilliform dense clusters of acicular subsacciform spinelets.** The abactinal plates are mostly irregular on the abactinal surface of the arms, tending to be in rows perpendicular to the margin in the interradial area. The marginal plates are subequal, in regular series with dense tufts of spinelets. The inferomarginals project beyond the superomarginals. The actinal interradial plates are in oblique series, the superambulacral plates are absent and the superactinal plates are present as multiple struts.

Type species: *Paranepanthia platydisca* (Fisher 1913) (Indonesia, the Philippines)
Three species are referred to this genus by O'Loughlin and Waters (2004), only one of which, *P. grandis* (page 172), is found in Australia. *Paranepanthia rosea* was removed from *Paranepanthia* by O'Loughlin and Waters (2004) and placed in *Aquilonastra*. One other species, *P. brachiata*, is located in this genus in WoRMS (Mah, 2019).

Paranepanthia grandis, Hopetoun, Western Australia (C. Bryce).

GRAND SEASTAR
Paranepanthia grandis
(H.L. Clark, 1928)

DESCRIPTION

A fairly large asterinid with a stellate form, **usually five armed with more or less extended flat interradial area and arched disc and arms, actinal surface flat**; disc and median portion of arms covered with irregularly arranged plates of several sizes and shapes, round, elliptical or crescentic in longitudinal rows along arms, **densely covered with very delicate, glassy spinelets giving the surface a velvety appearance**; papulae in groups of up to seven on the disc, 2–3 on the base of arms, and singly near margin except interradially where there are none; actinal plates rhombic, each with a tuft of glassy spinelets; 6–11 furrow spines, 6–11 subambulacral spines with smaller spinelets covering the rest of the plate.

MAXIMUM SIZE

R/r = 57/25 mm (R = ~2.3r).

COLOUR

Pale pink to deep pink or orange.
Actinal surface usually cream.

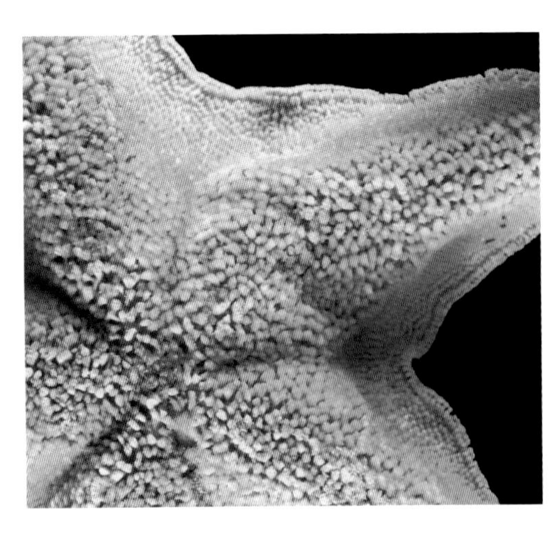

Abactinal plate characteristics,
South Australian Museum (SAMK3765).

FOOD

Small gastropods and compound ascidians.

HABITAT, DEPTH AND AUSTRALIAN DISTRIBUTION

Found under rocks and in rock pools, 0–40 m; uncommon. From northern NSW south to Tasmania and west to Cape Peron, near Fremantle, WA. It is rarely found north of Cape Naturaliste. Southern Australian endemic.

TYPE LOCALITY

'South Australian coast'.

FURTHER DISTRIBUTION

Australian endemic.

REFERENCES

Clark, H.L., 1928: 393–395; fig. 113a–d.

Livingstone, 1933: 14; pl. 3 figs 1–2; pl. 4 figs 2–3 (as *Asterinopsis praetermissa*).

O'Loughlin and Waters, 2004: 4–5 (key), 25–27; figs 5c, 15c.

Shepherd, 1968: 748.

GENUS *PARVULASTRA* O'Loughlin and Waters, 2004

Australian species of this genus have five arms, are pentagonal to subpentagonal in shape, flat actinally, and low and convex abactinally. They have a noticeable integument, and the carinal series of plates is variably present. **The abactinal plates are broadly notched for papulae, and crescentiform**. The papulate areas are extensive and the papulae are large with a few per space. There are few secondary plates per space. **The abactinal spinelets are clustered or spread, not paxilliform, and are short and columnar**. There are no pedicellariae. The superomarginal and inferomarginal plate series are subequal, with the inferomarginals projecting beyond the superomarginals. There is commonly an extensive non-plated actinal area distal to the oral plates. **The actinal interrradial plates are in oblique series, and usually have two spines** mid-arm to distally; these spines are pointed to short. Superambulacral plates are present to varying extent, superactinal plates are always present and of small to very small size.

Type species: *Parvulastra exigua* (Lamarck, 1816) (Indo-West Pacific)

The genus includes five species, all occurring in the southern hemisphere, but only three are found in Australian waters.

Parvulastra exigua, Lord Howe Island, Tasman Sea (N. Coleman).

KEY TO THE SPECIES OF *PARVULASTRA* IN SHALLOW AUSTRALIAN WATERS

1	Carinal plate series regular except proximally.	***P. exigua*** page 175
	Carinal plate series irregular.	**2**
2	Very small size, R ≤ 5mm; South Australia.	***P. parvivipara*** page 177
	Small size, R ≤ 15mm; Tasmania.	***P. vivipara*** page 179

Parvulastra exigua, Cape Paterson, Victoria (D. Paul).

SMALL CUSHION STAR
Parvulastra exigua
(Lamarck, 1816)

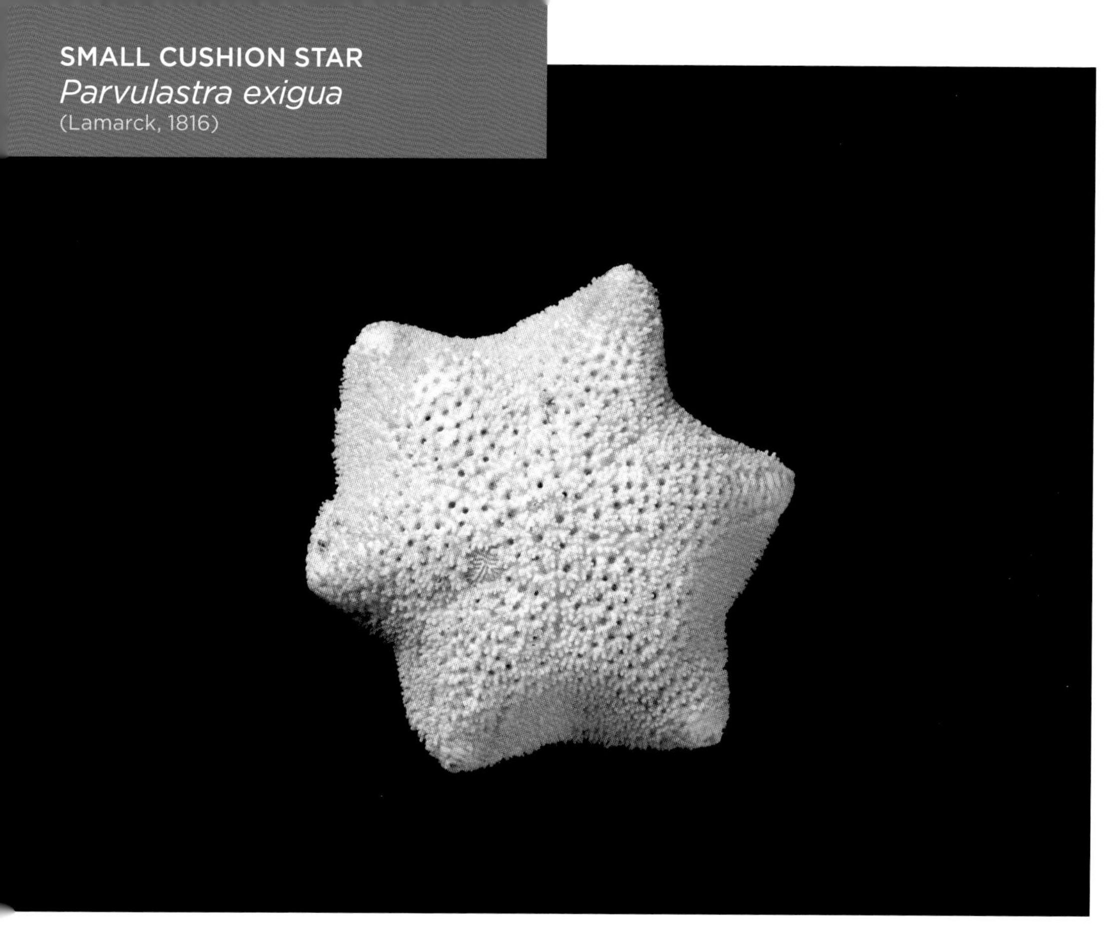

Museums Victoria (NMVF73058).

DESCRIPTION

A small five armed species of *Parvulastra,* nearly pentagonal, with a convex disc of imbricating abactinal plates with spinelets 200–300 μm long; carinal series regular, except proximally; plates crescentiform in regular longitudinal series on arms, broadly notched for papulae; papulate areas extensive with large papular spaces, a few papulae and secondary plates per space, papulae large; spinelets short, columnar; the inferomarginal plates form the margin of the body; their proximal spinelets are columnar or with a distal fringe of thinner, longer spinelets; actinal interradial plates with a single spine, two mid-radially to near the margin and none near the mouth where there is a 'floating' plate in an uncalcified area; there are two short, thin furrow spines and a single tall thick subambulacral spine; oral plates have 5–6 spines, 1–2 suboral spines; gonopores open near the third adambulacral plate on the actinal surface.

MAXIMUM SIZE

R/r = 15/10.7 mm (R = 1.2–1.4r).

COLOUR

Dull green or brown, sometimes with marginal and disc spinelets red, orange, purple, brown or pastel greens and browns. Actinal surface blue-green as are the tube feet with white suckers.

FOOD

Small animals on rocks and algae.

REPRODUCTION

This species is a protandrous hermaphrodite, becoming female at R = 6–11 mm, but some animals are simultaneous hermaphrodites.

Eggs are large (400 µm), the gonopores open on the actinal surface and the eggs are deposited in masses cemented to the underside of boulders. There is no parental care of the eggs. The fertilisation membrane hardens to become an egg case and the larva remains there, using its yolk reserves until it hatches as a brachiolaria with three well developed arms forming a tripod for locomotion. When suitable substrate is located an adhesive disc keeps it in contact with the bottom until metamorphosis at 15–17 days. Newly laid egg masses were found from August to October, but some breeding activity takes place for about nine months of the year.

HABITAT, DEPTH AND AUSTRALIAN DISTRIBUTION

In shallow pools on intertidal rock platforms, often associated with the brown alga *Hormosira*. Abundant in suitable habitats in the intertidal. From southern Queensland to southern Tasmania and the Eyre Peninsula, SA and eastward to Lord Howe Island.

Merimbula Bay, New South Wales [top], Blue Point, New South Wales [bottom] (S. Burrows).

TYPE LOCALITY

False Bay, South Africa.

FURTHER DISTRIBUTION

Southern latitudes including South Africa and Amsterdam and St Paul islands in the southern Indian Ocean, St Helena in the southern Atlantic, and from tropical localities including Mozambique and Chagos.

REFERENCES

Byrne, 1991: 502–505; figs 1c, 2g–k (as *Patiriella exigua*).

Byrne, 1992: 297–316 (reproduction, as *Patiriella exigua*).

Byrne, 1995: 293–305 (larval morphology, as *Patiriella exigua*).

Clark, H.L., 1946: 136 (as *Patiriella exigua*).

Dartnall, 1971: 40–43; fig. 1; pl. 4c, (as *Patiriella exigua*).

O'Loughlin and Waters, 2004: 4–5 (key), 27–28; figs 2l, 11b, 16a–d (transfer to new genus *Parvulastra*).

SOUTHERN SMALL CUSHION STAR
Parvulastra parvivipara
(Keough and Dartnall, 1978)

Smooth Pool, South Australia (G. Edgar).

DESCRIPTION

A very small pentagonal asterinid with a fairly flat body, lobed flat abactinal plates closely imbricated, few secondary plates present; four rows of papulae on each side of the radial midline, only the inner two reach the end of the arm; **spines of carinal plates in groups of 4–7, 3–5 on abactinal interradial spines that are 110–150 µm long and ~100 µm broad at the base**; superomarginal plates not distinct from abactinal plates; proximal inferomarginals with 4–5 spines, distal ones with two spines; **distal actinal plates imbricated, with one (occasionally two) spines 200 µm long**; there is an uncalcified area behind oral plates with one or more 'floating' plates; proximal adambulacral plates with two furrow spines, distal plates with one; subambulacral spines single; 4–5 oral spines per plate, suboral spines usually absent; living animals can be distinguished from *P. exigua* by their colour.

MAXIMUM SIZE

R/r = 5/3.4 mm (R = 1.1–1.4r).

COLOUR

Reddish yellow, pale.

FOOD

Biofilm.

REPRODUCTION

They are simultaneous hermaphrodites and fertilisation takes place within the gonad.

Sexual maturity is reached at R = 2 mm; all specimens over this size contain juveniles that are brooded within the gonad. Eggs are 135 µm diameter when fertilised, they develop into gastulae then into pear-shaped brachiolaria larvae (207 µm long) with reduced brachiolar arms. Metamorphosis takes place in the cavity of the gonad without attachment, newly metamorphosed juveniles have a diameter of 244 µm, the largest intragonadal juvenile has a radius of 1.2 mm. Premetamorphic development appears to depend on reserves in the small egg, post metamorphosis cannibalism supports further growth.

Juveniles have R = ~1.5 mm when released through rupture of the abactinal surface of the parent's disc, apparently causing death of the parent. Each adult carries up to 10 juveniles. Release of juveniles is believed to take place in spring and early summer because, by February, the population consists mostly of juveniles that reach maturity (at R = 2 mm) by September.

Population density varies from one animal per m^2 of suitable habitat to 2,000 per m^2 and the population is believed to be replaced annually.

HABITAT, DEPTH AND AUSTRALIAN DISTRIBUTION

Under rocks in mid to low intertidal rock pools in granite, in sheltered parts of exposed rocky shores, locally abundant in a few localities. It is separated geographcially from *P. vivipara*. Known only from seven localities between Ceduna and Cape Carnot on the west coast of Eyre Peninsula, SA. Southern Australian endemic.

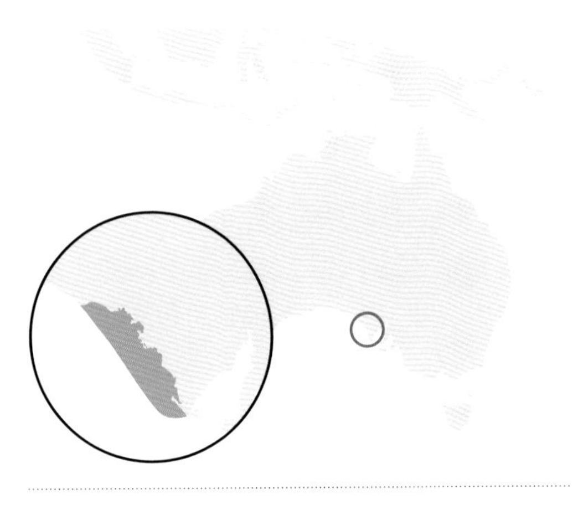

TYPE LOCALITY

Smooth Pool, South of Point Westall, SA.

FURTHER DISTRIBUTION

Australian endemic.

REFERENCES

Byrne, 1996: 551–567.

Byrne and Cerra, 1996: 17–26; figs 2–5 (as *Patiriella parvivipara*).

Byrne and O'Hara, 2017: 127.

Keough and Dartnall, 1978: 407–416; figs 1–7 (as *Patiriella parvivipara*).

O'Loughlin and Waters, 2004: 4–5 (key), 27–28 (transfer to new genus *Parvulastra*).

Roediger and Bolton, 2008: 205–213.

REMARKS

Parvulastra parvivipara is one of the smallest known seastar species in the world (Byrne and O'Hara, 2017).

TASMANIAN CUSHION STAR
Parvulastra vivipara
(Dartnall, 1969b)

Tasman Peninsula, Tasmania (G. Edgar).

DESCRIPTION

A very small asterinid with a **slightly rounded pentagonal shape; the abactinal radial area is covered with imbricating crescentic plates**; papulae are in five rows on each side of the radial area; **the abactinal plates carry 4–13 cylindrical, slightly capitate granular spinelets**; the inferomarginal plates carry 4–5 spinelets on the outer margin and form a conspicuous edge to the body; **actinal plates carry a single spine except near the disc margin where there may be two**; the adambulacral plates bear two furrow and one subambulacral spine; each oral plate carries six oral spines and one suboral spine; an uncalcified area behind the oral plates has one or two 'floating' plates in it; living animals can be distinguished from *P. exigua* by their colour.

MAXIMUM SIZE

R/r = 15/11.2 mm (R = ~1.25r).

COLOUR

Orange-yellow.

FOOD

Algal biofilm on the surface of rocks, emerging at night or in dim light.

REPRODUCTION

This species is a simultaneous hermaphrodite with male, female and hermaphroditic gonads in one animal. Eggs are ~120–150 μm in diameter and are apparently self fertilised. Eggs are brooded in sacs derived from the gonad, not attached to the ovary wall. There are 10 gonads with up to 20 embryos per gonad, at different stages, present at one time. There is a reduced brachiolaria larva that before birth develops into a crawl away juvenile with an $R \leq 2.5$ mm, exiting through the abactinal gonopores.

Reproduction takes place throughout the year, with a period of enhanced reproduction from November to January. The advanced size of juveniles is attained through intragonadal cannibalism. Sexual maturity is attained after two years.

HABITAT, DEPTH AND AUSTRALIAN DISTRIBUTION

Found on the under surface of rocks on gently sloping sheltered to exposed rocky shores, and in the *Galeolaria* fanworm zone and tide pools on rock platforms. Locally abundant in the intertidal. Known only from south-eastern Tasmania between Bruny Island and the Tasman Peninsula. It is separated geographically from *P. parvivipara*. Tasmanian endemic.

TYPE LOCALITY

Pittwater, south-eastern Tasmania.

FURTHER DISTRIBUTION

Australian endemic.

REFERENCES

Byrne, 1991: 505; figs 1d–f (as *Patiriella vivipara*).
Byrne, 1996: 551–567 (as *Patiriella vivipara*, reproduction).
Byrne and Cerra, 1996: 17–26 (as *Patiriella vivipara*).
Chia, 1976: 181 (as *Patiriella vivipara,* reproduction).
Dartnall, 1969b: 294–296; fig. 1; pl. 29a–f.
Dartnall, 1971: 45, 47 (as *Patiriella vivipara*).
O'Loughlin and Waters, 2004: 4–5 (key), 27–28.
Prestedge, 1998: 161–170 (biology, as *Patiriella vivipara*).

REMARKS

Juveniles of *Parvulastra parvivipara* and *P. vivipara* have been observed preying on their siblings once released from their parent (Byrne, 1996).

GENUS *PATIRIELLA* Verrill, 1913

Seastars of this genus have **5–7 arms with the interradial margin straight to incurved. The arms are subpentagonal to short**, with their ends pointed or rounded. There is a noticeable integument, and **the plates on the arms are irregularly arranged**. The abactinal spinelets are granuliform to digitiform and not webbed, either distributed close together or more widely spaced over the projecting surface of the plates. A regular series of superomarginal and inferomarginal plates are covered with granuliform or digitiform spinelets, the actinal plates are in oblique series, the actinal adradial spines in incomplete series. **The actinal spines are digitiform to short and conical with not more than three per plate**. Superambulacral plates are present distally only and the superactinal plates are present as single and multiple plate supports.

Type species: *Patiriella regularis* (Verrill, 1867) (New Zealand, Australia)

Patiriella inornata is only known from the type specimen from an unknown locality in Western Australia. Despite extensive and intensive collecting in Western Australia no further specimens of this species have been found, throwing some doubt on its occurrence in Western Australia. Thus it is included with reservations in this work. *P. littoralis* was previously doubtfully assigned to the genus *Marginaster*. Recently it has been reassigned to *Patiriella*, and the species, commonly known as the Derwent River seastar, is considered to be extinct (O'Hara et al., 2018).

Five species are included in the genus, four of which, *P. regularis, P. inornata, P. littoralis* and *P. oliveri* (with junior synonym *P. nigra*), are reported in Australian waters.

The introduction of both *Asterias amurensis* (front) from the northern Pacific and *Patiriella regularis* (back) from New Zealand in Tasmania's Derwent River are believed to have contributed to the disappearance of *P. littoralis*. Derwent River estuary, Tasmania (G. Edgar).

KEY TO THE SPECIES OF *PATIRIELLA* IN SHALLOW AUSTRALIAN WATERS

1	Inferomarginal spines present.	2
	Inferomarginal spines absent.	***P. inornata*** page 183
2	Abactinal plates more or less regular.	3
	Abactinal plates more or less irregular.	***P. littoralis*** page 184
3	Inferomarginal spines short, columnar to subsacciform.	***P. regularis*** page 187
	Inferomarginal spines digitiform.	***P. oliveri*** page 186

Patiriella regularis, Derwent River, Tasmania (G. Edgar).

UNADORNED SEASTAR
Patiriella inornata
Livingstone, 1933

DESCRIPTION
Body stellate with five arms, interbrachial angles acute; body moderately elevated; abactinal plates on disc not regularly arranged and scarcely imbricated; on top of the arms the plates are crescentic, imbricated and regularly arranged; in the interradial areas the plates are almost circular in outline; **all the abactinal plates are covered by coarse and fairly widely spaced granules**; the superomarginal plates are nearly rectangular, almost twice the size of nearby abactinal plates, and they are bare; there are about 45 superomarginals on the side of each arm; the inferomarginals project slightly beyond the superomarginals but are less than half the size of the superomarginals, and they are bare; papular pores are small and scattered randomly on the disc but radially they form 12 or more irregular series, decreasing distally; the actinal interradial plates are in 9–11 chevrons, plates larger near the mouth, each bears 1–3 short conical spinelets; there are 2–3 very short furrow spines, when three the middle one is longest, outside the furrow are 1–3 subambulacral spinelets; the oral plates bear four marginal spines, the innermost largest, flat, and almost forked at the tip; 2–4 suboral spines.

MAXIMUM SIZE
R/r = 27/15 mm (R = 1.8r).

COLOUR
Not recorded.

HABITAT, DEPTH AND AUSTRALIAN DISTRIBUTION
The only known specimen is from an unknown depth, habitat and locality in WA.

TYPE LOCALITY
'Western Australia'.

REMARKS
No further specimens of this species have been found despite extensive collecting around WA.

REFERENCES
Livingstone, 1933: 17–18;
pl. 1 figs 2, 4; pl. 2 figs 1, 4, 7.
O'Loughlin and Waters, 2004: 29–31.

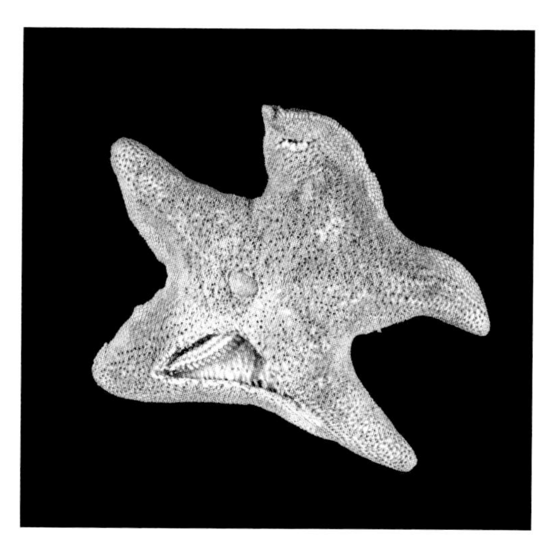

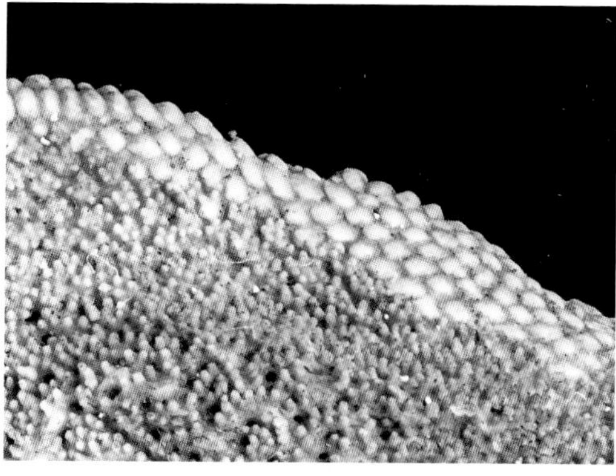

Holotype, Australian Museum (AMJ3198), superomarginal and abactinal plate characteristics [above].

DERWENT RIVER SEASTAR
Patiriella littoralis
(Dartnall, 1970)

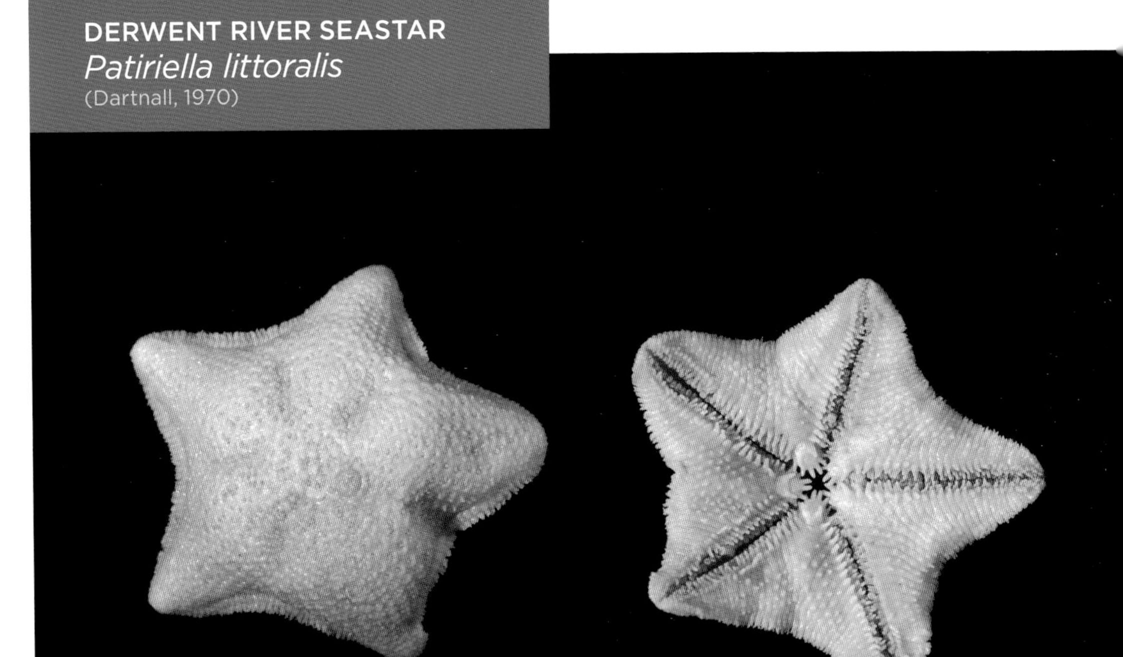

Paratypes, Australian Museum (AMJ7734), juvenile (bottom).

DESCRIPTION

A bluntly stellate seastar; abactinal surface convex, interradial areas slightly depressed, arm bases slightly swollen; **the body is covered with a thick skin that conceals the skeletal plates; the abactinal plates form an open reticulum and carry 4–7 granule-like spinelets on a raised knob**; the spinelets are up to 500 μm long, slightly broader at the base than at the tip; papulae single except for a group near the arm tip in large specimens; the terminal plate is distinct and may carry up to 10 spinelets; inferomarginal plates form the margin of the body and carry 2–5 small pointed tubercles on the upper surface and a fan of 2–4 webbed spinelets on the margin; **actinal interradial plates are slightly imbricate with small secondary ossicles near the margin, each plate has one spine; the skin covering the interradial actinal areas is marked with**

grooves running from the adambulacral plates to the margin; the adambulacral plates bear two furrow and one subambulacral spine; the oral plates have 2–3 oral spines and may or may not carry a suboral spine.

MAXIMUM SIZE

R/r = 17/11.3 mm (R = 1.5r).

COLOUR

Greenish brown, bordered by an off-white band outlining the body. Actinal surface off-white.

FOOD

Probably a particulate feeder or algal grazer, with feeding assisted by mucus in the skin grooves of the actinal surface.

HABITAT, DEPTH AND AUSTRALIAN DISTRIBUTION

Rocky intertidal, under stones on mud on rock platforms. Only known from the Derwent River estuary, from Cornelian Bay to Powder Bay, Tasmania; very rare, possibly extinct. Tasmanian endemic.

TYPE LOCALITY

Derwent River estuary, near Hobart, Tasmania.

FURTHER DISTRIBUTION

Australian endemic.

REFERENCES

Dartnall, 1970c: 207–211; figs 1–2; pl. 13a–c.

Clark, A.M., 1962: 99 (as *Marginaster* sp. cf. *paucispinus*).

Clark, A.M., 1984: 27.

Byrne and O'Hara, 2017: 113, 126, 272–273; fig 4.12.

O'Hara et al., 2019: 1–8, figs 1–3.

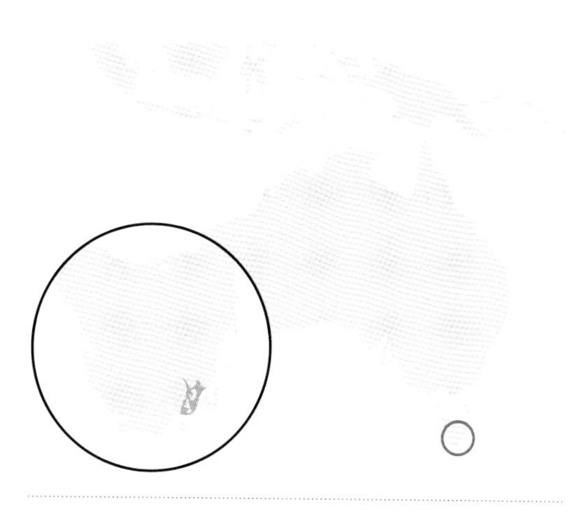

REMARKS

Byrne and O'Hara (2017) discussed this species, noting there was confusion over the identity of some specimens assigned to the species, that the species decline may be related to the appearance of *Patiriella regularis* or the introduced seastar, *Asterias amurensis*, or environmental factors, and that it may be an introduced species. More recently O'Hara et al. (2019) confirmed the identity of the species, and reassigned it to *Patiriella*. They concluded that it is likely to be endemic and may be extinct, which would be the first recorded extinction for an echinoderm species. Targeted surveys in 1993 and 2010 failed to find the species.

Other non-holotype specimens assigned to this species do belong in the genus *Marginaster*, and are thus Poraniidae (Mah, personal communication).

OLIVER'S SEASTAR
Patiriella oliveri
(Benham, 1911)

DESCRIPTION

Pentagonal to nearly stellate with broad, blunt-ended arms; abactinal plates slightly crescentic, the carinal row conspicuous with much narrower plates on either side decreasing in size interradially; papulate and non-papulate areas extensive; **slender cylindrical spinelets are in up to four rows on each plate**, about 14 in each row on the radial plates, 10 in each row laterally (up to a total of 50), up to 20 on superomarginal plates; **most actinal plates have one spine**, stouter than the abactinal spinelets, **near the margin smaller plates of five or six rows have two spines**; furrow spines nearly equal, paired; subambulacral spines single, longer and stouter than the furrow spines, blunt-pointed and somewhat flattened; adradial actinal plates bare or with reduced spines; crystal bodies present below elevations on abactinal plates; no pedicilleriae.

MAXIMUM SIZE

R/r = 35/23 mm (R = 1.5r).

Lord Howe Island, Tasman Sea (N. Coleman).

COLOUR

Dark green black, which dries to dark blue-grey. Actinal surface bluish green; actinal spines greenish with white tips and those near the margin also have a white base.

HABITAT, DEPTH AND AUSTRALIAN DISTRIBUTION

Intertidal, on under surface of rock fragments, outer edge of reef flat or in rock pools. In Australia it is only recorded from Lord Howe Island, Tasman Sea.

TYPE LOCALITY

Sunday Island, Kermadec Islands.

FURTHER DISTRIBUTION

Kermadec Islands, New Zealand.

REFERENCES

Benham 1911: 147; fig. 6 (as *Asterina oliveri*).

Clark, H.L., 1938: 167; pl. 21 figs 3–4 (as *Patiriella nigra*).

Clark, H.L., 1946: 136 (as *Patiriella nigra*).

Dartnall, 1971: 45 (as *Patiriella nigra*).

O'Loughlin and Waters, 2004: 4–5 (key), 29–31; fig. 16e–f.

Mah et al., 2009: 395.

NEW ZEALAND CUSHION STAR
Patiriella regularis
(Verrill, 1867)

Bruny Island, Tasmania (G. Vallee).

DESCRIPTION

A stellate to nearly pentagonal species, usually five, occasionally 6–7 arms; body domed abactinally, flat actinally with an acute angle at the margin; madreporite large; pedicellariae absent; abactinal surface with papulate areas more extensive than non-papulate areas; **radial and proximal interradial plates crescentic, openly imbricate, numerous small secondary plates**, very variable in size and shape; papular spaces large with up to 12 secondary plates and 12 papulae in interradial areas near the disc, up to five secondary plates and papulae in proximal carinal papular spaces, secondary plates rare in distal papular spaces; plates and papulae usually in regular longitudinal series along sides of arms, often irregularly arranged along crest of arm, regular series of up to six carinal plates sometimes present on proximal half of arms, **carinal plates crescentic or with a double notch for papulae**; distal interradial plates closely imbricate, in regular series, without papulae; **abactinal plates with glassy convexities**, raised spine bearing ridge or dome variably evident; **abactinal spinelets digitate, slender, up to 600 µm long; tapering to cylindrical with spinose ends**, spinelets **10–30 in single to double irregular rows across proximal plates**; distal interradial plates with clusters of 2–8 spinelets; superomarginal plates are larger than the inferomarginals and have 6–10 spinelets but the latter project beyond them and have up to 10 spinelets; furrow spines 2–3, usually two, unequal; one subambulacral spine per plate up to 2 mm long, stout, bluntly pointed; adradial plates usually bare; **actinal plates bear a single spine except near the margin where they have 1–2 very small spines**; actinal plates in regular series curving from furrow to margin; some proximal actinal areas not calcified.

MAXIMUM SIZE

R/r = 42/23 mm (R = 1.4–1.8r).

COLOUR

Mottled, predominantly olive green, with yellow, orange, mauve and brown hues. Actinal surface off-white.

FOOD

Small molluscs, microalgae and detritus. An omnivorous scavenger, it is unable to maintain body weight or reproduce on a diet solely of substrate biofilm.

REPRODUCTION

Spawning occurs in summer (January to March) in New Zealand. Gonopores are abactinal and sexes are separate. Eggs are small (150 μm) with indirect development through planktotrophic bipinnaria and brachiolaria larvae that attach to the substrate by an adhesive disc and brachiolar arms, and metamorphose 9–10 weeks after fertilisation. Five to six days later they take up independent existence.

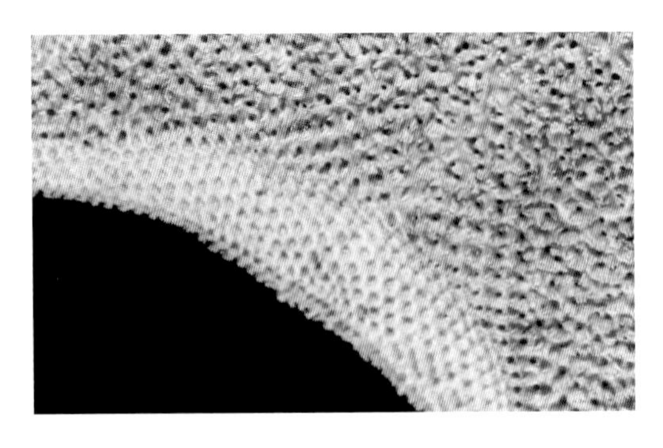

Abactinal interradial plate characteristics, South Australian Museum (SAMAK3926).

HABITAT, DEPTH AND AUSTRALIAN DISTRIBUTION

On rock, sand or mud, 0–100 m. Newcastle Bight, NSW and Tasmania. Abundant near Hobart, Tasmania (from D'Entrecasteaux Channel to the Tasman Peninsula) and in New Zealand from where it is believed to have been accidentally introduced to Tasmania, possibly on New Zealand oysters in the 1930s.

TYPE LOCALITY

Auckland, New Zealand.

FURTHER DISTRIBUTION

North and South Islands, Stewart and Chatham Islands, New Zealand, not reported from the Kermadec Islands.

REFERENCES

Byrne, 1991: 501–502; figs 2a–b, 2d–e.
Byrne and Barker, 1991: 332–345 (reproduction).
Crump, 1971: 137–162.
Dartnall, 1969a: 53–54 (distribution).
Dartnall, 1970b: 73–77.
Fell, 1959: 127–142.
Livingstone, 1933: 16–17; pl. 1 figs 6–7; pl. 2 figs 8, 10–11. (as *Patiriella mimica*).
O'Loughlin et al., 2002: 697–711.
O'Loughlin and Waters, 2004: 4–5 (key), 29–31; fig. 3b.
Mah et al., 2009: 395.

GENUS *PSEUDONEPANTHIA* A.H. Clark, 1916

Seastars of this genus usually have five (4–10) arms that are subcylindrical, not flat actinally, **and lack a marginal edge**. The integument is variably noticeable, and secondary plates are absent. The abactinal spinelets are thick or thin, or subsacciform, either covering plates or in tufts. Pedicellariae are absent, there are no crystal bodies on the plates. Inferomarginal plates are not projecting, furrow spines are up to five per plate. Actinal interradial spines are digitiform, 7–30 per plate. Superambulacral plates are present, transactinal plates are rare, and superactinal plates are lacking. Two species are found in depths of less than 30 m.

Type species: *Pseudonepanthia gotoi* A.H. Clark 1916 (Japan)

Seven species are included in the genus, of which four, *Pseudonepanthia gracilis* (100–540 m depth), *P. briareus* (100 m depth), *P. nigrobrunnea* (page 190) and *P. troughtoni* (page 191) are found in Australian waters. These species were formerly included in *Nepanthia* (Rowe and Marsh, 1982).

The genus is distributed from Japan and the Mariana Islands to the Philippines, eastern Indonesia, eastern and southern Australia and New Zealand.

Pseudonepanthia nigrobrunnea, Byron Bay, New South Wales (N. Coleman).

BLACK AND BROWN SEASTAR
Pseudonepanthia nigrobrunnea
(Rowe and Marsh, 1982)

DESCRIPTION

Small disc, five subcylindrical to slightly tapering blunt-ended arms, not fissiparous; no distinct margin to arms; abactinal plates rather irregular in shape, often double notched with 2–3 (occasionally 1–4) papulae per space; **abactinal plates covered with up to 45 short, stout columnar spinelets with 12–16 points**, the central spinelet somewhat enlarged; furrow spines in a comb of 3–4 subequal spines with 6–7 subambulacral spines in two rows; **crystal bodies and pedicellariae absent**; superambulacral plates contiguous with abactinal plates.

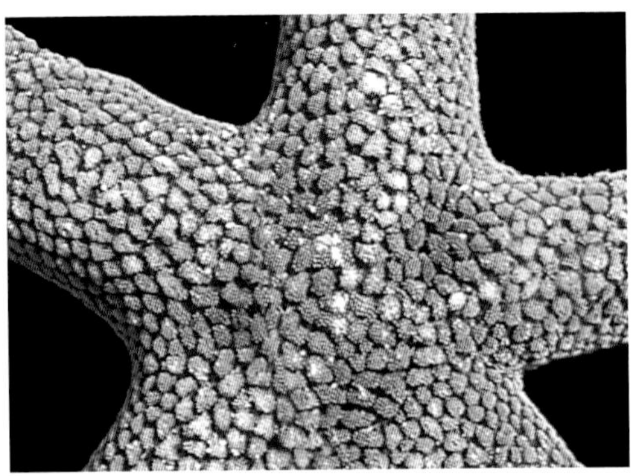

Paratype, Australian Museum (AMJ10920), abactinal plate characteristics [bottom].

MAXIMUM SIZE

R/r = 75/15 mm (R = 4.3–5r).

COLOUR

Very dark brown with black papulae.

HABITAT, DEPTH AND AUSTRALIAN DISTRIBUTION

On rocky substrates, 10–30 m depth. From Double Island Point (25°56'S), Queensland to Solitary Islands, NSW. Eastern Australian endemic.

TYPE LOCALITY

Groper Island, Coffs Harbour, NSW.

FURTHER DISTRIBUTION

Australian endemic.

REFERENCES

Rowe and Marsh, 1982: 95–97; figs 1, 2a–c, 5i, 6j, 6k (as *Nepanthia nigrobrunnea*).

O'Loughlin and Waters, 2004: 4–5 (key), 32–33; fig. 14e.

TROUGHTON'S SEASTAR
Pseudonepanthia troughtoni
(Livingstone, 1934)

Pearson Island, South Australia (P. Southwood).

DESCRIPTION

Small disc with five long cylindrical to slightly tapering, blunt-ended arms; non-fissiparous; no distinct margin to arms; abactinal plates rhomboidal, slightly convex; **abactinal spinelets cover plates, they are short and stout, with 10–20 points surrounding a prominent hemispherical to slightly pointed knob**; furrow spines in overlapping slightly angled combs of 4–5 subequal spines, 6–10 subambulacral spines in 2–3 rows; there are no secondary abactinal

Cape Paterson, Victoria (J. Finn).

plates, glassy convexities or pedicellariae; papulae single; superambulacral plates in a single series along the ambulacrum, sometimes contiguous with the inferomarginal plates.

MAXIMUM SIZE

R/r = 80/17 mm (R = 3.3–5.9r).

COLOUR

Pinkish white or salmon pink to rose, with red papulae and scarlet skin visible between the plates.

FOOD

Encrusting organisms such as bryozoans, sponges and compound ascidians.

REPRODUCTION

Gonopores are on the actinal surface, suggesting benthic development.

HABITAT, DEPTH AND AUSTRALIAN DISTRIBUTION

On open coasts, on rock, often under reef ledges, intertidal to 73 m. From Wilsons Promontory, Victoria to Green Head (30°05'S), WA. Southern Australian endemic.

TYPE LOCALITY

Albany, WA.

FURTHER DISTRIBUTION

Australian endemic.

REFERENCES

Clark, H.L., 1938: 180; pl. 21 (as *Parasterina occidentalis*).

Clark, H.L., 1946: 143 (as *Parasterina troughtoni* and *P. occidentalis*).

Clark, A.M., 1966: 320; fig. 3a–b; pl. 3 figs 4–6 (as *Nepanthia hadracantha*).

Livingstone, 1934: 179; pl. 18 figs 1–6 (as *Parasterina troughtoni*).

O'Loughlin and Waters, 2004: 4–5 (key), 32–33; fig 4e, 14d.

Rowe and Marsh, 1982: 94–95; figs 1, 2d–e, 5k, 6l (as *Nepantia troughtoni*).

Shepherd, 1968: 748 (food).

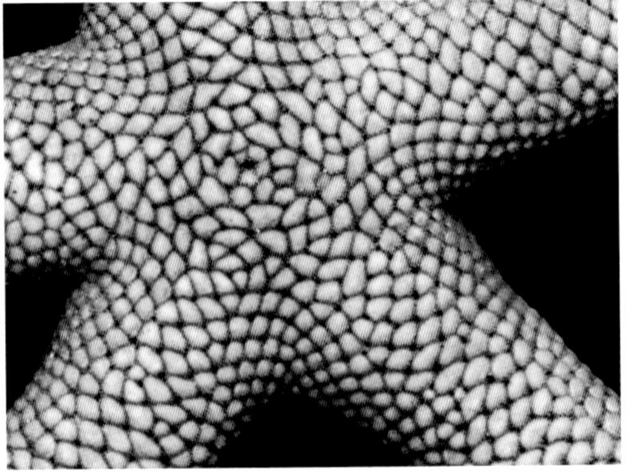

Abactinal plate characteristics, Cape Paterson, Victoria [left] (J. Finn), Western Australian Museum [right] (WAMZ9080).

GENUS *TEGULASTER* Livingstone, 1933

Seastars of this genus are of small size, have five arms, and a thick body covered by thin integument. They form medium to long-armed stellate forms, with elevated arms. They are flat actinally and narrow or broad basally. The abactinal plates are thick, imbricating, shallow notched and crescentiform. These plates are irregular in form and arrangement on a narrow median band on the upper arms. Plates and papulae are in a few series along the sides of the arms, the papulae are mostly single per space, and secondary plates are rare. Abactinal spinelets are few or absent, glassy, sacciform, conical or acicular. Crystal bodies are present, superomarginal plates are in regular series. The inferomarginals have up to seven short spinelets or a tuft of acicular spinelets. The actinal interradial plates are in oblique series, each with 1–2 long or short sacciform spines. The superambulacral plates are rare or not in complete series, and the superactinal plates are present as single plate struts.

Type species: *Tegulaster emburyi* Livingstone, 1933 (Australia)

Apart from the type species, *Tegulaster emburyi,* three species were transferred to *Tegulaster* from other genera and one, *T. ceylanica*, is removed to *Disasterina* (O'Loughlin and Waters 2004). Currently recognised in *Tegulaster* are *T. alba* (from *Asterina*), *T. leptalacantha* (from *Disasterina*) and *T. praesignis* (with synonym *Disasterina spinulifera*) from *Disasterina*. These all occur in Australian waters in depths of less than 30 m.

Tegulaster praesignis, Heron Island, Queensland (N. Coleman).

KEY TO THE SPECIES OF *TEGULASTER* IN SHALLOW AUSTRALIAN WATERS

1	Abactinal plates loosely arranged, not well imbricated.	***T. leptalacantha*** page 197
	Abactinal plates imbricated.	**2**
2	Circle of large crescentic plates in centre of disc.	***T. emburyi*** page 196
	No large crescentic plates in centre of disc.	**3**
3	Superomarginal plates squarish with central group of minute spinelets; body small (R ≤ 9 mm); Tasman Sea.	***T. alba*** page 195
	Superomarginal plates large with spinelets rare; body large (R ≤ 16 mm); Queensland.	***T. praesignis*** page 198

Tegulastra alba brooding juveniles, Lord Howe Island, Tasman Sea (N. Coleman).

WHITE SEASTAR
Tegulaster alba
(H.L. Clark, 1938)

DESCRIPTION
Arms five, with rounded tips; form stellate with rather acute interbrachial arcs; abactinal plates imbricated radially, less so or not at all interradially; **secondary plates absent; each plate carries two or more minute spinelets (visible with a hand lens),** usually in a single linear series of 2–7 (up to 12), but sometimes a double series on large plates; crystal bodies on plates; superomarginals squarish, with a central group of 3–4 minute spinelets; the margin is formed by the inferomarginals, which have projecting tufts of seven spinelets; **actinal interradial plates have one (sometimes two) central spines,** smaller plates have two smaller spines; there are four slender, blunt furrow spines and three subambulacral spines set obliquely on the plate; the oral plates have 6–9 marginal spines and a single large suboral spine on the surface of the plate; larger specimens have one or more uncalcified spots in the actinal interradial areas.

MAXIMUM SIZE
R/r = 9/5 mm (R = ~2r).

COLOUR
All white, larger individuals have minute traces of orange-yellow and a few scattered patches of grey or purple.

REPRODUCTION
Gonopores are actinal and the young are brooded under the discs with up to 150 juveniles released.

HABITAT, DEPTH AND AUSTRALIAN DISTRIBUTION
On under surface of rocks, 0–20 m. Known only from Norfolk and Lord Howe Islands, Tasman Sea, common at the type locality. Tasman Sea endemic.

TYPE LOCALITY
Neds Beach, Lord Howe Island, Tasman Sea.

FURTHER DISTRIBUTION
Australian endemic.

REFERENCES
Clark, H.L., 1938: 150; pl. 22 fig. 7 (as *Asterina alba*).

Clark, H.L., 1946: 132 (as *Asterina alba*).

Coleman, 2007: 19 (reproduction).

O'Loughlin and Waters, 2004: 4–5 (key), 35–36; fig. 18a–b.

EMBURY'S SEASTAR
Tegulaster emburyi
Livingstone, 1933

DESCRIPTION
Arms five, stellate with distinct, strongly convex to keeled arms, narrow at base; the abactinal plates are in regular order except near the ends of the arms, where they become irregular; all are imbricated, strongly so in the radial areas where the plates are largest; **abactinal plates are bare except for erect, forficiform pedicellariae with two conical valves** and a few small, centrally placed spinelets on the disc and near the margin; secondary plates are rare; **the centre of the disc is strongly demarcated by a circle of five large crescentic plates**; the surface of the body is covered by a thin transparent skin; glassy convexities (crystal bodies) are present on the plates; the superomarginals are small, circular and bare; the inferomarginals project beyond the superomarginals and bear a row of 3–6 very short spinelets; **each actinal interradial plate carries a straight series of two (occasionally 1–4) webbed spinelets**; 6–7 furrow spines in a fan, and 2–4 subambulacral spines, slightly longer than the actinal spines; 8–9 oral spines and one large and four short suboral spines.

Holotype, Australian Museum (AMJ5605).

MAXIMUM SIZE
R/r = 19/7.6 mm (R = 2.5r).

COLOUR
The circular area in the centre of the disc is bright orange as are some interradial plates, the remainder is deep cream, blotched with light and dark magenta.

REPRODUCTION
Gonopores open on the abactinal surface, suggesting planktonic development.

HABITAT, DEPTH AND AUSTRALIAN DISTRIBUTION
Under dead coral boulders on the reef crest. Known only from the type locality, intertidal. Queensland endemic.

TYPE LOCALITY
North West Island, Capricorn Group, Queensland.

FURTHER DISTRIBUTION
Australian endemic.

REFERENCES
Clark, H.L., 1946: 143–144.

Clark and Rowe, 1971: 67.

Livingstone, 1933: 11–13; pl. 1 figs 1, 3; pl. 2 figs 2–3, 6, 9.

FINE-EDGED SEASTAR
Tegulaster leptalacantha
(H.L. Clark, 1916)

DESCRIPTION
Stellate; five arms narrowly rounded; disc slightly elevated; **abactinal plates are loosely arranged, not well imbricated**, grading from large to small; the upper edge of each is domed and deeply notched for a papula; **abactinal disc plates bare except for a fringe of small needle-like spinelets around the anus and madreporite**; interradial plates bear 6–10 spinelets particularly in small specimens, more evident distally; the superomarginals are small and bare, the inferomarginals each carry a tuft of 12–15 slender needle-like spinelets forming a well defined flange; the body is covered by skin, which partly obscures the abactinal plates; crystal bodies are present on the plates; the **actinal interradial plates each have a long sacculate spine**; near the mouth is a conspicuous patch of uncalcified membrane, sometimes bearing a 'floating' plate; **furrow spines are in combs of 4–6 webbed spinelets arranged in a fan**, there is a single long sacculate subambulacral spine; 6–9 oral spines and a single long sacculate suboral spine.

MAXIMUM SIZE
R = 25 mm (R = 2–2.5r).

Holotype, Australian Museum (AMJ3082).

COLOUR
Orange-yellow, with greenish grey marks on the upper side. Actinal surface yellowish.

REPRODUCTION
Gonopores open on the abactinal surface, suggesting planktonic development.

HABITAT, DEPTH AND AUSTRALIAN DISTRIBUTION
Under dead coral boulders in pools at low tide, 0–366 m; uncommon. Eclipse Island (13°52'S) in the northern Kimberley, WA and Great Barrier Reef, from Orpheus Island to Heron Island (23°30'S), Queensland.

TYPE LOCALITY
Masthead Island, Queensland.

FURTHER DISTRIBUTION
Mauritius, Réunion Island, Madagascar and east Africa.

REFERENCES

Clark, H.L., 1946: 139 (as *Disasterina leptalacantha*).

Clark and Rowe, 1971: 67 (as *Disasterina leptalacantha*).

Livingstone, 1933: 8–10; pl. 3 figs 5–6; pl. 4 figs 1, 4 (as *Disasterina leptalacantha*).

O'Loughlin and Waters, 2004: 4–5 (key), 35–36.

O'Loughlin, 2009: 210–212, fig. 5.

MARKED SEASTAR
Tegulaster praesignis
(Livingstone, 1933)

DESCRIPTION

Arms five, body stellate, arms convex, narrowly rounded and flattened at their ends, interradial area marked by a deep groove; **abactinal plates strongly imbricated**, covered by a thin skin; the plates are notched for papulae on their upper edges; the abactinal plates are of varying sizes and shapes, not regularly arranged; **except for part of the carinal series, the abactinal plates are bare apart from small granule-like sacciform spinelets around the anus and madreporite**; crystal bodies are present on the plates; the superomarginal plates are large and prominent, heart-shaped and twice as large as adjacent plates; the inferomarginals are smaller than the superomarginals and carry a tuft of 4–6 small, short spinelets, not webbed; the inferomarginals project beyond the superomarginals; **actinal interradial plates carry a single long sacculate spine; the furrow spines are long, slender and webbed and in combs of 3–4**; subambulacrals bear a single long sacculate spine, sharply pointed; the oral plates carry 6–9 partly webbed marginal spines and a single long sacculate suboral spine; there is an uncalcified area distal to the oral plates with 1–3 'floating' plates.

MAXIMUM SIZE

R = 16 mm (R = 1.7r).

COLOUR

Pale cream to light brown with lighter edges.

Abactinal plate characteristics, holotype, Australian Museum (AMJ5059).

REPRODUCTION

Gonopores open on the abactinal surface suggesting planktonic development.

HABITAT, DEPTH AND AUSTRALIAN DISTRIBUTION

Muddy sand with dead and living coral, and under coral rock, 0–7 m. Murray Islands, Torres Strait, Curtis and Heron Islands (23°30'S), Queensland. Queensland endemic.

TYPE LOCALITY

Off Curtis Island, North of Port Curtis, Queensland.

FURTHER DISTRIBUTION

Australian endemic.

REFERENCES

Clark, H.L., 1946: 139 (as *Disasterina praesignis*).
Clark and Rowe, 1971: 67 (as *Disasterina praesignis*).
Livingstone, 1933: 10–11; pl. 1 figs 5, 8; pl. 2 fig. 5 (as *Disasterina praesignis*).
O'Loughlin and Waters, 2004: 4–5 (key), 35–36.

REMARKS

O'Loughlin and Waters (2004) have synonymised *Disasterina spinulifera* with *Tegulaster praesignis*.

ASTERODISCIDIDAE

Asterodiscididae is distinguished from other Valvatids by having a stellate to pentagonal shape and a large inflated disc area. The seastars are of moderate size, and have a reticulate abactinal skeleton with papulae grouped between the plates. There are few, well spaced marginal plates, usually inconspicuous in adults but **with large smooth distalmost superomarginal plates**. The actinal papulae are absent, the interbrachial septum is membranous, and the tube feet have a terminal disc.

The family contains three recent genera and 20 species. Two genera, *Paulia* and *Amphiaster*, are found in the eastern tropical Pacific, and *Asterodiscides* occurs in the Indo-West Pacific region (Lane and Rowe, 2009). Eighteen species belong to the genus *Asterodiscides*. Separated from the Oreasteridae by Rowe (1977), the family is represented in Australia by one genus (*Asterodiscides*), and seven species (six from less than 30 m depth) one known only from north Queensland, one from southern Australia and four found on the west or north-west coasts of Western Australia, one of which has also been found in New Caledonia. Three species have been found sympatrically in passages between the Houtman Abrolhos and another is known only from Shark Bay. *Asterodiscides grayi* Rowe, 1977 is recorded from 70 m depth in Moreton Bay and from 70–90 m depth in northern New Zealand.

Asterodiscides truncatus, Jervis Bay, New South Wales (F. Bavendam).

GENUS *ASTERODISCIDES* A.M. Clark, 1974

Seastars of this genus have **a** pentagonal to stellate body form with an R = 1.1–2r. There are 3–4 superomarginal plates of which the distalmost is larger than the 2–3 proximal plates. They are bare and prominent (except in one species, *A. culcitulus*, in which the distalmost superomarginal is as small as the three proximal plates). The abactinal tubercles are variously shaped, no higher than 4 mm. The actinal granulation and tuberculation is varied, and the intermarginal papulae are abundant. There are 3–7 furrow spines, and the subambulacral spines are in one or two, rarely three, series.

Type species: *Asterodiscides elegans* (Gray, 1847) (West Pacific, South China Sea)

Eighteen species are recognised worldwide, six of which are found in Australian waters in depths of less than 30 m.

Asterodiscides culcitulus, Houtman Abrolhos, Western Australia (C. Bryce).

KEY TO THE SPECIES OF *ASTERODISCIDES* IN SHALLOW AUSTRALIAN WATERS

1	Body more or less pentagonal; R = 1.1–1.5r.	**2**
	Body stellate; R = 1.5–2r.	**4**
2	Actinal plates with more than 10 subequal granules on each plate; distalmost superomarginal plates small, ovate (height nearly equal to diameter); abactinal tubercles subspherical.	***A. culcitulus*** page 202
	Actinal plates have one (occasionally 2–3) prominent subspherical granules on each plate; distalmost superomarginal plates large, elongate (height greater than diameter); abactinal tubercles usually conical or pointed.	**3**
3	Two subambulacral spines in a double row; north-western Australia.	***A. macroplax*** page 203
	Three (occasionally four) subambulacral spines in single row; Queensland.	***A. multispinus*** page 204
4	No ring of granules around base of abactinal tubercles, which are cylindrical to short inverted cone-shaped; distalmost superomarginal plates very convex, ovate meeting abactinally (height nearly equal to diameter); southern Australia (from northern NSW to the Great Australian Bight).	***A. truncatus*** page 208
	A ring of granules around the base of abactinal tubercles; distalmost superomarginal plates elongate or squarish, flat to moderately convex.	**5**
5	Distalmost superomarginal plates very large, squarish, flat, meeting broadly above the terminal plate; abactinal tubercles with the form of an inverted cone.	***A. soelae*** page 206
	Distalmost superomarginal plates elongate (height greater than diameter), moderately convex; abactinal tubercles pointed to subspherical.	***A. pinguiculus*** page 205

(Modified from Rowe, 1985.)

DIAMANTINA'S SEASTAR
Asterodiscides culcitulus
Rowe, 1977

DESCRIPTION
Pentagonal, cushion-like; abactinal and marginal plates obscured by subspherical granules of 3–4 sizes, largest ~2 mm high, 2.5 mm wide; **four superomarginal plates, distalmost one not enlarged**, separated from terminal plate by 1–2 small plates; actinal plates with up to seven subequal tubercles; 5–7 furrow spines, two subambulacral spines and one spine of a second row; pedicellariae, when present, slender, pincer-like.

MAXIMUM SIZE
R/r = 143/85 mm (R = 1.1–1.6r).

COLOUR
Skin and smaller granules yellow-brown, darker in the centre of the disc and radially. Small specimens paler, sometimes disc centre and arms are pinkish red, the remainder lighter in colour.

HABITAT, DEPTH AND AUSTRALIAN DISTRIBUTION
On silty sand and gravelly substrates, 30–180 m. Mandurah to Shark Bay including the Houtman Abrolhos. Western Australian, endemic to the mid-west coast.

TYPE LOCALITY
Off Dongara, WA.

FURTHER DISTRIBUTION
Australian endemic.

REFERENCES
Rowe, 1977: 193–197; fig. 1a–c.

Rowe, 1985: 550–551.

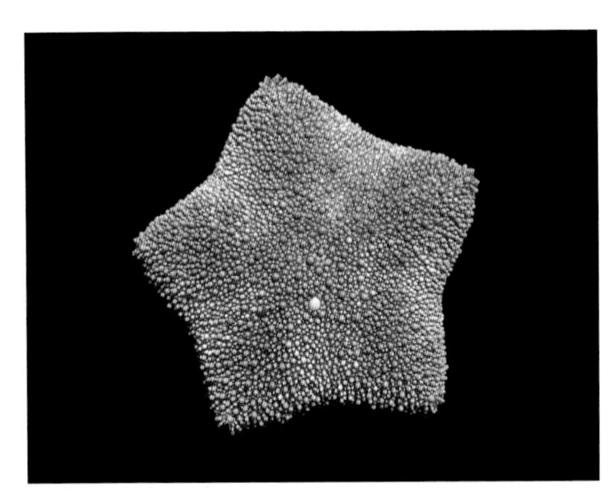

Abactinal tubercle characteristics [above], Dongara, Western Australia [top left] (L. Marsh); abactinal arm tip characteristics [bottom left], Western Australian Museum (WAMZ15642).

LONG-PLATED SEASTAR
Asterodiscides macroplax
Rowe, 1985

DESCRIPTION
Body pentagonal, substellate in juveniles; **abactinal tubercles of three sizes, largest are 2.4 mm high and 1.9 mm wide, conical to bluntly pointed**, spaced or crowded, larger tubercles with or without a basal ring of granules, interstitial granules present; three **superomarginal** plates on each side of arm, **distalmost bare, flat, elongate tear-shaped (8 mm by 4 mm)** the two proximal superomarginals only discernible in juveniles; 4–5 furrow spines, two subambulacral spines, one outside the other; actinal plates usually with a single prominent tubercle (but up to four), pedicellariae usually present.

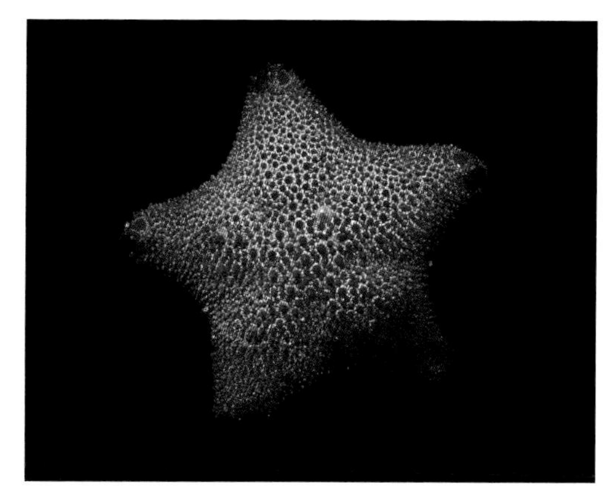

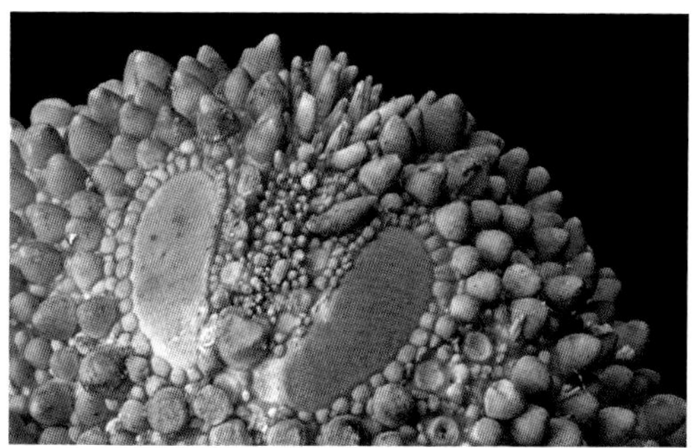

North West Shelf, Western Australia [top] (L. Marsh); abactinal arm tip characteristics [bottom], Western Australian Museum (WAMZ17851).

MAXIMUM SIZE
R/r = 88/60 mm (R = 1.28–1.66r).

COLOUR
Dark red.

HABITAT, DEPTH AND AUSTRALIAN DISTRIBUTION
On silty sand and gravelly substrates, 7–100 m. From the Houtman Abrolhos, Exmouth Gulf and the North West Shelf to Port Walcott, near Roebourne. Western Australian endemic.

TYPE LOCALITY
Houtman Abrolhos, WA.

FURTHER DISTRIBUTION
Australian endemic.

REFERENCES
Rowe, 1977: 204–206
(as *Asterodiscides pinguiculus*).
Rowe, 1985: 536–540; fig. 1; pls 2d–f, 3a–d.

REMARKS
This species is very similar to *Asterodiscides pinguiculus* but the latter has differently shaped distalmost superomarginal plates.

MULTI-SPINE SEASTAR
Asterodiscides multispinus
Rowe, 1985

DESCRIPTION
Body more or less pentagonal; abactinal tuberculation and granulation as in *A. macroplax;* two proximal superomarginal plates not discernible even after denuding arm; **distalmost superomarginals flat, tear-shaped, nearly or twice as long as wide**; four (occasionally 3–5) furrow spines, **3–4 (up to six) subambulacral spines** across the width of the adambulacral plate; actinal plates with single, rarely 2–3 prominent tubercles; no pedicellariae on the abactinal surface, few on the actinal surface.

MAXIMUM SIZE
R/r = 97/75 mm (R = 1.3r).

COLOUR
Cream with dark red tubercles and orange tube feet.

HABITAT, DEPTH AND AUSTRALIAN DISTRIBUTION
Sand and mud, 23–28 m. So far only known from north-east of Townsville, Queensland on the inner continental shelf.

TYPE LOCALITY
Off Townsville, Queensland.

FURTHER DISTRIBUTION
Australian endemic.

REFERENCES
Rowe, 1985: 540–541; fig. 1; pl. 3e–f.

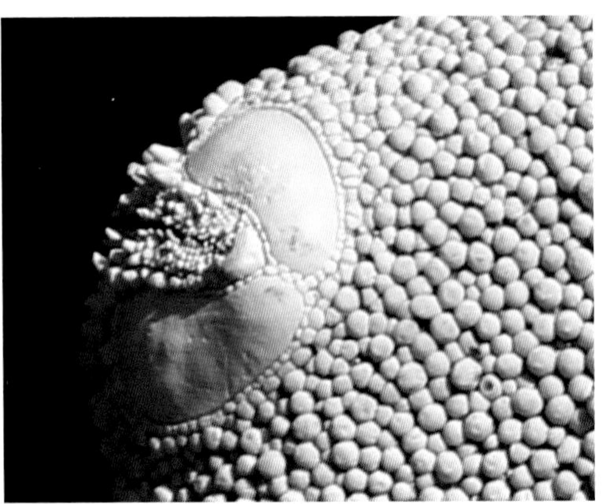

Abactinal tubercle [above] and abactinal arm tip characteristics [bottom left], Australian Museum (AMJ1686).

ROTUND SEASTAR
Asterodiscides pinguiculus
Rowe, 1977

DESCRIPTION
Stellate; **the only marginal plates visible are the flat, pear-shaped distalmost superomarginals** and the underlying 4–6 distal inferomarginals; abactinal tubercles small, conical, closely packed, of two sizes, the largest (1.75 mm) sharply conical, with a ring of granules at the base; actinal plates each with one large (2.75 mm) central conical tubercle with a peripheral ring of variously sized squarish granules; 3–5 furrow spines, **two subambulacral spines, one outside the other; pedicellariae not present on the holotype**.

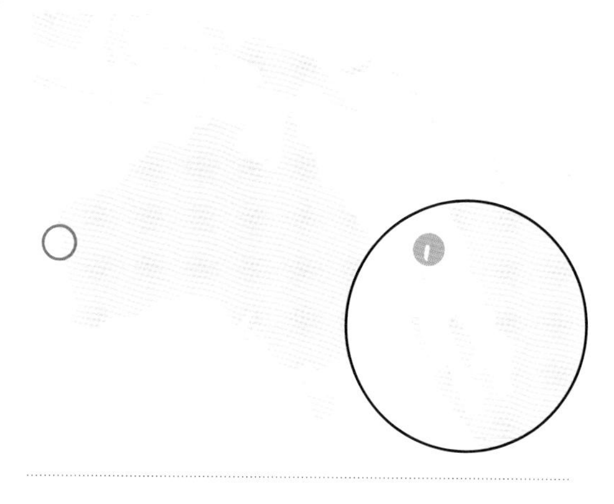

MAXIMUM SIZE
R/r = 92.5/-58.5 mm (R = 1.6r).

COLOUR
Uniform deep maroon.

HABITAT, DEPTH AND AUSTRALIAN DISTRIBUTION
On rubble and sand substrates, 23–27 m. Known only from the holotype from near Bernier Island, Shark Bay, WA. Western Australian endemic.

TYPE LOCALITY
Off Bernier Island, Shark Bay, WA.

FURTHER DISTRIBUTION
Australian endemic.

REFERENCES
Rowe, 1977: 204–206; fig. 4h–j.

Rowe, 1985: 537 (Exmouth Gulf specimen referred to *Asterodiscides macroplax*).

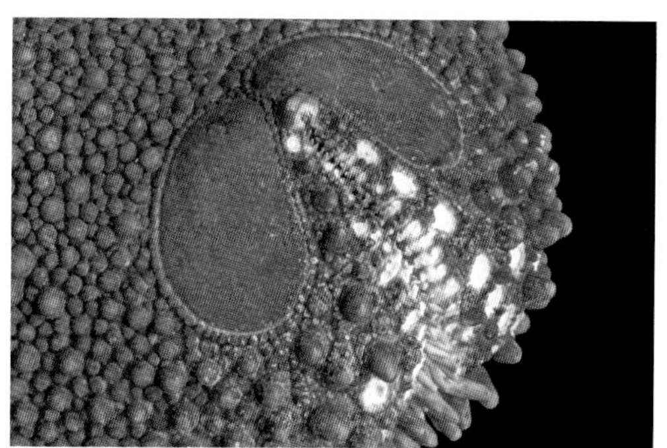

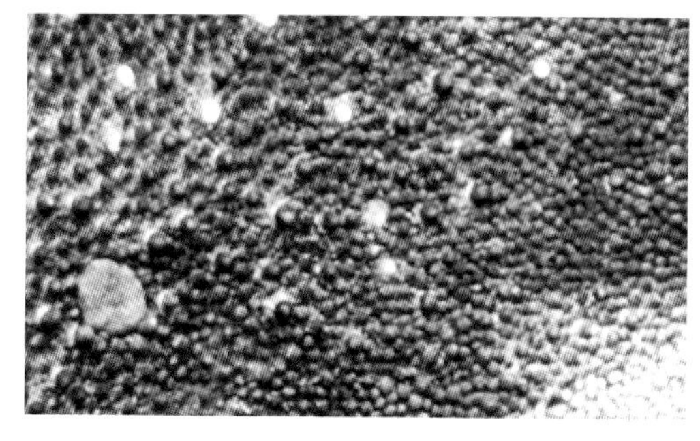

Abactinal arm tip [above] and abactinal tubercle characteristics [right], holotype, Western Australian Museum (WAMZ1731).

SOELA'S SEASTAR
Asterodiscides soelae
Rowe, 1985

DESCRIPTION
Stellate; abactinal tuberculation coarse, tubercles cone-shaped, up to 3.6 mm diameter, top slightly convex, with a basal ring of granules, interstitial granules present; **four superomarginal plates, only the distalmost discernable, squarish (13.4 mm by 10.2 mm), flat (with a peripheral ring of small granules)**, broadly in contact above terminal plate; 3–4 furrow spines, **usually a single subambulacral spine**, a second (outer) spine irregularly present; actinal plates with a spatulate or subspherical tubercle; pedicellariae present or absent.

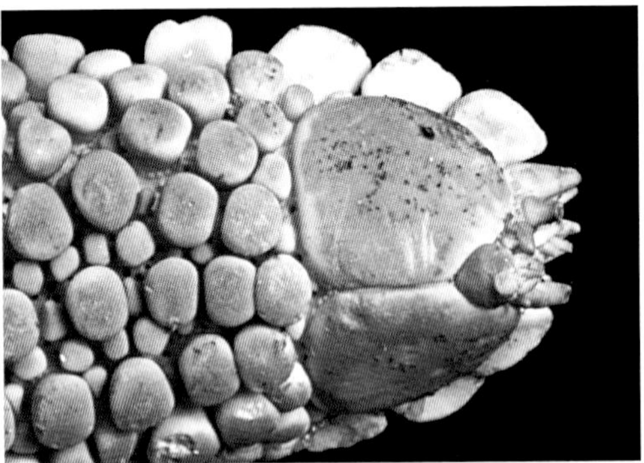

Abactinal tubercle characteristics [top], Houtman Abrolhos, Western Australia (L. Marsh); abactinal arm tip characteristics [bottom], Western Australian Museum (WAMZ17826).

MAXIMUM SIZE
R/r = 135/68 mm (R = 2–2.2r).

COLOUR
Tan to pinkish red, distal superomarginals sometimes paler.

HABITAT, DEPTH AND AUSTRALIAN DISTRIBUTION
On silty sand and gravelly substrates, 20–80 m. From the Houtman Abrolhos to the North West shelf north of the Dampier Archipelago, WA.

TYPE LOCALITY
Goss Passage, Wallabi Group, Houtman Abrolhos, WA.

FURTHER DISTRIBUTION
New Caledonia.

REFERENCES
Rowe, 1985: 547–550; fig. 1; pls 4a–f, 5a.

REMARKS
The name honours the research vessel RV *Soela* from which many specimens were collected.

Houtman Abrolhos, Western Australia (L. Marsh).

Houtman Abrolhos, Western Australia (C. Bryce).

FIREBRICK SEASTAR
Asterodiscides truncatus
(Coleman, 1911)

DESCRIPTION

Stellate, with **four superomarginal plates**, the first three from the interradius small, the **last one on each side large, bare, strongly convex** and meeting the opposite one above the terminal plate; **large abactinal tubercles** more or less cylindrical to shaped like a truncated inverted cone with a convex lower surface, smaller ones cylindrical with a convex upper surface; there are scattered granules among the tubercles; actinal plates each with a prominent nearly cylindrical tubercle surrounded by unequal granules; 3–4 furrow spines, a double row of subambulacral spines the outer one small, angular; pedicellariae slender, curved, pincer-like, on marginal, actinal and some abactinal plates.

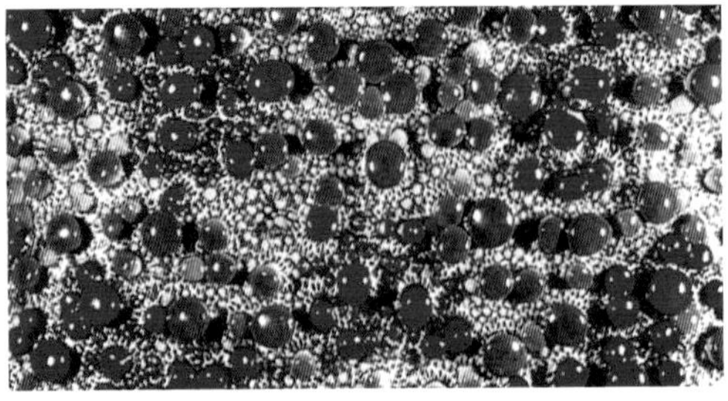

Australian Museum (AMJ24600), abactinal tubercle characteristics [bottom].

MAXIMUM SIZE

R/r = 166/83 mm, R = 1.5–2r.

COLOUR

Variable, typically purple and red or mauve, orange and yellow.

REPRODUCTION

Observed spawning at Poor Knights Islands, New Zealand during April (Green, personal communication).

HABITAT, DEPTH AND AUSTRALIAN DISTRIBUTION

On reef or silt, 14–804 m. Off North Solitary Island, New South Wales to Albany, WA.

TYPE LOCALITY

Off Botany Bay, NSW and off the east coast of Victoria.

FURTHER DISTRIBUTION

Kermadec Islands and North Island, New Zealand.

REFERENCES

Clark, H.L., 1946: 108 (as *Asterodiscus*).
Coleman, 1911: 699 (as *Asterodiscus*).
Mah et al., 2009: 395.
Rowe, 1977: 200–201; fig. 3e–g.
Rowe, 1985: 532–534 (tabular key).

Jimmies Island, New South Wales [top] (A. Green);
King George Sound, Western Australia [bottom] (S. Morrison).

ASTEROPSEIDAE

Asteropseids have five arms **and a stellate body covered by a more or less thickened skin, usually obscuring the skeleton**. The abactinal skeletal plates are tessellate in juveniles, and imbricating or reticulate in adults (changing during growth in some species). The plates bear single or numerous spines or none. The pedicellariae are bivalved, granuliform or elongate tong-shaped. The marginal plates are well developed, and the actinal plates are in longitudinal series parallel to the ambulacral groove.

The family Asteropseidae has five genera found in tropical and warm temperate seas; three genera and three species are found in Australian seas, all in shallow water.

Asteropsis carinifera, Réunion Island (P. Bourjon).

Asteropsis carinifera, Jicarita Island, Panama (C. Bryce).

KEY TO THE GENERA OF ASTEROPSEIDAE IN SHALLOW AUSTRALIAN WATERS

1	Body covered by a thick, smooth, rubbery skin obscuring the skeleton except in dried specimens; a pair of large bivalved pedicellaria interradially on the abactinal surface.	***Petricia*** page 214
	Body covered by skin bearing one small spine per plate or large marginal and carinal spines only.	2
2	Prominent fringe of large, horizontal marginal spines and an irregular series of carinal spines.	***Asteropsis*** page 211
	Small spines on all abactinal plates; very prominent superomarginal plates each with an obliquely placed giant bivalved pedicellariae surrounded by short conical spines.	***Valvaster*** page 214

GENUS *ASTEROPSIS* Müller and Troschel, 1840

Type species: *Asteropsis carinifera* (Lamarck, 1816) (Indo-Pacific)

Monotypic genus. Diagnosis as for the species (page 212).

HORNED SEASTAR
Asteropsis carinifera
(Lamarck, 1816)

Muséum National d'Histoire Naturelle (MNHNIE201431).

DESCRIPTION

Stellate, except for very small specimens that are almost pentagonal; body covered by smooth skin more or less obscuring the close fitting oval or circular plates (small specimens) or more widely spaced plates (large specimens); the arms are triangular in cross section; **the carinal plates form a radial ridge emphasised by a series of spines**, small specimens are flatter and lack radial spines; the edge of the body is formed by a flange of superomarginal plates each bearing a prominent conical, pointed spine; in very small specimens the inferomarginals bear small spines; actinal surface bare, actinal plates in rows parallel to the furrow; furrow spines five, the three middle ones longest, a single very stout subambulacral spine per plate; papulae single in small specimens, in groups in large ones, confined to the abactinal surface; a single straight, erect pedicellaria only on adambulacral plates next to the furrow spines.

MAXIMUM SIZE

R/r = 88/25 mm (R = 2.3–3.5r).

COLOUR

Variable, often mottled shades of grey-green or brown sometimes blotched with darker brown, black, white, blue, yellow or red matching the varied colours on the underside of boulders.

FOOD

Scavenges dead animals. Has been observed eating cone shells in an aquarium.

REPRODUCTION

In the Red Sea reproduction takes place during April. The eggs are 200 µm in diameter and develop into a planktotrophic bipinnaria larva that goes through a brachiolaria stage before metamorphosis after 30 days. The larvae swim close to the surface.

Great Barrier Reef, Queensland [top]; Papua New Guinea [bottom] (R. Steene).

HABITAT, DEPTH AND AUSTRALIAN DISTRIBUTION

On coral reefs and rocky tropical shores, under rocks or coral, 0–13 m. In northern Australia it is only known from the offshore reefs, Rowley Shoals and Scott Reef, WA; Ashmore Reef, Timor Sea and from Torres Strait to the Capricorn Group, Great Barrier Reef, Queensland.

TYPE LOCALITY

Not recorded.

FURTHER DISTRIBUTION

This is one of the most widely distributed Indo-West Pacific species, found from east Africa to India, South-East Asia, Taiwan and the Ryukyu Islands to Hawaii, in the eastern Pacific at Mexico, Panama and the Galapagos Islands.

REFERENCES

Clark, H.L., 1946: 109–110 (as *Asterope carinifera*).

Clark, A.M. and Rowe, 1971: 65; pl. 9 fig. 9.

Maluf, 1988: 120 (distribution).

Mortensen, 1937: 44; pl. 10 fig. 4 (reproduction).

Yamaguchi, 1975: 20 (food).

GENUS *PETRICIA* Gray, 1847

Type species: *Petricia vernicina* (Lamarck, 1816) (Australia, New Zealand)

One species is found in shallow Australian waters. Diagnosis as for the species (page 215).

GENUS *VALVASTER* Perrier, 1875

Type species: *Valvaster striatus* (Lamarck, 1816) (Indo-Pacific)

Monotypic genus. Diagnosis as for the species (page 217).

Petricia vernicina, Port Jackson, New South Wales (J. Turnbull).

VELVET SEASTAR
Petricia vernicina
(Lamarck, 1816)

South Australian Museum (SAMAK3962).

DESCRIPTION

Disc large with five short, thick arms; **the body is covered with a smooth, tough, rubbery skin, which obscures the skeleton**; in dried specimens the reticulate skeleton and prominent marginal plates are easily seen; in small specimens the skeleton is composed of close fitting, slightly lobed plates that become separated during growth and joined by narrow connecting ossicles giving the reticulate skeleton of adults; there are no tubercles, but small spines sometimes

Houtman Abrolhos, Western Australia (C. Bryce).

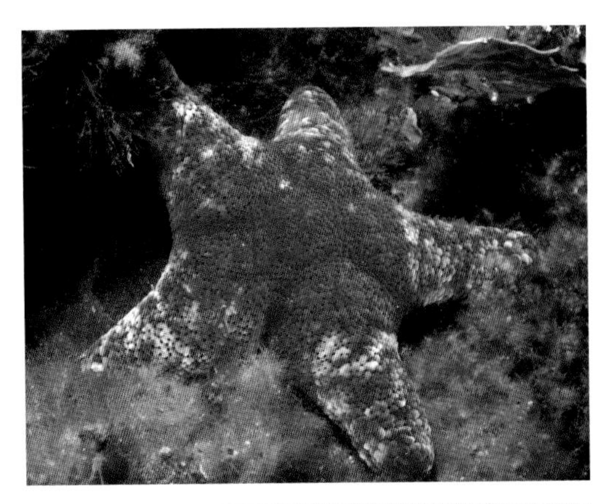

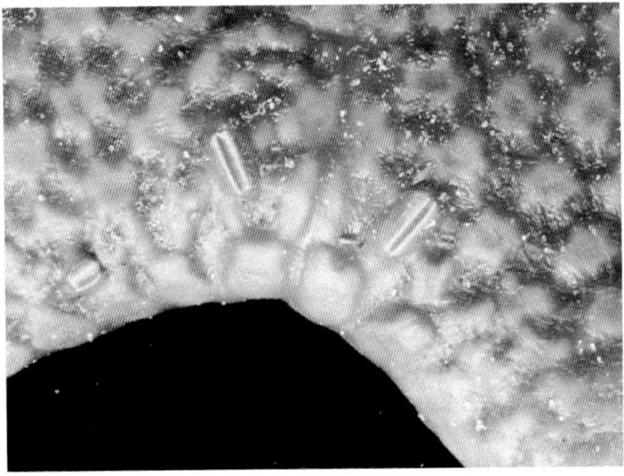

Port Phillip, Victoria [top] (J. Finn); abactinal interradial characteristics [bottom], South Australian Museum (SAMAK3962) .

occur on the distal inferomarginal plates; **two large bivalved pedicellariae are usually conspicuous on the abactinal surface in each interradial area** and sometimes on the actinal surface near the mouth; the furrow spines are paired, followed by 1–2 subambulacral spines.

MAXIMUM SIZE

R/r = 104/55 mm (R = 1.8–2.8r).

COLOUR

Variable, often red, orange, dull green, brown, pink or purple, sometimes mottled.

FOOD

Encrusting sponges (76% of diet), bryozoans, colonial ascidians and coralline algae.

HABITAT, DEPTH AND AUSTRALIAN DISTRIBUTION

Under ledges of shallow rock platforms and on rocky substrates on sheltered to moderately exposed coasts, 0–70 m. Found from Caloundra, Queensland, around southern Australia and Tasmania, also from Lord Howe and Norfolk Islands, to the Houtman Abrolhos, WA.

TYPE LOCALITY

'Australian Seas'.

FURTHER DISTRIBUTION

Kermadec Islands, New Zealand.

REFERENCES

Clark, H.L., 1946: 110 (as *Petricia vernicina* and *P. obesa* H.L. Clark, 1923).

Mah et al., 2009: 395.

Shepherd, 1968: 743–744.

Sloan, 1980: 71 (food).

STRIATED SEASTAR
Valvaster striatus
(Lamarck, 1816)

DESCRIPTION

Stellate, with fairly short, flat, broad arms; abactinal plates connected by ossicles that elongate with growth to give a regular reticulate skeleton in large animals; **superomarginal plates prominent, each with a very large bivalved pedicellaria except in small specimens** in which pedicellariae are few or absent; abactinal plates and connecting ossicles bear conical spines up to 3 mm long giving a slightly spaced, spiny covering to the disc and arms; there are erect pincer-like pedicellariae among the spines; actinal plates have a mixture of large and small spines; furrow spines in a fan of 4–5 with a single blunt subambulacral spine per plate.

MAXIMUM SIZE

R/r = 85/30 mm (R = 2.8–3.5r).

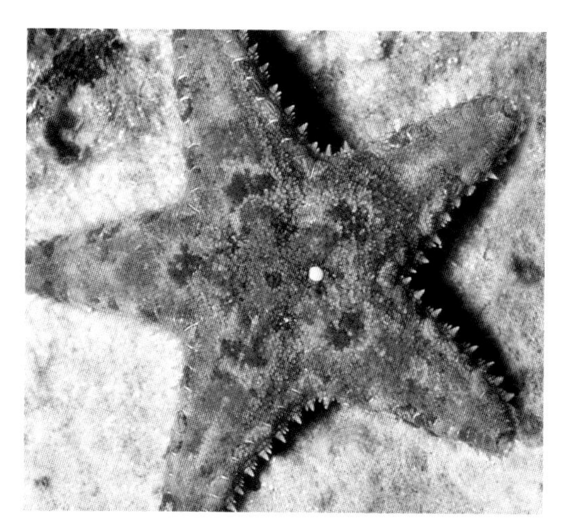

Madang, Papua New Guinea, superomarginal characteristics [bottom] (R. Steene).

COLOUR

Variable, often brown and cream, green, banded or blotched with purple, pink or red.

HABITAT, DEPTH AND AUSTRALIAN DISTRIBUTION

Under dead coral near the outer edge of coral reef platforms and on outer reef slopes on sand or under rocks, 0–25 m. Off WA it has only been found once, on the outer slope of Cartier Island reef in the Timor Sea. It is recorded from Torres Strait (as *V. spinifera*), and the northern Great Barrier Reef.

TYPE LOCALITY

Mauritius.

FURTHER DISTRIBUTION

Found from Mauritius, the Bay of Bengal, the Philippines, Papua New Guinea, New Caledonia, Fiji to Hawaii and may be expected wherever intensive collecting is undertaken on tropical reefs. It seems to be a genuinely rare species.

REFERENCES

Blake, 1980: 167–173, 178–180 (transfer of *Valvaster* from Valvasteridae to Asteropseidae).

Clark, A.M. and Rowe, 1971: 71; pl. 9 figs 10–11.

Clark, H.L., 1921: 102; pl. 6 fig. 6; pl. 36 figs 8–9 (as *Valvaster spinifera*).

Clark, H.L., 1946: 151 (as *Valvaster spinifera*).

CHAETASTERIDAE

The single genus in the family, *Chaetaster* Müller and Troschel, 1840, has **paxilliform plates with slender, glassy spinelets**. The seastars have five, slender, cylindrical arms. The marginal plates are small with an **odd interradial marginal plate in each series**. Superambulacral plates are present and **interbrachial septa are calcareous**. One species is known from Australia.

Chaetaster moorei, New Caledonia (P. Laboute).

GENUS *CHAETASTER* Müller and Troschel, 1840

Type species: *Chaetaster moorei* Bell, 1894 (Indo-West Pacific)

Four species are recognised worldwide, only the type species (page 219) is found in shallow Australian waters.

MOORE'S SEASTAR
Chaetaster moorei
Bell, 1894

DESCRIPTION
Disc very small, arms cylindrical, long and very slender; the skeletal plates, except the adambulacrals, are paxilliform, oval on the carinal plates, rounded on the remainder; paxillae are covered by about 60 short, fine, glassy spinelets, some also bear a sharp spine in their centre; the peripheral spinelets are slightly longer than the others; abactinal plates are in 9–13 (size dependent) rows; **marginal plates small with an odd interradial marginal plate in each series**; there are 5–8 furrow spines in a fan followed by four upright stouter spines; papulae are only found on the abactinal surface where they occur singly, between the paxillae; there are no pedicellariae.

MAXIMUM SIZE
R/r/br = 87.5/9.5/7.5 mm (R = 9.2r).

COLOUR
Reddish brown with yellow arm tips. Actinal surface orange-brown.

FOOD
Probably feeds on substrate biofilm.

Interradial and carinal plate characteristics, Australian Museum (AMJ10958).

HABITAT, DEPTH AND AUSTRALIAN DISTRIBUTION
On dead coral on outer reef slopes, 20–73 m; rare. In Australia it is only known from Middleton Reef and Lord Howe Island, Tasman Sea.

TYPE LOCALITY
Macclesfield Bank, South China Sea.

FURTHER DISTRIBUTION
Andaman Islands, Bay of Bengal, New Caledonia and Ogasawara Islands, Japan.

REFERENCES
Bell, 1894: 404.
Clark, A.M., 1951: 1256.
Clark, A.M. and Rowe, 1971: 65.
Guille et al., 1986: 142–143.

GONIASTERIDAE

Goniasterids commonly have five arms and a large, often pentagonal disc. The interradial arcs are rounded, and the arms are usually short and blunt or long, slender and pointed. The marginal plates often form a conspicuous edge to the disc and arms. Plates may be tabulate, with granule-covered skin, or bare and ringed by granules. The skeleton often bears spines. The plates are tessellate or connected by short connecting ossicles. Pedicellariae are often present and are large and valvate. The papulae are generally single, and are usually only on the abactinal surface. The tube feet have sucking discs, and **the interbrachial septum is membranous**.

This is a large cosmopolitan family of 72 genera of which 24, with 52 species, are recorded from Australian waters. Twelve genera with 30 species are from depths of less than 30 m.

As a result of genetic studies two genera formerly in the family Ophidiasteridae, *Fromia* and *Neoferdina* (Mah and Foltz, 2011a) and one (*Nectria*) from the Oreasteridae (Mah, 2005), have been transferred to the Goniasteridae.

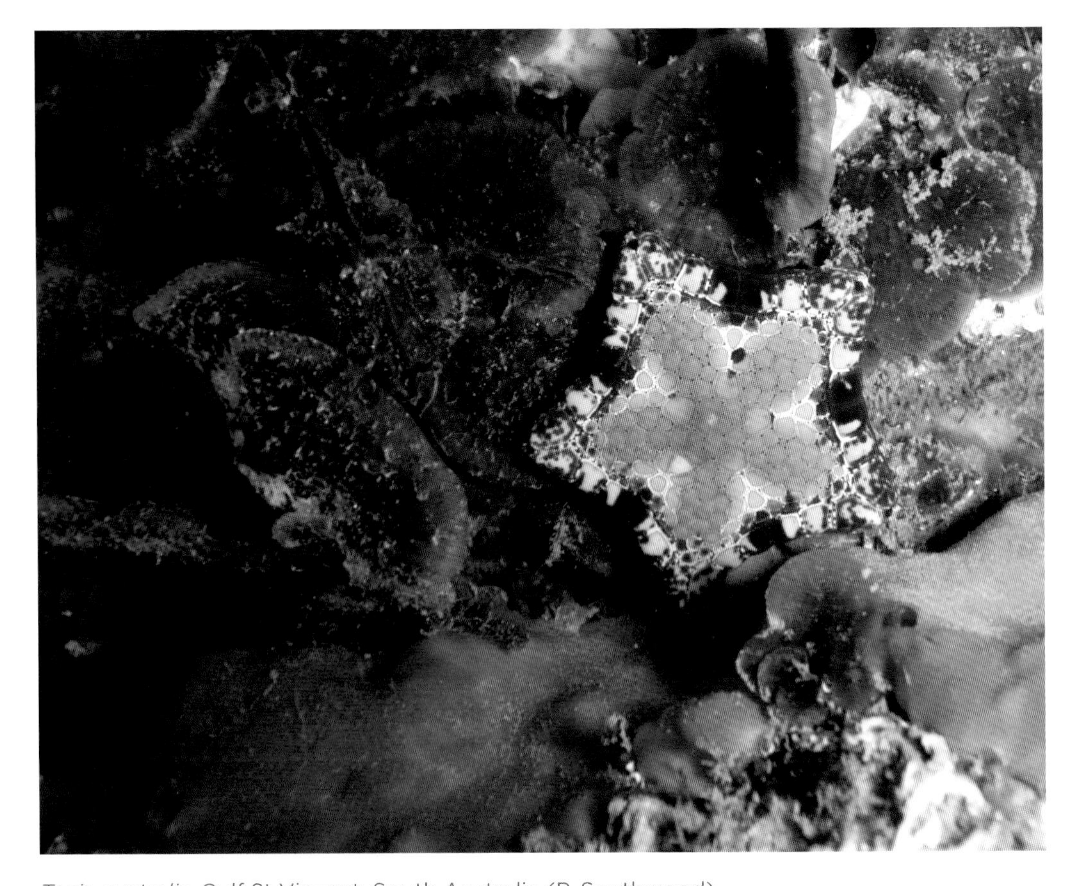

Tosia australis, Gulf St Vincent, South Australia (P. Southwood).

KEY TO THE GENERA OF GONIASTERIDAE IN SHALLOW AUSTRALIAN WATERS

1	Abactinal plates paxilliform, tabulate or convex, not flat or covered with membrane; marginal plates granulose.	2
	Abactinal plates flat or slightly convex, polygonal or stellate.	3
2	Abactinal plates paxilliform, small and very numerous.	***Mediaster*** page 245
	Abactinal plates tabulate, very large, crowned with more or less flattened, often enlarged granules surrounded by a sharply defined circle of short, wide spinelets.	***Nectria*** page 247
3	Abactinal plates bare, not covered by smooth or granulose membrane.	4
	Abactinal surface covered by smooth or granulose membrane which may obscure the outline of the underlying plates.	7
4	Abactinal plates bare, extending to the arm end or nearly so; plates outlined by small granules often contrasting in colour.	5
	Abactinal plates bare, confined to disc; superomarginals meet medially for the whole length of the arm; granules between plates prominent radially on disc.	***Iconaster*** page 242
5	Body form pentagonal to slightly stellate.	***Tosia*** page 278
	Body form stellate or substellate.	6
6	Body form substellate; terminal plate larger than adjacent marginal; north-eastern Australia.	***Anchitosia*** page 223
	Body form substellate to stellate; terminal plate smaller than adjacent marginal plates; southern Australia.	***Pentagonaster*** page 262
7	Disc pentagonal to stellate with rapidly tapering arms, surface granulose; $R < 3r$.	8
	Disc small, arms elongate, gradually tapering, surface granulose; $R > 3r$.	11

8	Granules small to minute.	9
	Granules coarse.	10
9	Disc small; granules minute, spaced apart; abactinal plates extending almost to arm tips; marginal plates without spines; $R < 3r$.	***Anthenoides*** page 223
	Disc large; granules very small, crowded; abactinal plates extending to tips of short pointed arms; inferomarginals usually bear one or more spines; R usually $< 3r$.	***Stellaster*** page 267
10	Disc, arms and actinal surface densely covered with fairly coarse, convex granules; tubercles (if any) small; arms broad with blunt ends; pedicellariae absent.	***Styphlaster*** page 267
	Disc and arms closely covered by slightly convex polygonal granules and five very large rounded tubercles, one at the base of each radial series of plates; numerous very large bivalved pedicellariae on actinal plates, small bivalved pedicellariae abactinally.	***Pseudogoniodiscaster*** page 262
11	Furrow spines only, no subambulacral spines; actinal granulation smooth, continuous up to and partly covering furrow spines; no actinal pores.	***Neoferdina*** page 257
	Subambulacral spines and actinal pores usually present; intermarginal plates may also be present.	***Fromia*** page 226

Pentagonaster duebeni, King George Sound, Western Australia (C. Bryce).

GENUS *ANCHITOSIA* Mah, 2007

Type species: *Anchitosia queenslandensis* (Livingstone, 1932a) (West Pacific)
Monotypic genus. Diagnosis as for the species (page 224).

GENUS *ANTHENOIDES* Perrier, 1881

Species of this genus have a body that is covered by a thin membrane or by granules. The actinal interradial areas are large. The arms taper to an acute tip, and the abactinal plates are polygonal to stellate. There are no spines on the abactinal surface, and rarely a few short tubercles are present. The furrow spines are in a comb, followed by subambulacral granules or an enlarged flattened spine. The actinal plates have hemispherical granules and slit-like bivalved pedicellariae, or two-jawed pincer-like pedicellariae. The marginal plates are without conspicuous granules, spinelets or pedicellariae. Abactinally there is a row of small secondary plates on either side of the median radial series.

Type species: *Anthenoides peircei* Perrier, 1881 (Gulf of Mexico, Caribbean Sea)
A single species, *A. dubius* (page 225), is recorded from shallow Australian waters, described by H.L. Clark (1938) from a single specimen found off north-western Australia.

Anchitosia queenslandensis, Milne Bay, Papua New Guinea (N. Coleman).

QUEENSLAND BISCUIT STAR
Anchitosia queenslandensis
(Livingstone, 1932a)

DESCRIPTION

Body form weakly stellate, arm tips rounded; abactinal plates flat to convex, round to polygonal in outline; primary circlet plates enlarged, distinctive; papular pores single or double, absent interradially; granules surrounding plates are polygonal to quadrate; madreporite triangular; marginal plates bare, smooth, slightly convex, surrounded by granules, wider than long; superomarginals aligned with inferomarginals, 8–12 between arm tips; **terminal plate enlarged, 1–2 times size of adjacent superomarginals**; actinal plates angular to polygonal, smooth, bare with peripheral granules; furrow spines 2–5 compressed but thickened; 2–3 subambulacral spines granular, thickened; pedicellariae when present in pits with tong-shaped valves; *Anchitosia* differs from *Tosia australis* in having more superomarginal plates, differs from *T. magnifica* in lacking granules on the actinal plates, and differs from *Tosia* and *Pentagonaster* in having **the terminal plate enlarged.**

MAXIMUM SIZE

R/r = 20/10 mm (R = 1.7–2r).

Samaurez Reef, Coral Sea (N. Coleman).

COLOUR

Crimson, including the granules between the plates, or red to orange with yellow arm tips and yellow granules between the plates.

FOOD

Probably a benthic grazer.

REPRODUCTION

Presumed to have lecithotrophic development as a specimen collected in November extruded two very large eggs between the rows of marginal plates in the arm angle.

HABITAT, DEPTH AND AUSTRALIAN DISTRIBUTION

Reef crest, reef front and reef cavities, 0–58 m; rare. In Australia it is confined to the Great Barrier Reef, Queensland, from Lizard Island to the Capricorn Group, and Lady Elliott Island and reefs in the Coral Sea.

TYPE LOCALITY

Pixie Reef, Great Barrier Reef, Queensland.

FURTHER DISTRIBUTION

New Caledonia, Papua New Guinea and the Ryukyu Islands, Japan.

REFERENCES

Clark, H.L., 1946: 94–95 (as *Tosia*).

Livingstone, 1932a: 243–244; pl. 5 figs 1–2, 7 (as *Tosia*).

Mah, 2007: 322–323; figs 4d–f (new genus *Anchitosia*).

DOUBTFUL SEASTAR
Anthenoides dubius
H.L. Clark, 1938

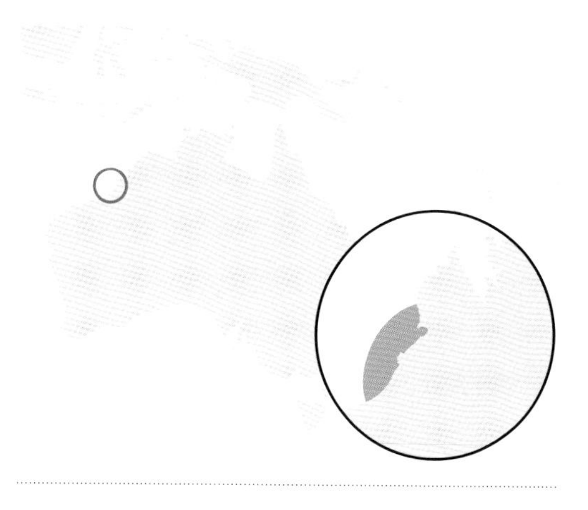

DESCRIPTION
A small species with disc fairly flat, five short narrow, pointed arms; abactinal surface covered by a membrane, with numerous minute, well spaced granules obscuring the outlines of the underlying plates except on arms where a single median series of squarish plates extends nearly to the tip; single small papulae occur at each corner of the radial plates and each corner of the first three superomarginals; on the first carinal plate is a rounded tubercle, higher than thick; a few low, oblong pedicellariae are scattered on distal parts of the disc; superomarginals squarish, the distal three are in contact on the abactinal surface of the arm, covered with bare skin apart from a few very minute granules; terminal plate with four horizontally projecting short thick spinelets; interradial actinal areas small, actinal plates with scattered minute granules, some have a pedicellaria or a larger granule; five subequal flattened furrow spines reducing to two distally; 3–4 minute granules in a line on the outer part of the plate.

MAXIMUM SIZE
R/r of holotype = 18/6 mm (R = 3r, 3.6br).

COLOUR
Pale olive grey, variegated with a darker shade, terminal plates and a band on distal half of arm dark grey. Actinal surface white.

HABITAT, DEPTH AND AUSTRALIAN DISTRIBUTION
On soft substrates, 10–37 m. Endemic to north-western Australia between Lagrange Bay and Broome. North-western Australian endemic.

TYPE LOCALITY
Broome, WA.

FURTHER DISTRIBUTION
Australian endemic.

REFERENCES
Clark, H.L., 1938: 91; pl. 17 figs 5–6.
Clark, H.L., 1946: 95.

REMARKS
This small species is only known from the type specimens collected near Broome, WA. It bears little resemblance to other species of *Anthenoides* but it has not been determined where it should be placed, it appears to be closer to *Stellaster*. Mah (2019) suggests questioning the validity of this species, noting that the holotype is a juvenile of another goniasterid.

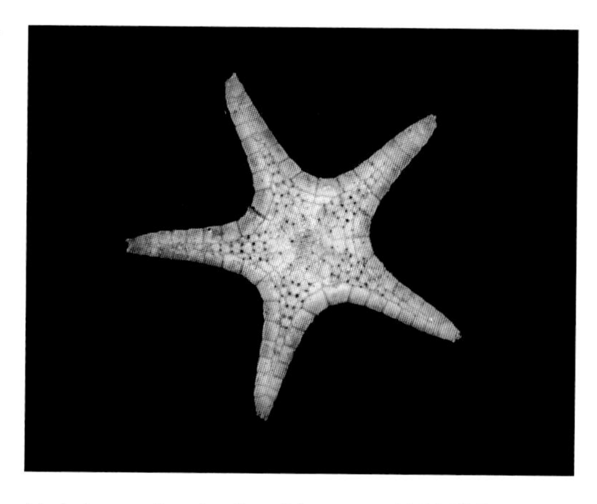

Holotype, Australian Museum (AMJ6113).

GENUS *FROMIA* Gray, 1840

Species of this genus are small to medium sized seastars, normally with five fairly flat to slightly convex arms. The plates are covered with fine granules of uniform size or sometimes with a few larger central granules. The marginal plates are prominent or not. The papulae are single and on abactinal intermarginal and actinal surfaces. Furrow and subambulacral spines are present but few, and not in fans or combs. The abactinal plates on the arms are not usually regularly arranged. Fresh specimens of most species can be readily distinguished by colour.

Type species: *Fromia milleporella* (Lamarck, 1816) (Indo-Pacific)

Eight species are found in Australian waters, all in depths of less than 30 m. Most are found on tropical reefs, some with restricted distributions to the north of Australia while several have a wide Indo-West Pacific distribution. One species is endemic to southern Australia. One undescribed species is known from north-western Australia.

Fromia heffernani was originally described as a species of *Ferdina* but was placed in a new monotypic genus *Celerina* by A.M. Clark (1967). Mah and Foltz (2011a) showed that *Celerina heffernani* and *Fromia monilis* were members of the same clade within the Goniasteridae and Mah (2019) formally synonymised *Celerina* with *Fromia*.

Fromia monilis, Houtman Abrolhos, Western Australia (C. Bryce).

KEY TO THE SPECIES OF *FROMIA* IN SHALLOW AUSTRALIAN WATERS

1	Superomarginal plates prominent; disc and arm tips often red; arms slender, fairly flat.	2
	Superomarginal plates not prominent.	3
2	Superomarginal plates alternate large and small, at least on outer half of arms.	4
	No alternation of superomarginals which decrease in size evenly along arm.	5
3	Disc and arms more or less convex in adults.	7
	Disc and arms very convex, arms nearly round in section; abactinal plates irregular in size and shape, enlarged plates may be more convex than the remainder and may contrast in colour; colour variable, uniform rust red or dark red-brown with enlarged plates paler.	***F. indica*** page 233
4	Intermarginal plates present.	***F. heffernani*** page 230
	Intermarginal plates absent.	***F. monilis*** page 237
5	Abactinal plates in fairly regular series; numerous split-granule pedicellariae on actinal surface particularly on disc; abactinal granulation generally smooth; colour pattern similar to *F. monilis*.	***F. eusticha*** page 229
	Abactinal plates not in regular series; granulation not smooth.	6
6	Most marginal plates with one or more enlarged central tubercular granules; arms slender, about three times width at base.	***F. hemiopla*** page 232
	Marginal plates without enlarged central granules; arms about 2–2.5 times width at base.	***F. milleporella*** page 236

7

Arms flattish to sub-cylindrical, slightly tapering. — ***F. pacifica*** page 239

Arms usually broad based (in adults), arms taper strongly; abactinal plates small irregular in arrangement; granulation coarse; colour uniform yellow to light orange in western and southern Australia, light to dark red on the east coast, papulae black; southern Australia, subtropical to temperate waters. — ***F. polypora*** page 240

Fromia milleporella, Réunion Island (P. Bourjon).

WELL-LINED SEASTAR
Fromia eusticha
Fisher, 1913

DESCRIPTION

Superomarginal plates in regular series, not alternating large and small, mostly longer than wide; abactinal plates fairly regularly arranged at base of arm, plates roundish or hexagonal; actinal plates in three regular series at base of arm; furrow spines three at base of arm, two distally; two subambulacral spines and 1–3 subambulacral pedicellariae; **the most characteristic feature is an abundance of split-granule pedicellariae on the actinal plates**; papulae are single at the corners of the plates; plates of both surfaces are covered in close set polygonal granules, about 8–10 across a carinal plate.

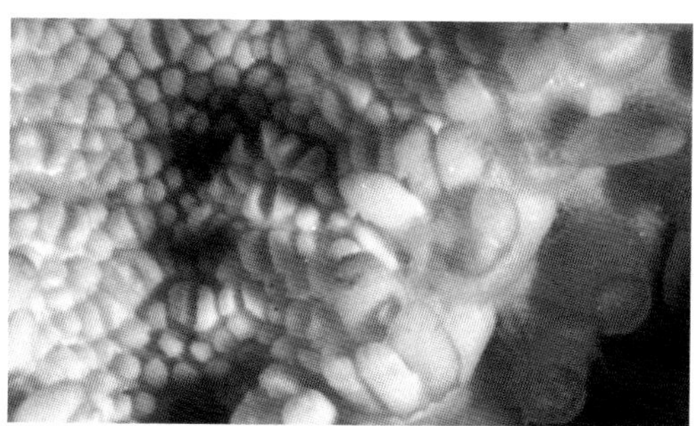

Rowley Shoals, Western Australia [top] (L. Marsh); actinal pedicellariae [bottom], Western Australian Museum (WAMZ15770).

MAXIMUM SIZE

R/r = 45/10 mm (R = 4.5r).

COLOUR

Disc and arm tips red, the remaining body is cream with red between the plates, outlining them.

HABITAT, DEPTH AND AUSTRALIAN DISTRIBUTION

On coral reef slopes, or sandy substrate with rocks and sponges, 18–82 m. In Australia it is only recorded from the Rowley Shoals and Browse Island off north-western Australia.

TYPE LOCALITY

Near Siasi, Sulu Archipelago, the Philippines.

FURTHER DISTRIBUTION

Philippines, eastern Indonesia, and the Marshall Islands.

REFERENCES

Fisher, 1919: 375; pl. 95 fig. 2; pl. 105 fig. 1; pl. 106 fig. 1; pl. 107 figs 3, 5.

REMARKS

Fromia monilis sometimes has split-granule pedicellariae on the actinal plates, but the presence of even superomarginal plates with no alternation in size distinguishes *F. eusticha*.

HEFFERNAN'S SEASTAR

Fromia heffernani

(Livingstone, 1931)

Papua New Guinea (L. Marsh).

DESCRIPTION

Resembles *Fromia monilis*; disc flat, five flat to convex arms tapering to a sharp tip; abactinal plates polygonal, often irregularly arranged, with single papular pores where the plates meet; the surface is covered in fine granules, sometimes larger in the centre of plates; some of the carinal plates are swollen, consecutively near the disc and alternating, often irregularly, with flat plates distally; swollen superomarginal plates alternate, often irregularly, with small flat plates; **intermarginal plates lie between the marginal series** in the arm angle in small specimens but along the arms in larger ones; there are no papulae below the inferomarginals and no pedicellariae; **the furrow spines are paired (sometimes three per plate), usually sharply pointed and curved, claw-like over the furrow**; subambulacral spines absent or represented by an enlarged granule; the oral spines (except the inner one) are smaller than the furrow spines.

MAXIMUM SIZE

R/r = 50/8 mm (R = 3.3–6.6r), R is usually < 40 mm.

COLOUR

Usually the disc, flat plates of the outer half of the arm and joints between the plates of the inner half of the arm red, dark red or purplish black; convex plates and inner half of the arm cream, pale pink or beige. Actinal surface cream.

HABITAT, DEPTH AND AUSTRALIAN DISTRIBUTION

On coral reefs, on lagoon slopes or more usually outer reef slopes, 0–55 m; uncommon. In Australia it has only been recorded from Ashmore Reef, Seringapatam Reef and the Rowley Shoals, off the North West Shelf, WA and Wheeler Reef to 19°20'S on the Great Barrier Reef, Queensland.

TYPE LOCALITY

Santa Cruz Islands, West Pacific.

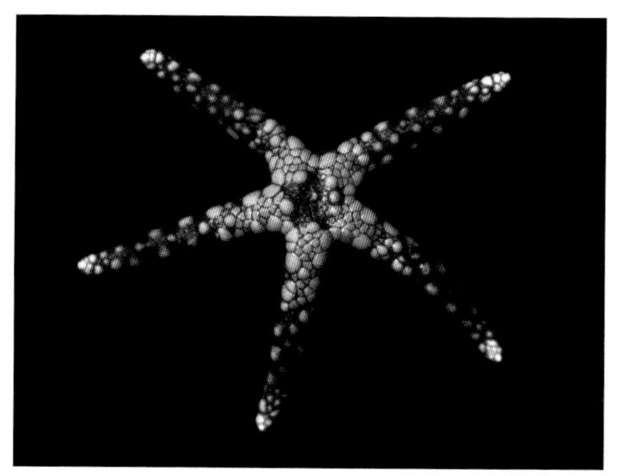

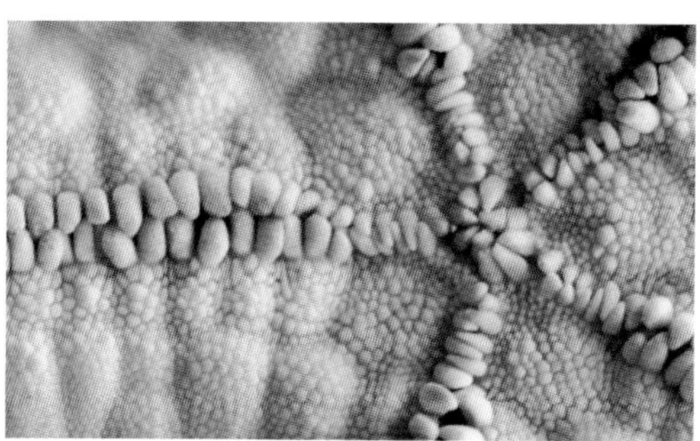

Papua New Guinea [top] (L. Marsh); furrow characteristics [bottom], Western Australian Museum (WAMZ15735).

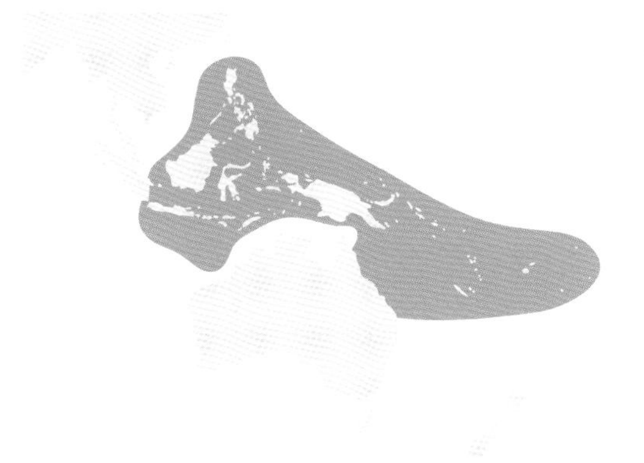

FURTHER DISTRIBUTION

Fromia heffernani has a limited tropical Indo-West Pacific distribution, from the Cocos (Keeling) and Christmas islands (Indian Ocean), Indonesia, Papua New Guinea, Vanuatu, Palau to New Caledonia and Samoa in the Pacific and north to Macclesfield Bank (South China Sea).

REFERENCES

Clark, A.M., 1967: 193–194; pl. 6 figs 4–5 (new genus *Celerina heffernani*).

Clark, A.M. and Rowe, 1971: 65.

Livingstone, 1931: 306; pl. 24 figs 1–5 (as *Ferdina heffernani*).

Mah, 2017: 8; fig. 1.

Marsh, 1977: 255.

REMARKS

This species looks like the much more common *Fromia monilis* when viewed from above but can be readily distinguished by its single row of claw-like furrow spines, lack of subambulacral spines, lack of actinal papular pores and the presence of intermarginal plates. It was previously placed in a monotypic genus *Celerina*.

HALF-NODULED SEASTAR
Fromia hemiopla
Fisher, 1913

DESCRIPTION
Arms fairly slender, superomarginal plates are in a regular series, not alternating large and small; abactinal plates irregular in arrangement; one or more tubercular granules in the centre of the superomarginal plates on the outer half or three-quarters of the arm; 3–4 furrow spines (two distally), flat, spatulate, in overlapping fans; 2–3 subambulacral spines, shorter, thicker than the furrow spines, followed by 2–3 enlarged granules.

MAXIMUM SIZE
R/r/br = 36/9/10 mm (R = 4r).

COLOUR
Dark red, superomarginal plates darker.

FOOD
Encrusting sponges and the substrate biofilm.

HABITAT, DEPTH AND AUSTRALIAN DISTRIBUTION
On coral reefs, 3–5 m. In Australia it is only recorded from Lizard Island, Queensland.

TYPE LOCALITY
Tonquil Island, Mindanao, the Philippines.

FURTHER DISTRIBUTION
Andaman Sea, Papua New Guinea, the Philippines and Marshall Islands, West Pacific.

REFERENCES
Clark, A.M. and Rowe, 1971: 63.

Fisher, 1919: 377–378; pl. 95 fig. 3; pl. 105 fig 2; pl. 106 fig. 3; pl. 107 figs 2, 4.

REMARKS
Differs from *Fromia milleporella* in having one or more tubercular granules in the centre of the prominent marginal plates and in having more slender arms and broad, thin, furrow spines. In view of the variability of species of *Fromia, F. hemiopla* may be a variant of *F. milleporella.*

Marginal plate characteristics [top], Milne Bay, Papua New Guinea (N. Coleman).

INDIAN SEASTAR
Fromia indica
(Perrier, 1869)

Heron Island, Queensland (F. Michonneau).

DESCRIPTION

Disc and arms are convex; the marginal plates less prominent than in other species, not alternating large and small; **abactinal plates rather irregular in size and shape, usually wider than long, sometimes some are enlarged and more convex than the remainder**; the surface of all plates is granular, covered with slightly larger granules outlining the margin of the abactinal plates; papular pores scattered singly between the plates; there are two furrow spines per plate, backed by two subambulacral spines and one or two granules; pedicellariae of the split-granule type may be numerous.

MAXIMUM SIZE

R/r = 55/12 mm (R = 3.4–4.6r, 3.6–4.6br).

COLOUR

Usually rust red to dark red-brown. Outside Australia the colour may be mottled with the raised arm plates contrasting with the remainder, either lighter or darker.

FOOD

Algae and bottom detritus.

Spawning posture and sperm release, Great Barrier Reef, Queensland [top] (A. Hoggett); Houtman Abrolhos, Western Australia [bottom] (C. Bryce).

REPRODUCTION

Observed spawning concurrently with the mass spawning of corals in late March at the Houtman Abrolhos, WA. During spawning they adopt a tip-toe posture, resting on their arm tips. Gonopores open abactinally in the interradial area just above each first superomarginal plate; of seven specimens examined four were female, two were hermaphrodites (with mature sperm and immature ova) and one was immature. Larvae were observed for four days to the development of brachiolar arms. The eggs are very large, 1.15 mm, and develop through a non feeding planktonic brachiolaria larva that sinks to the seabed after four days.

HABITAT, DEPTH AND AUSTRALIAN DISTRIBUTION

On coral reefs, under dead coral slabs or among corals, 0–55 m; common. From Houtman Abrolhos, WA and Ashmore Reef, Timor Sea to Lady Elliott Island, Queensland.

Raja Ampat, Indonesia (S. Morrison).

TYPE LOCALITY

'Indian Ocean'.

FURTHER DISTRIBUTION

From the Maldive Islands, India to Indonesia, the Philippines and Southern Japan to Fiji in the West Pacific.

REFERENCES

Clark, H.L., 1921: 43; pl. 7 fig. 3; pl. 29 figs 5–6 (as *Fromia elegans*).

Clark, H.L., 1946: 113 (as *Fromia andamanensis* and *F. elegans*).

Clark, A.M. and Rowe, 1971: 62 (as *Fromia indica* and *F. elegans*).

Coleman, 2007: 31 (food).

Marsh, 1988: 187–192 (reproduction).

Sastry, 2007: 70.

REMARKS

The three names previously given to this species indicate its variability, which may be found within one population. The three were synonymised by Hayashi (1938).

THOUSAND-PORED SEASTAR
Fromia milleporella
(Lamarck, 1816)

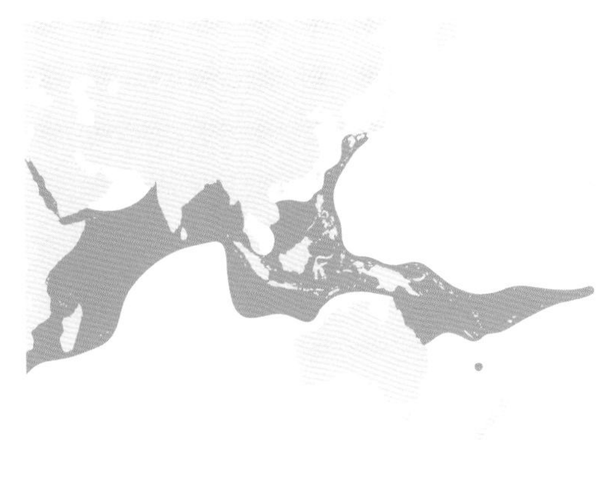

DESCRIPTION
Usually five armed; disc and arms slightly convex but **marginal plates more prominent than in *F. indica*, not alternating large and small**; abactinal plates irregularly arranged to fairly regular; arms usually 2–2.5 times as long as the width at their base; granulation moderately fine; three furrow spines, followed by two shorter subambulacral spines; papulae single, uniformly distributed over the abactinal surface and in 1–2 rows actinally, they appear as black dots on the living animal; usually no pedicellariae.

MAXIMUM SIZE
R/r = 36/9 mm (R = 4r).

COLOUR
Dark plum red, paler on the actinal surface, papulae appear black.

Marginal plate characteristics [top], Hibernia Reef, Timor Sea (C. Bryce).

FOOD
Encrusting sponges and the substrate biofilm.

HABITAT, DEPTH AND AUSTRALIAN DISTRIBUTION
In the open on coral reefs or algal covered hard bottom, 0–30 m; common. Scott Reef, WA, Ashmore Reef, Timor Sea and Murray Islands to Bunker Group, Great Barrier Reef, Queensland also Swain Reefs and Norfolk Island, Tasman Sea.

TYPE LOCALITY
'European seas' (erroneous).

FURTHER DISTRIBUTION
From east Africa and the Red Sea, Cocos (Keeling) Islands, through South-East Asia to the Ryukyu Islands (Japan) and the West Pacific to New Caledonia, Fiji, Samoa and Palmyra (Line Islands).

REFERENCES

Clark, H.L., 1946: 112–113.

Clark, A.M. and Rowe, 1971: 63; pl. 8 fig. 10.

REMARKS
F. milleporella and *F. hemiopla* are very similar and may prove to be one species.

NECKLACE SEASTAR

Fromia monilis

Perrier, 1869

Browse Island, Western Australia (C. Bryce).

DESCRIPTION

Disc and arms flat, arms slender, elongate; **superomarginal plates are prominent, alternating large and small on the distal half of arms**; carinal plates on the arms usually larger and more convex than the lateral series, plates fairly regularly arranged, usually five plates across the base of the arm, three in small specimens; plates covered with fine granules; the largest specimens may have four furrow spines basally, usually 2–3; usually two subambulacral spines and two series of 4–6 granules; pedicellariae sometimes occur among the granules; papulae single, at the corners of the plates on the abactinal, marginal and actinal surfaces.

MAXIMUM SIZE

R/r = 60/9 mm (R = 6.6r), usually much smaller.

COLOUR

The disc is all or partly red, the outer half or only the ends of arms red, the remainder is cream or beige with red between the plates, outlining them.

FOOD

Algae and the substrate biofilm.

REPRODUCTION

F. monilis was observed spawning concurrently with the mass spawning of corals at the Houtman Abrolhos, WA in late March. During spawning they adopt a tip-toe posture resting on the arm tips. The gonopores open abactinally; of five specimens examined two were female and three were hermaphrodites with mature testes and immature oocytes. The eggs are large and a non feeding brachiolaria larva is inferred.

HABITAT, DEPTH AND AUSTRALIAN DISTRIBUTION

In the open on offshore coral reefs, 0–100 m; common. From the Houtman Abrolhos, WA to Ashmore Reef, Timor Sea and the Great Barrier Reef, Queensland to Elizabeth Reef, Tasman Sea.

TYPE LOCALITY

Ambon, Indonesia.

FURTHER DISTRIBUTION

Red Sea, Maldive Islands, Indonesia to New Caledonia, Samoa and Fiji in the West Pacific and north to Taiwan and the Ryukyu Islands, Japan.

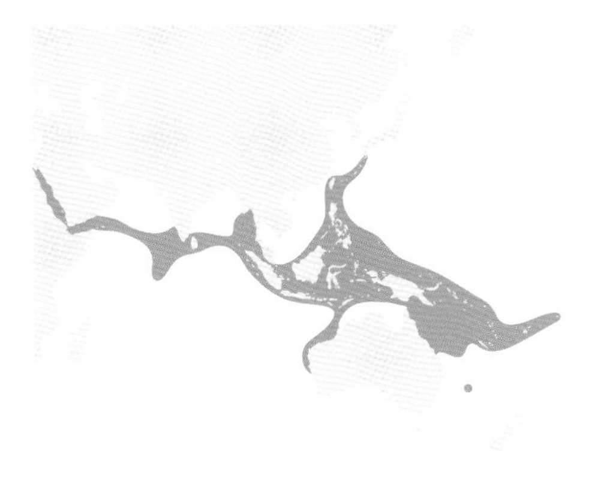

REFERENCES

Clark, A.M. and Rowe, 1971: 62.

Fisher, 1919: 373; pl. 105 fig. 4; pl. 106 fig. 2; pl. 107 figs 1–7 (as *Fromia japonica*).

Marsh, 1988: 187–191 (reproduction).

REMARKS

Fromia monilis resembles *F. heffernani* which is distinguished by having a single row of furrow spines (no subambulacrals), intermarginal plates and no actinal papulae.

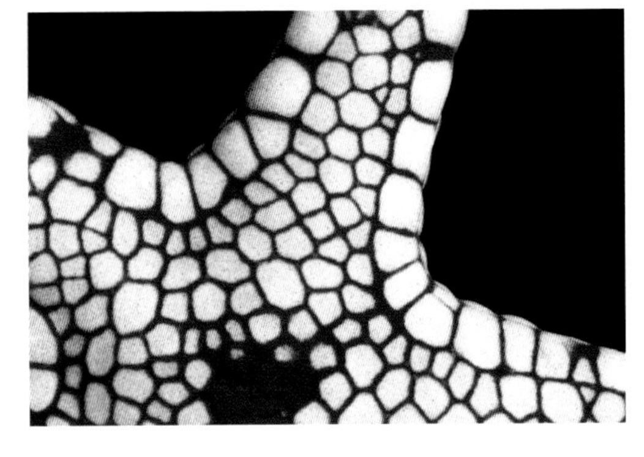

Komodo, Indonesia [top left] (S. Morrison); Houtman Abrolhos, Western Australia [left], abactinal plate characteristics [above], (C. Bryce).

PACIFIC SEASTAR
Fromia pacifica
H.L. Clark, 1921

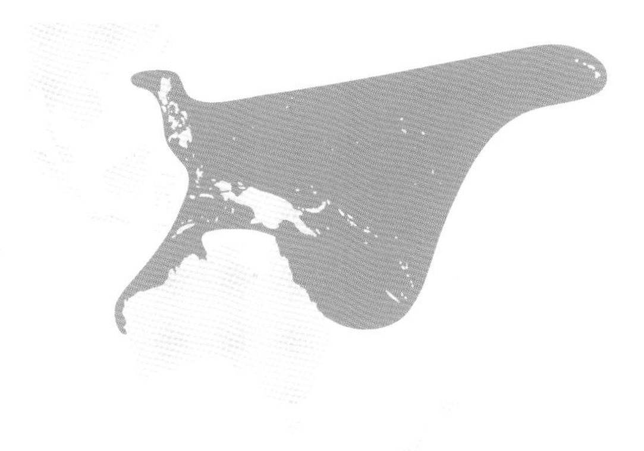

DESCRIPTION

Disc small and flat to convex; arms moderately long, flat to convex, tapering to the tip; abactinal plates slightly convex; the carinal row may be distinct with one row of dorsolateral plates on either side, sometimes two rows at the base of the arm while at the arm tip the superomarginals touch the carinal row in small specimens but may be separated from it by one row of dorsolateral plates in larger specimens; sometimes the abactinal plates are rather irregular in arrangement; the whole surface has a close set uniform cover of medium sized granules; the superomarginal plates are wider than long basally but longer than wide on the mid-arm and there may be small plates between the larger ones irregularly, on the outer half of the arm; papulae occur singly around the abactinal plates, intermarginally and in two rows on the actinal surface; 2–4 thick, blunt furrow spines, 2–3 subambulacral spines, shorter and stouter than the furrow spines, and several enlarged granules.

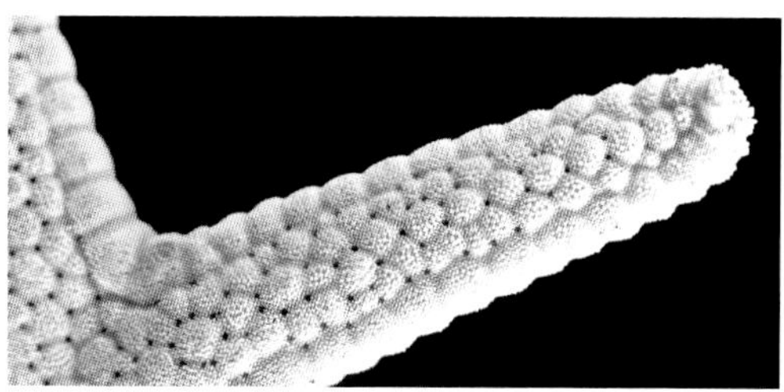

Houtman Abrolhos, Western Australia [top] (L. Marsh); abactinal plate characteristics [bottom], Western Australian Museum (WAMZ99827).

MAXIMUM SIZE

R/r/br = 36/9/8 mm (R = 4r, 4.5br).

COLOUR

Scarlet, arm tips and madreporite sometimes yellow.

FOOD

Everts the stomach on the epibenthic biofilm (Birtles, personal communication).

HABITAT, DEPTH AND AUSTRALIAN DISTRIBUTION

Associated with coral reefs or rubble bottom, 13–90 m. Houtman Abrolhos and North West Shelf, WA to Swain Reefs, Great Barrier Reef, Queensland.

TYPE LOCALITY

Hawaiian Islands.

FURTHER DISTRIBUTION

Indonesia (Moluccas), the Philippines to Hawaii.

REFERENCES

Clark, A.M. and Rowe, 1971: 63.

MULTI-PORED SEASTAR
Fromia polypora
H.L. Clark, 1916

DESCRIPTION
The largest of the *Fromia* genus and the only species confined to temperate seas; it attains its maximum size in the southern part of its distribution, rarely R > 60 mm on the west coast; the arms are variable in length and width from slender to very plump; **the plates are obscured by a dense coat of coarse granules**, most of a uniform size but sometimes those in the centre of the plates are enlarged; marginal plates prominent in small specimens, less so in large ones; there are 2–3 furrow spines backed by 2–3 short subambulacral spines followed by 4–6 granules; pedicellariae are of the split-granule type, few in number; the numerous papular pores are conspicuous because of the dark papulae, found on both the abactinal and actinal surfaces.

MAXIMUM SIZE
R/r = 112/28 mm (R = 3–5.4r, 3.3–5.1br).

COLOUR
Bright yellow to orange with contrasting dark, almost black papulae. A red colour form is found from the Queensland border to Jervis Bay, NSW and at Norfolk Island.

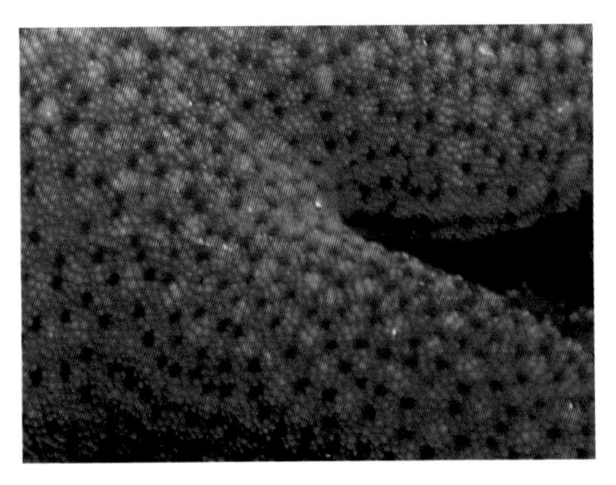

Abactinal plate characteristics, Sorrento Reef, Western Australia (B. Wilson).

FOOD
Encrusting sponges, bryozoans and compound ascidians.

HABITAT, DEPTH AND AUSTRALIAN DISTRIBUTION
In the open on algal covered temperate rocky reefs of the open coast, 0–160 m; common. From Sunshine Beach, Queensland to Houtman Abrolhos, WA, also Norfolk Island, Tasman Sea. Southern Australian endemic.

TYPE LOCALITY
East of Maria Island, Tasmania.

FURTHER DISTRIBUTION
Australian endemic.

REFERENCES
Clark, H.L., 1946: 114 (as *Austrofromia polypora*).

Shepherd, 1968: 744 (as *Austrofromia polypora*).

REMARKS
This species was originally placed in the genus *Fromia* but was made the type of a new genus *Austrofromia* by H.L. Clark (1921) then referred back to *Fromia* by Rowe (1989).

Recherche Archipelago, Western Australia [top];
Oyster Harbour, Western Australia [bottom] (C. Bryce).

GENUS *ICONASTER* Sladen, 1889

Species of this genus have flat, rigid bodies and large stellate central discs. The disc and marginal plate surfaces are flush. The arms are relatively elongate with superomarginals abutted across the midpoint. The primary circlet and interradial plates are enlarged. There are papular plates on the radial regions. The abactinal disc plates and marginals are outlined with small close fitting granules between the plates, abactinal plate corners have enlarged accessories. Actinal plates are bare. Furrow spines are short and squat, round and polygonal in cross section, and crowded. Subambulacrals are round and polygonal in cross section. A subambulacral furrow spine space is absent. There are narrow flaps on the ambulacral base. Superomarginal plates with glassy tubercles occur in two species.

Type species: *Iconaster longimanus* (Möbius, 1859) (Indo-West Pacific)

Four species occur worldwide, only the type species (page 243) is found in Australian waters in depths of less than 30 m.

Iconaster longimanus, Dampier Archipelago, Western Australia (C. Bryce).

ICON SEASTAR
Iconaster longimanus
(Möbius, 1859)

Dampier Archipelago, Western Australia (S. Morrison).

DESCRIPTION

Body flat, rigid, interradial arcs rounded; arms narrow and elongate with the marginal plates meeting for the whole length of the arm; no granules on the surface of the abactinal or actinal plates but the abactinal disc plates and marginals are outlined with small close fitting granules between the plates; **the radial plates (and in large specimens a lateral row of plates) have granules only along their sides, so that two (or four) radial white lines of granules are conspicuous** on the outer part of the disc; plates of the actinal surface are conspicuously outlined by granules; furrow spines are short, blunt, four to a plate, outside of which are two rows of 3–4 granules each; pedicellariae, when present, are of the sugar-tongs form with curved blades, no teeth; glassy tubercles on abactinal and superomarginal plates.

MAXIMUM SIZE

R/r = 110/25.6 mm (R = 4.3r).

COLOUR

Variable, for example yellow with orange radial areas on disc; yellow with light orange-brown arms (irregularly banded with brown) and marginal plates; reddish-brown disc with white granules outlining the radial plates; dark brown or grey-brown abactinal plates and some marginal plates; or orange-red arms and a red disc.

FOOD

Stomach is everted on encrusting bryozoans, sponge and the epibenthic biofilm. Also a scavenger (Birtles, personal communication).

Disc colour and pattern variations, Bintan Island, Indonesia [top], Cassini Island, Western Australia [bottom] (C. Bryce).

REPRODUCTION

In Singapore reproduction takes place between June and September. Sexes are separate, the eggs are very large (~1 mm diameter). The ciliated larva develops into a lecithotrophic pear-shaped brachiolaria with three blunt brachiolar arms. After 10 days in the plankton the larvae settle, metamorphosis is complete by 18 days, juveniles measure 1.8 mm disc diameter.

HABITAT, DEPTH AND AUSTRALIAN DISTRIBUTION

On sand, rubble or rock, 5–84 m. Found from Exmouth Gulf around northern Australia to the Capricorn Group of the Great Barrier Reef, Queensland.

TYPE LOCALITY

Not recorded.

FURTHER DISTRIBUTION

Saudi Arabia, Oman, Malaysia, Indonesia and the Philippines.

REFERENCES

Fisher, 1919: 303–306; pl. 77 fig. 2; pl. 83 fig 5; pl. 93 figs 2, 2a; pl. 104 fig. 3.

Lane and Hu, 1994: 343–346; figs 1–6 (development).

GENUS *MEDIASTER* Stimpson, 1857

Species of this genus have long slender arms with several series of abactinal ossicles separating the superomarginls, one series may reach the tips of the arms. The abactinal ossicles are tabulate. Rudimentary superambulacral ossicles are present.

Type species: *Mediaster aequalis* Stimpson, 1857 (north-eastern Pacific)

Of the three species recorded from Australian waters, only *M. australiensis* (page 246) is known from depths of less than 30 m.

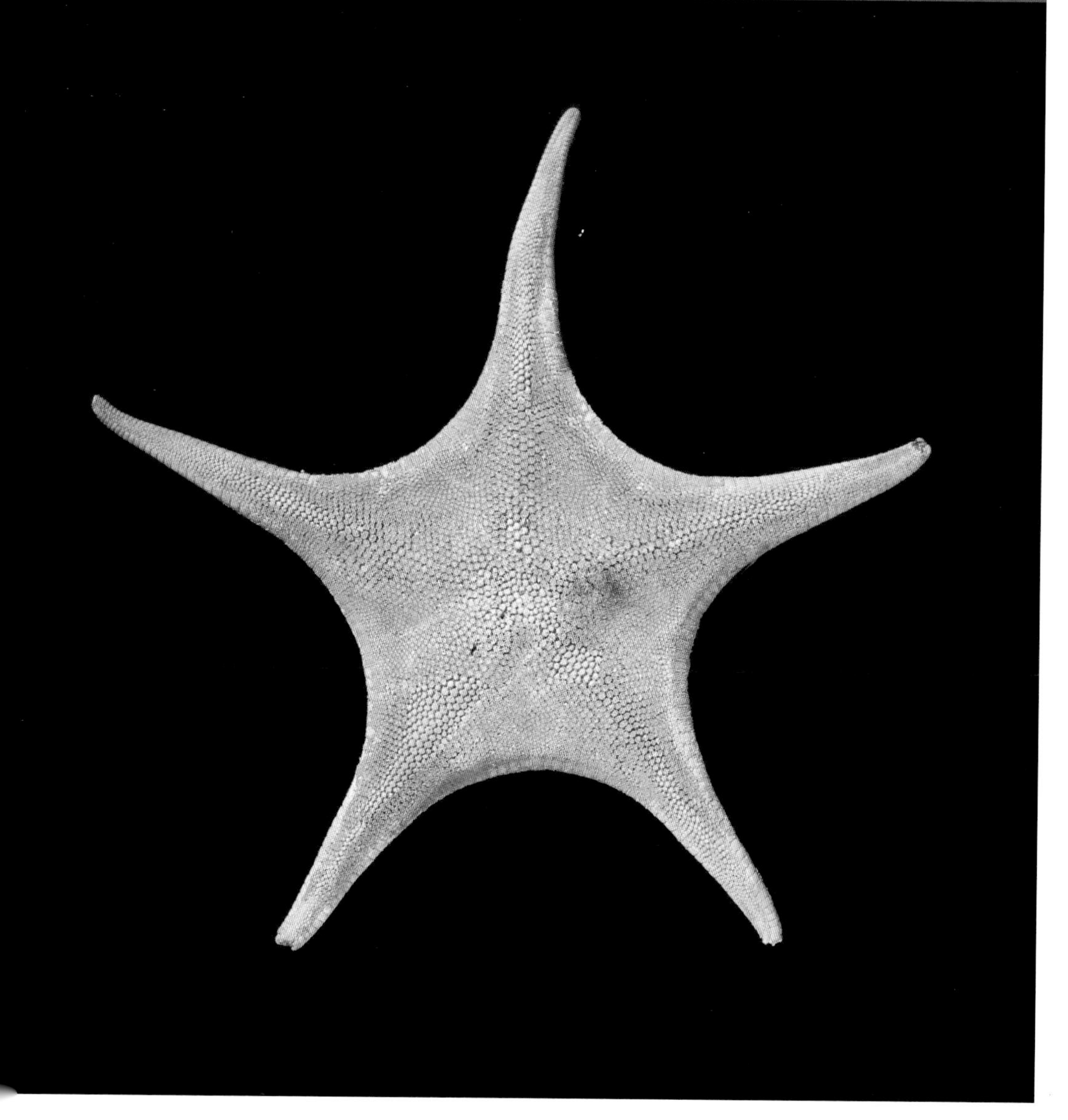

Mediaster australiensis, holotype, Australian Museum (AMJ1535).

MYRIAD SEASTAR
Mediaster australiensis
H.L. Clark, 1916

DESCRIPTION
Disc large and flat; arms wide at base but tapering rapidly to a nearly cylindrical tip; **abactinal plates tabulate, more or less paxilliform, small and numerous**; the largest abactinal plates on disc centre and carinal series, have tabulae with a marginal series of 12–15 angular blunt spinelets (or coarse granules) surrounding 3–8 similar or smaller granules (number of granules highly variable), sometimes replaced by two-jawed pedicellariae; papulae numerous, small, single, scattered irregularly among the plates; superomarginal plates up to 48 on each side of each arm, closely covered with granules that are coarser and more rounded than those on the disc; granules sometimes replaced by a pedicellaria; inferomarginals similar to superomarginals; actinal plates in 8–9 series with coarse, angular granules, no pedicellariae; 5–6 flat, blunt furrow spines; subambulacral spines in two rows, the inner with four short thick angular spines, the outer with 3–4 even shorter and thicker spines, there is a bare space between these two rows.

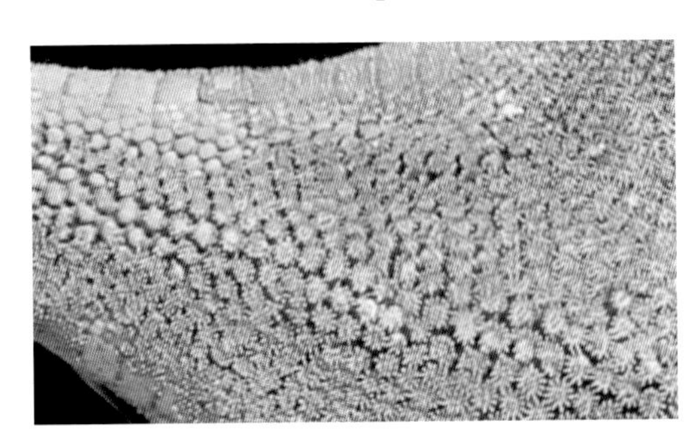

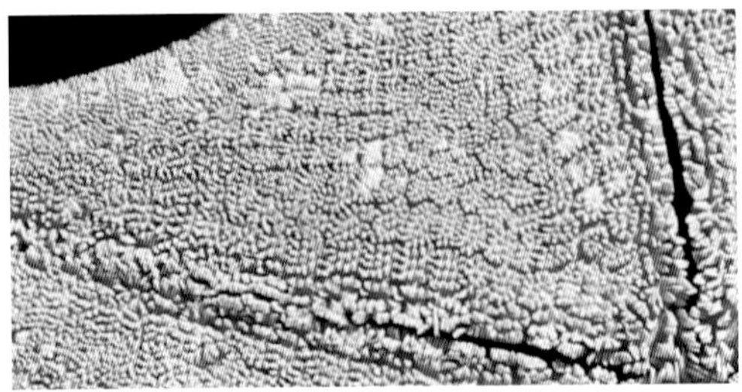

Abactinal [top] and actinal plate characteristics [bottom], holotype, Australian Museum (AMJ1535).

MAXIMUM SIZE
R/r = 100/40 mm (R = 2.5r).

HABITAT, DEPTH AND AUSTRALIAN DISTRIBUTION
On soft substrates, 11–503 m. From off Brush Island, NSW to Oyster Bay, Tasmania; there is one record from off Point D'Entrecasteaux on the south coast of WA at 378 m. Southern Australian endemic.

TYPE LOCALITY
Off Flinders Island, Bass Strait.

FURTHER DISTRIBUTION
Australian endemic.

REFERENCES
Clark, H.L., 1916: 39; figs 1–3; pl. 9 figs 1–2.
Clark, H.L., 1946: 83.

REMARKS
Two other species of *Mediaster* are found in Australian waters, one from deep water off NSW (*M. arcuatus*) and one off North Queensland from an unknown depth (*M. praestans*).

GENUS *NECTRIA* Gray, 1840

Species of this genus have five tapering arms and a relatively large disc. The abactinal surface is convex, and the actinal surface is flat. **The abactinal skeleton comprises distinctive raised tabulae** each supported by six radiating ossicles at the base, forming a network. The disc tabulae are crowned with a peripheral ring of usually prominent granules encircling the central granules, while other tabulae have granules of varying shape and size. The tabulae are distinguishable on the disc, decreasing in size along the arm to about halfway, after which they become unrecognisable as tabulae. The marginal plates are obvious, flat and squarish. The infero- and superomarginals are of similar size and number decreasing regularly in size to the arm tip, and covered with even-sized granules. The actinal plates are covered with coarse granules. The adambulacral plates have furrow spines slightly larger than the adjacent granules on the actinal plates. The papulae are in groups between ossicles of the abactinal skeleton, sometimes between the marginal plates, rarely on the actinal surface. The interradial septae are partially calcified. Pedicellariae are often present. Superambulacral plates are present.

Type species: *Nectria ocellata* Perrier, 1875 (Australia)

Nectria is an endemic southern Australian genus, believed to have evolved in south-western Australia, although lack of fossil material prevents assessment of its evolutionary origin (Zeidler and Rowe 1986). Eight species are recognised, of which seven are found in south-western Australia. The eighth, *N. humilis*, is from deeper water in Bass Strait. *N. ocellifera* is excluded from this book as it has not been found in less than 45 m. Most of the species can be readily identified from the illustrations but a key is included to cover the variability of some species, which can be confusing. Originally described in the Goniasteridae this genus was transferred to the Oreasteridae by Zeidler and Rowe (1986), but as a result of genetic data was returned to the Goniasteridae by Mah and Foltz (2011a). Following Zeidler (1995) *N. ocellata* is recognised here as the type species.

Nectria ocellata, Denmark, Western Australia (D. Lane).

KEY TO THE SPECIES OF *NECTRIA* IN SHALLOW AUSTRALIAN WATERS

1	Intermarginal papulae usually present, when absent many distal arm plates enlarged.	2
	Intermarginal papulae absent, distal arm plates never enlarged.	5
2	Tabulae with concave aspect, with central granules lower in profile and smaller than peripherals.	***N. macrobrachia*** page 250
	Tabulae with convex aspect, with central granules higher in profile and larger than peripherals.	3
3	Abactinal arm plates decrease regularly in size towards arm tip; pedicellariae with slender valves.	***N. pedicelligera*** page 254
	Abactinal arm plates of mixed sizes towards arm tip, some almost as large as disc plates; pedicellariae with slender or broad valves.	4
4	Furrow spines 4–6 (rarely three or seven); pedicellariae common with broad valves; intermarginal papulae always present.	***N. multispina*** page 251
	Furrow spines 3–4 (rarely two); pedicellariae rare with slender valves; intermarginal papulae sometimes absent.	***N. saoria*** page 255
5	Tabulae with peripheral and central granules few in number (usually each less than 20), irregular in size and shape, round in cross section, peripherals radiating; abactinal arm plates always indistinct distally with enlarged central granules.	***N. wilsoni*** page 256
	Tabulae with numerous peripheral and central granules (usually more than 20, often more than 30), usually regular in size and shape, peripherals forming compact ring around centrals, rarely radiating; central granules convex, usually crowded, larger than peripherals, abactinal arm plates with more or less distinct limits distally and with central granules not prominent ('form 1'); central granules irregular in size, abactinal arm plates indistinct distally with central granules or granules prominent (as in *N. wilsoni*) ('form 2'); central granules lower than peripherals, becoming flat in extreme cases, abactinal arm plates with distinct limits distally and with even granulation ('form 3').	***N. ocellata*** page 252

(Modified from Zeidler and Rowe, 1986.)

Nectria multispina (front) and *N. ocellata* (back), Wilsons Promontory, Victoria [top] (J. Turnbull); *Nectria ocellata*, Haycock Point, South Australia [bottom] (P. Southwood).

LARGE-ARMED SEASTAR

Nectria macrobrachia

H.L. Clark, 1923

DESCRIPTION

Disc convex; arms relatively long; **tabulae low, crowded**, occurring beyond the disc until the last third of the arm; **tabulae have larger, raised peripheral granules than central ones giving them a concave appearance**; abactinal arm plates flat, decreasing regularly in size towards the arm tips; intermarginal papulae present; 2–3 furrow spines with two rows of 2–3 subambulacral spines; pedicellariae with 3–4 broad valves; the form and granulation of the tabulae and the distribution of the papulae easily separates this species from others of the genus.

MAXIMUM SIZE

R/r = 75/21 mm (R = 3.5r).

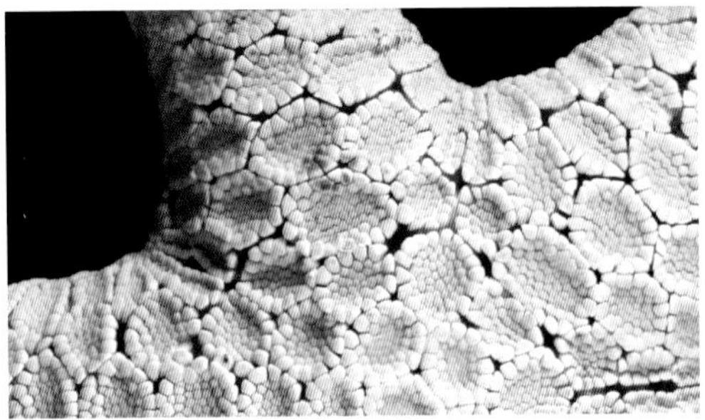

Two Peoples Bay, Western Australia [top] (G. Edgar); tabulae characteristics [bottom], South Australian Museum (SAMAK3479).

COLOUR

Light yellow to orange, brown or pink with arm tips darker than rest of body, usually pinkish brown.

FOOD

Encrusting bryozoans and sponges.

HABITAT, DEPTH AND AUSTRALIAN DISTRIBUTION

On rock with algae or encrusting animals, 0–350 m. From Wilsons Promontory, Victoria and Flinders Island, Bass Strait to the Houtman Abrolhos and off Port Gregory, WA. Southern Australian endemic.

TYPE LOCALITY

Pelsart Group, Houtman Abrolhos, WA.

FURTHER DISTRIBUTION

Australian endemic.

REFERENCES

Shepherd, 1967a: 475 (food).

Zeidler and Rowe, 1986: 121; figs 2b, 8b–c.

MULTISPINA SEASTAR
Nectria multispina
H.L. Clark, 1928

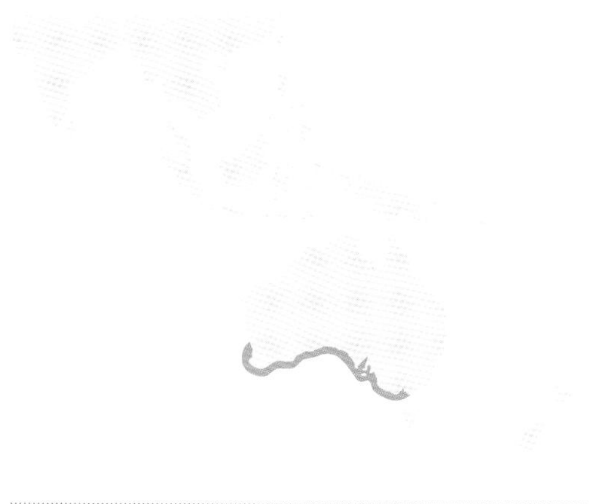

DESCRIPTION
Body fairly large; arms narrow tapering to a blunt tip; **tabulae convex and crowded, polygonal, with central granules irregular in size and larger than those of the periphery, granules very compact giving a smooth appearance**; peripheral granules similar to each other in size forming a compact ring around central granules; tabulae appear to extend to arm tips but become convex plates on outer third of arm and are interspersed with smaller, flatter plates; 4–6 furrow spines, subambulacral spines in two rows, three (sometimes four) in first row, 3–8 in second row; **papulae** extend to arm tips and **are present intermarginally**; large pedicellariae with 3–4 broad valves.

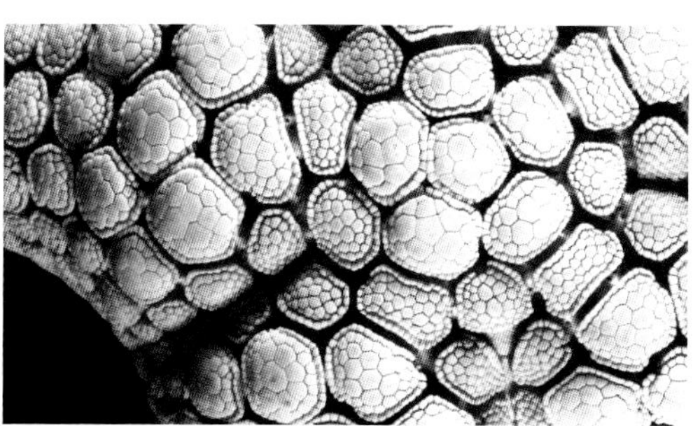

Denmark, Western Australia [top] (N. Coleman); tabulae characteristics [bottom], South Australian Museum (SAMAK3500).

MAXIMUM SIZE
R/r = 102/34 mm (R = 2.6–3r).

COLOUR
Yellow or orange to brick red, mottled with darker tabulae, paler below.

FOOD
Rock-encrusting bryozoans, sponges, ascidians, small bivalves and red algae.

HABITAT, DEPTH AND AUSTRALIAN DISTRIBUTION
On rock, intertidal to 20 m. Wilsons Promontory, Victoria to Cape Peron south of Fremantle, WA. Southern and south-western Australian endemic.

TYPE LOCALITY
Not recorded. Probably Gulf St Vincent or Spencer Gulf, SA.

FURTHER DISTRIBUTION
Australian endemic.

REFERENCES
Shepherd, 1967a: 472, (food).
Zeidler and Rowe, 1986: 122–123; figs 2d, 10c, 11a.

OCELLATE SEASTAR

Nectria ocellata

Perrier, 1875

Tasman Peninsula, Tasmania (D. Lane).

DESCRIPTION

Disc convex; **tabulae with central granules of similar size and shape, close but not compact**, rarely lower than the peripheral granules, which are similar to each other in size, smaller than the central granules and wedge-shaped forming a compact ring around the central granules, sometimes tending to radiate; tabulae flat to convex, occurring beyond the disc until the last third of the arm; abactinal arm plates convex to flat, decreasing in size to arm tip; central granules prominent, peripherals reduced; 2–4 (occasionally five)

Cape Paterson, Victoria (J. Finn).

furrow spines, subambulacrals in two rows of 1–3; **papulae absent intermarginally**; pedicellariae with 3–4 elongate valves.

MAXIMUM SIZE

R/r = 125/45 mm (R = 2.78r).

COLOUR

Brick red, orange to orange-brown sometimes mottled with lighter and darker coloured tabulae.

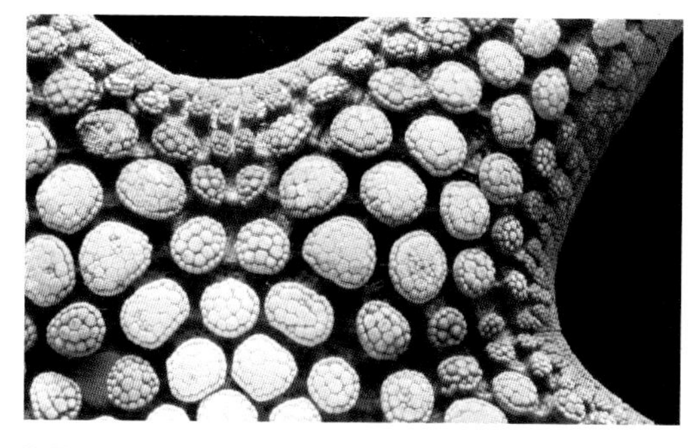

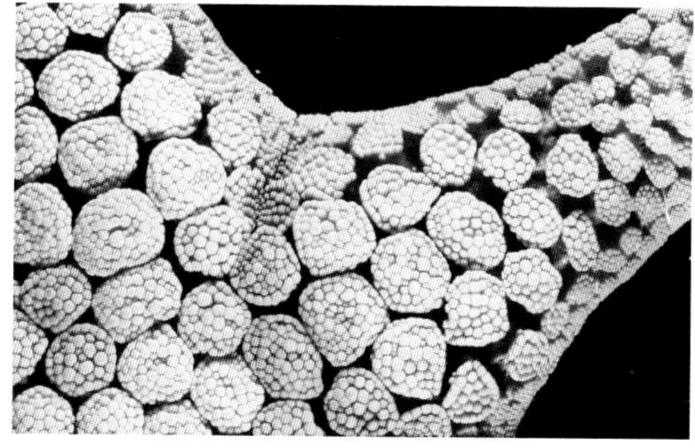

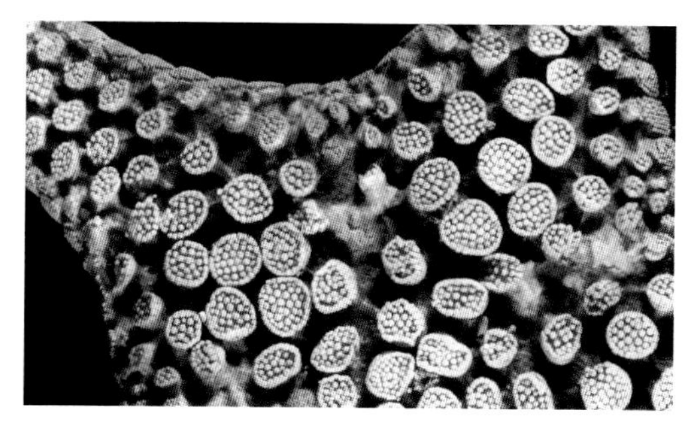

Tabulae, with varying granule characteristics, 'form 1' [top], 'form 2' [middle] and 'form 3' [bottom], South Australian Museum (SAMAK3517, K3591, K3586).

FOOD

Encrusting sponges, algae, bryozoans, compound ascidians and detritus.

HABITAT, DEPTH AND AUSTRALIAN DISTRIBUTION

On rock, 0–230 m. From Broughton Island, NSW, around southern Australia and Tasmania to the western end of the Great Australian Bight (125°30'E). Rare in south-western Australia and usually in deep water. Southern Australian endemic.

TYPE LOCALITY

Tasmania.

FURTHER DISTRIBUTION

Australian endemic.

REFERENCES

Coleman, 1979: 155 (food).

Zeidler and Rowe, 1986: 119–121; figs 2a, 4b–8a.

REMARKS

Nectria ocellata is easily confused with *N. pedicelligera* but the absence of intermarginal pedicellariae in *N. ocellata* separates them.

PEDUNCULATE SEASTAR
Nectria pedicelligera
Mortensen, 1925

DESCRIPTION
Tabulae with central granules of similar size and shape, close or compact, larger than peripheral granules; peripheral granules similar in size to each other, wedge-shaped, forming a compact ring around the central granules, rarely radiating; tabulae slightly convex, occurring beyond the disc until halfway along the arm; abactinal arm plates more or less flat, decreasing in size towards arm tip, central granules sometimes prominent giving plates a rough appearance; 3–4 (up to six) furrow spines, 2–3 subambulacrals in first row, 3–4 (occasionally six) in second row; **papulae present intermarginally, rarely actinally**; pedicellariae with 3–5 (up to six) very slender valves.

MAXIMUM SIZE
R/r = 120/36 mm (R = 2.9–3.3r).

COLOUR
Uniformly brick red to orange often mottled with lighter coloured tabulae, especially at the base of the arms.

Port Lincoln, South Australia [top] (J. Keesing); tabulae characteristics [bottom], Museums Victoria (NMVF74003).

FOOD
Probably feeds on encrusting invertebrates.

HABITAT, DEPTH AND AUSTRALIAN DISTRIBUTION
Found on rocky substrates, 0–25 m. From Eden, NSW, around southern Australia, including north-western Tasmania to Denmark, WA. Southern Australian endemic.

TYPE LOCALITY
Gisborne, east coast of North Island, New Zealand (may be erroneous as there are no further records of any *Nectria* species from New Zealand).

FURTHER DISTRIBUTION
Australian endemic.

REFERENCES
Zeidler and Rowe, 1986: 121–122; figs 2c, 9, 10a–b.

REMARKS
Nectria pedicelligera is easily confused with *N. ocellata* and *N. multispina* but differs from *N. ocellata* in having intermarginal papulae and more regular granulation, while *N. multispina*, which has enlarged plates extending along the arms, has a greater number of furrow spines, and broader valved pedicellariae.

SAORI SEASTAR
Nectria saoria
Shepherd, 1967

DESCRIPTION

Disc moderately convex, arms long, tapering to blunt tip; **tabulae crowded, with central granules usually similar in size, compact, polygonal, giving the tabulae a smooth convex appearance**; central granules larger than peripheral granules, which are slightly irregular in size, forming an irregular peripheral ring; tabulae occur beyond the disc until the last third of the arm, but similar convex plates extend to the arm tips, mixed with smaller plates giving the arms a slightly knobbly appearance; furrow spines blunt, prismatic, 3–4 (occasionally two), subambulacrals prismatic, 2–3 in first row 2–4 in second row; **papulae** extend to arm tips abactinally, **usually present intermarginally**, absent actinally; pedicellariae with 3–5 (sometimes two or six) elongate valves only on the abactinal surface.

MAXIMUM SIZE

R/r = 90/26 mm (R = ~3.5r).

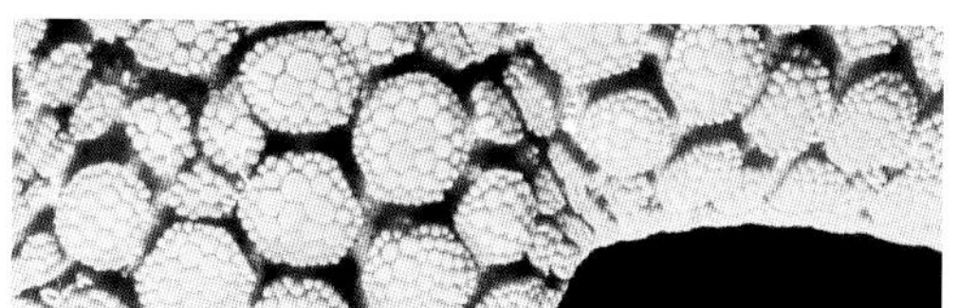

Hopetoun, Western Australia [top] (C. Bryce); tabulae characteristics [bottom], South Australian Museum (SAMAK3663).

COLOUR

Various shades of red; papulae black.

FOOD

Encrusting colonial ascidians and red algae.

HABITAT, DEPTH AND AUSTRALIAN DISTRIBUTION

On rock, 0–25 m. Port Phillip, Victoria to Sorrento, WA. This is the most common species of *Nectria* on the south coast of WA. Southern Australian endemic.

TYPE LOCALITY

Between Wright Island and the Bluff, Encounter Bay, SA.

FURTHER DISTRIBUTION

Australian endemic.

REFERENCES

Shepherd, 1967a: 477 (food).

Zeidler and Rowe, 1986: 123–124; figs 2f, 13a.

REMARKS

Nectria saoria is closest to *N. multispina* but the former has a smaller disc, fewer furrow spines, narrower pedicellariae, and irregular sized convex plates distally on the arms.

WILSON'S SEASTAR
Nectria wilsoni
Shepherd and Hodgkin, 1965

DESCRIPTION

Disc large convex; arms thick at base and slightly tapering; **tabulae crowded or well spaced, with central granules irregular in shape and size, both larger and smaller than peripheral granules**, markedly convex, usually well spaced; peripheral granules radiating, irregular in size and shape but forming a distinct peripheral ring; **tabulae more or less confined to disc**; abactinal arm plates flat, decreasing in size regularly towards arm tip, with coarse granulation, often 1–2 central granules on each plate enlarged and prominent; furrow spines 3–4 (sometimes two), subambulacral spines in two rows of 2–3 spines; **papulae numerous abactinally**, occurring on the disc and rarely beyond halfway along the arm, **absent intermarginally and actinally**; pedicellariae with typically 4–5 (occasionally 3–7) slender valves, sometimes absent.

Esperance, Western Australia [top] (S. Morrison); tabulae characteristics [bottom], South Australian Museum (SAMAK3742).

MAXIMUM SIZE

R/r = 120/40 mm (R = ~3r).

COLOUR

Deep orange to dark red sometimes blotched with darker colour.

FOOD

Algae and sponges.

HABITAT, DEPTH AND AUSTRALIAN DISTRIBUTION

On rock, jetty piles and in seagrass beds, 0–44 m. Off Lakes Entrance, Victoria to Beagle Island (29°50'S), WA. Southern Australian endemic.

TYPE LOCALITY

Sorrento, WA.

FURTHER DISTRIBUTION

Australian endemic.

REFERENCES

Coleman, 2007: 51 (food).

Shepherd and Hodgkin, 1965: 119–121; fig. 1.

Zeidler and Rowe, 1986: 123; figs 2e, 11b, 12.

GENUS *NEOFERDINA* Livingstone, 1931

A genus of seastars in which **at least some of the superomarginal and sometimes some of the abactinal and/or inferomarginal plates are bare, with a margin of fine granules**; the other plates are completely covered with fine granules. The abactinal plates of the arms are usually not in regular rows; **the adambulacral armature is in a single row of furrow spines, partly covered by granulated skin**. Papulae occur singly, only on the abactinal surface.

Type species: *Neoferdina cumingi* (Gray, 1840) (Indo-Pacific)

The genus is found from the Seychelles in the western Indian Ocean to Japan and east to the Tuamotu Islands. Two of the currently recognised twelve species are found in Australia; *N. cumingi* (page 258) and *N. insolita* (page 260).

(Jangoux, 1973.)

Neoferdina cumingi, Great Barrier Reef, Queensland (N. Coleman).

CUMING'S SEASTAR
Neoferdina cumingi
(Gray, 1840)

DESCRIPTION
The species is characterised by the **alternation of bare purple or cerise superomarginal plates with granule-covered plates**; additional bare plates may form transverse rows on the arms or as a radial series, or the transverse series may be of raised granule-covered plates; inferomarginal plates may or may not be bare; often the pattern of bare plates varies among arms of the same individual; **paired furrow spines are in a single row, their outer face partly covered by granulated skin**; papulae occur singly, only on the abactinal surface; this species is very variable and was known under three different species names.

MAXIMUM SIZE
R/r = 45/12 mm (R = 3–4.2r).

COLOUR
Abactinal plates vary from pale to deep beige, outlined with darker granules. Centre of disc is often rust or reddish coloured. The bare plates are purple or cerise.

FOOD
Substrate biofilm of algae and microorganisms.

HABITAT, DEPTH AND AUSTRALIAN DISTRIBUTION
On reef flats under boulders and on or under ledges, or on sand and rubble on reef slopes, 0–30 m; uncommon. Rowley Shoals and Scott Reef, WA, Ashmore and Cartier Reefs, Timor Sea and on the Great Barrier Reef, Queensland from Torres Strait to the Bunker Group and Lord Howe Island, Tasman Sea.

TYPE LOCALITY
Philippines.

FURTHER DISTRIBUTION
From the Cocos (Keeling) and Christmas islands in the eastern Indian Ocean, Indonesia, north to Japan and east to the Tuamotu Archipelago and Henderson Island in the Pacific.

REFERENCES
Jangoux, 1973: 788–789, 791.
Mah, 2017: 47–49; fig. 15.

REMARKS
Jangoux (1973) included *Neoferdina cancellata* (Grube, 1857) and *N. ocellata* (H.L. Clark, 1921) in the synonymy of *N. cumingi*.

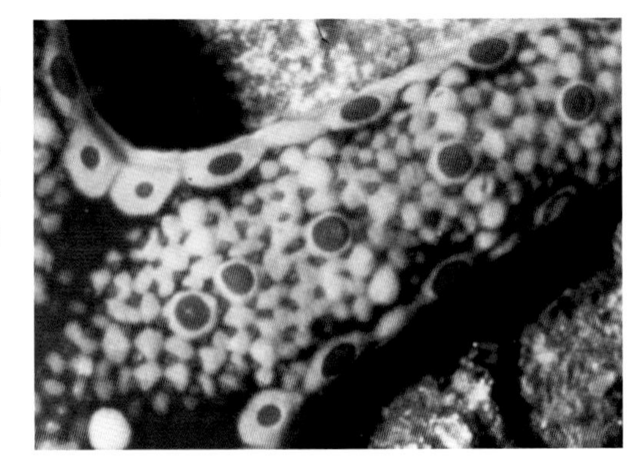

Superomarginal and abactinal plate characteristics, Madang, Papua New Guinea (R. Steene).

Madang, Papua New Guinea [top] (R. Steene);
Great Barrier Reef, Queensland [bottom] (F. Michonneau).

BARE-MARGINED SEASTAR
Neoferdina insolita
Livingstone, 1936

Cliff Head, Western Australia (L. Marsh).

DESCRIPTION
Disc and arms are flat, outlined by large superomarginal plates; the abactinal plates are irregularly arranged except for a carinal series, all are covered with fine granules; slightly larger granules lie between the plates and around the papular pores; **the superomarginals are all bare, not alternating**, although there are occasional irregularities, a border of minute granules encircles the bare area, which is very conspicuous because of the dark colour; some, or all, of the inferomarginals are also bare; the paired furrow spines are in a single row partly covered by granules on their outer side; papulae are only on the abactinal surface; no pedicellariae.

MAXIMUM SIZE
R/r = 33/17 mm (R = 2.5–3.9r).

COLOUR
Abactinal plates light grey-brown with yellow-brown granules between them. Centre of disc is bright yellow-brown with a central pink

plate next to the anus and several other pink plates on the disc, the bare superomarginals are nearly black. Actinal surface is cream, shading to buff near the margin.

HABITAT, DEPTH AND AUSTRALIAN DISTRIBUTION

Usually on coral reefs, reef flat, reef crest and slopes and on the continental shelf, 3–188 m; cryptic, rare. In WA it has been found rarely, dredged from the west coast shelf off Cliff Head (29°32'S) at 150 m, off Wedge Island (30°48'S) at 188 m and the North West Shelf at 100 m. It is found on the Great Barrier Reef, Queensland from Lodestone Reef (18°41'S) to Bushy-Redbill Reef (20°55'S), and the Solitary Islands, NSW.

TYPE LOCALITY

Samarai, Papua New Guinea.

FURTHER DISTRIBUTION

Papua New Guinea, Taiwan.

REFERENCES

Chao, 1999: 413–414; figs 23–24.
Clark and Rowe, 1971: 65.
Livingstone, 1936: 384–385; pl. 28 figs 2, 4, 6.
Mah, 2017: 50–53; fig. 17.

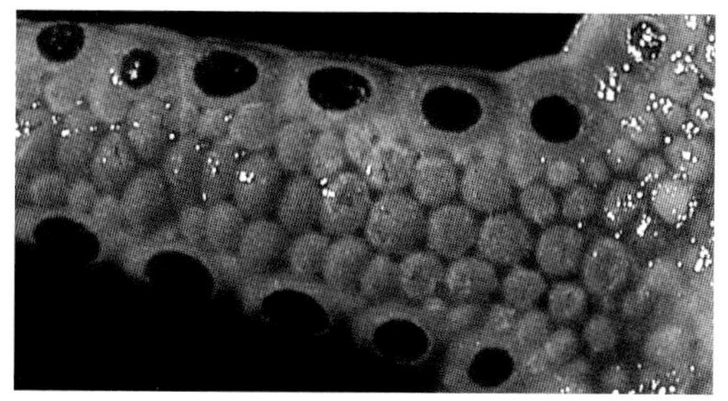

Timor-Leste [top] (N. Hobgood); superomarginal and abactinal plate characteristics [bottom], Cliff Head, Western Australia (L. Marsh).

REMARKS

Jangoux (1973) doubted the validity of this species but its characters show little variation and the bare plates are almost always red to purple, or dark grey to nearly black.

GENUS *PENTAGONASTER* Gray, 1840

Species of this genus are pentagonally shaped or have moderately long arms that have broad rounded ends. The marginal plates have a smooth central area, the distalmost plate, or one proximal to it, may be enlarged. The inferomarginals correspond with the superomarginals except at the extreme tip of the arms.

Type species: *Pentagonaster pulchellus* Gray, 1840 (New Zealand)

Only one species, *P. duebeni* (page 263), is found in shallow Australian waters.

GENUS *PSEUDOGONIODISCASTER* Livingstone, 1930

Type species: *Pseudogoniodiscaster wardi* Livingstone, 1930 (Australia)

Monotypic genus. Diagnosis as for the species (page 265).

Pentagonaster duebeni, Rottnest Island, Western Australia (C. Bryce).

VERMILLION SEASTAR
Pentagonaster duebeni
Gray, 1847

Fremantle, Western Australia (S. Morrison).

DESCRIPTION

Body flat; arms of variable length; the abactinal plates are flat to strongly convex, each is surrounded by a border of granules; **4–8 superomarginals on each side of each arm**, the distal ones are more or less swollen and may be very elongate, meeting abactinally; **the terminal plate is small**, pitted and perforated for the terminal sensory tentacle; three furrow spines (two distally), three subambulacral spines, the rest of the

Mullaloo, Western Australia (C. Bryce).

plate covered by granules; pedicellariae are tong-shaped with 2–3 tapered blades fitting into elongate grooves in the plates.

MAXIMUM SIZE

R/r = 70/28 mm (R = 2–2.5r).

COLOUR

Variable, often pale pink, yellow, orange or crimson, usually with white or beige granules outlining the plates.

FOOD

Sponges, bryozoans and compound ascidians.

HABITAT, DEPTH AND AUSTRALIAN DISTRIBUTION

On rock platforms, vertical rock faces and under ledges on sheltered and exposed coasts, 0–200 m. From Double Island Point (25°54'S), Queensland around southern Australia and Tasmania to Shark Bay, WA. Southern Australian endemic.

TYPE LOCALITY

'Western Australia'.

FURTHER DISTRIBUTION

Australian endemic.

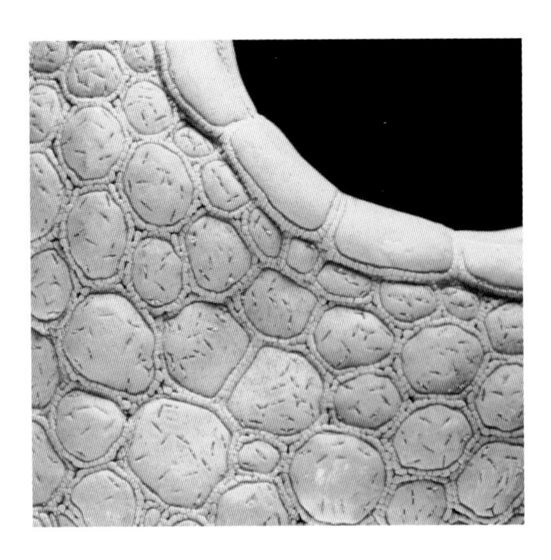

Abactinal plate characteristics, South Australian Museum (SAMAK3053).

REFERENCES

Clark, A.M., 1953b: 400–403; fig. 13b; pl. 43 figs 1–3; pl. 44 figs 1–3.

Clark, H.L., 1946: 88–90.

Mah, 2007: 324–326; fig. 6a–b (*Pentagonaster duebeni*); fig. 7a–b (*P. stibarus*).

REMARKS

H.L. Clark (1914) described *Pentagonaster stibarus* from 73–183 m between Fremantle and Geraldton. In 1946 he considered this to be a synonym of *P. crassimanus* (Möbius, 1859) but kept *P. crassimanus* separate from *P. duebeni.* Rowe and Gates (1995) include both *P. stibarus* and *P. crassimanus* in the synonymy of *P. duebeni* but Mah (2007), using cladistic analysis revived the name *P. stibarus* including a nearly pentagonal, probably juvenile form, from reef flats in the Houtman Abrolhos. The holotype of *P. stibarus* is beyond the depth range of this work. Genetic analysis should be used to resolve the status of the presumed juveniles that have been collected on many reefs in the Houtman Abrolhos between March and December.

WARD'S SEASTAR

Pseudogoniodiscaster wardi

Livingstone, 1930

Holotype, Australian Museum (AMJ5320).

DESCRIPTION

Stellate with rounded interradial arcs, five arms tapering to a blunt tip; arms flattened abactinally and actinally, disc large and raised; the abactinal skeleton is made up of more or less uniform sized plates except for the larger radial and median interradial series; plates, except those in the median interradial series, are connected by thin secondary plates that also connect them to the superomarginals; the abactinal plates are covered by fine granules and some plates bear bluntly pointed conical tubercles and 1–6 bivalved pedicellariae; **at the base of each carinal series is a very large tubercle, granulated basally with bare top**, the five tubercles form a pentagon on the disc; the madreporite is situated interradially between two of the large tubercles; papular spaces vary in size and shape with 8–38 papulae; there are 1–2 minute pincer-like pedicellariae in the granulated papular spaces; the prominent superomarginals are covered by very fine, close packed granules giving them a smooth appearance; many of the superomarginals bear a small to medium sized tubercle, tubercles are absent from the inferomarginal plates; small pedicellariae are found on both series, most numerous interradially; the inferomarginals project beyond the superomarginals except near the ends of the arms; actinal plates are closely granulated and most carry one very large bivalved pedicellaria 2–4.5 mm long, occasional pedicellaria are trivalved; there are 6–7 furrow spines, stout and flat-sided, in a fan-like comb; there are 3–4 broad, stubby subambulacral spines followed by groups of 2–3 enlarged granules; beyond the subambulacrals are enlarged prismatic granules.

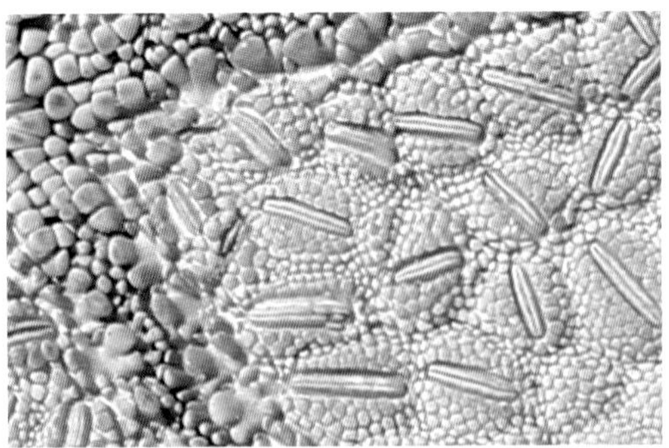

Abactinal surface characteristics including pedicellariae and carinal tubercle [top], actinal pedicellariae [bottom], holotype, Australian Museum (AMJ5320).

MAXIMUM SIZE

R/r/br of holotype = 77/37/42 mm (R = 2.08r).

COLOUR

Dark green.

HABITAT, DEPTH AND AUSTRALIAN DISTRIBUTION

Among weed, 0–73 m. Only known from Port Curtis and the Swain Reefs, Queensland. North-eastern Australian endemic.

TYPE LOCALITY

Rat Island, Port Curtis, Queensland.

FURTHER DISTRIBUTION

Australian endemic.

REFERENCES

Livingstone, 1930: 16–18; pl. 4 figs 1–2; pl. 5 figs 1–3.

REMARKS

Pseudogoniodiscaster wardi may need to be transferred to the Oreasteridae when it is determined whether the interbrachial septum is calcareous or membranous, but none of the few known specimens have been dissected.

GENUS *STELLASTER* Gray, 1840

Species of this genus have well developed tapering arms. The abactinal plates are polygonal or substellate, bearing very small crowded granules. The inferomarginal plates usually have a flattened mobile spine. Bivalved pedicellariae are flush with the surface, and erect tong-shaped pedicellariae may be present.

Type species: *Stellaster childreni* Gray, 1840 (Indo-Pacific)

Four species are found in Australian waters in depths of less than 30 m.

GENUS *STYPHLASTER* H.L. Clark, 1938

Type species: *Styphlaster notabilis* H.L. Clark, 1938 (Australia)

Monotypic genus. Diagnosis as for the species (page 276).

Stellaster inspinosus, Swan River estuary, Western Australia (G. Edgar).

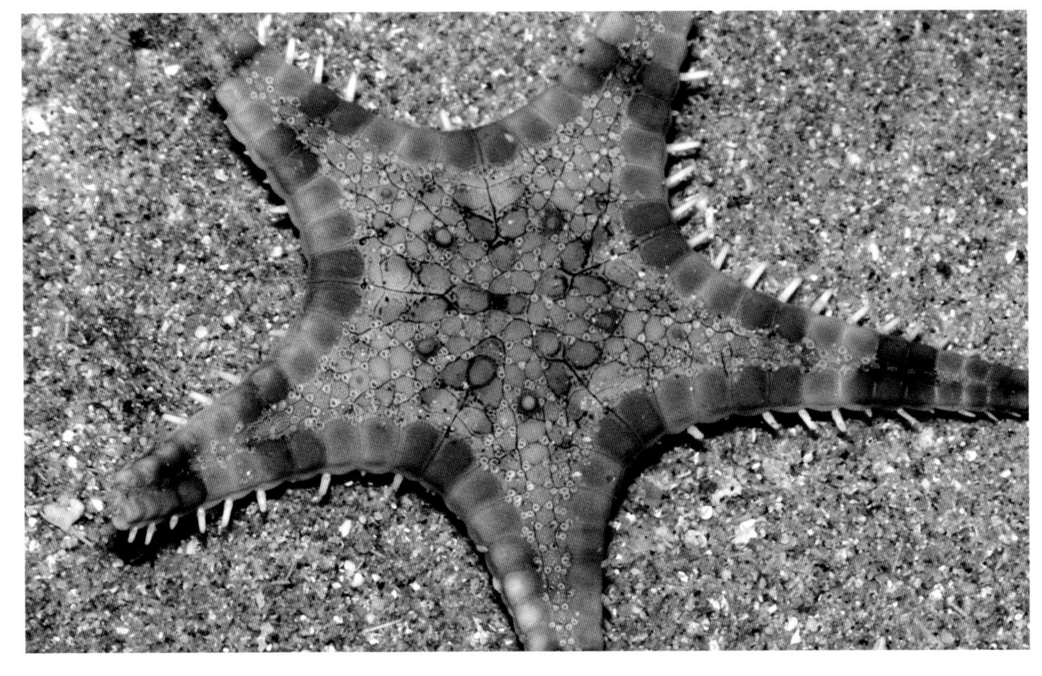

Stellaster childreni, Singapore (R. Tan).

KEY TO THE SPECIES OF *STELLASTER* IN SHALLOW AUSTRALIAN WATERS

1	Granules around papulae conspicuously swollen and scale-like; R ≤ 25 mm.	***S. squamulosus*** page 275
	Granules around papulae not swollen and scale-like.	**2**
2	Many spines on inferomarginal plates; actinal surface white, with a large, circular purple area around the mouth; R > 25 mm; northern Australia.	***S. princeps*** page 273
	Only one inferomarginal spine, or none, medium sized.	**3**
3	Inferomarginal spine usually present; larger plates of disc flat or slightly convex; many abactinal plates carry a bluntly pointed tubercle; sometimes a chisel-like spine at the base of the carinal series; abactinal surface usually brownish or greyish.	***S. childreni*** page 269
	Inferomarginal spine sometimes absent; larger plates of disc convex or swollen; few, if any, abactinal plates carry tubercles; abactinal surface pinkish brown; Western Australia.	***S. inspinosus*** page 271

CHILDREN'S SEASTAR

Stellaster childreni

Gray, 1840

Cassini Island, Western Australia (C. Bryce).

DESCRIPTION

Disc slightly convex to domed; abactinal plates nearly flat, covered by a coat of very fine granules; **denuded plates have few or many embedded crystal bodies**; a few small tubercles are often present, sometimes a large chisel-shaped tubercle is present at the base of the carinal series of plates; arms taper to a very narrow tip with a large terminal plate; superomarginal plates are swollen, finely granulated but lack crystal bodies; the last 3–5 superomarginals meet mid-radially; **inferomarginals match the superomarginals**

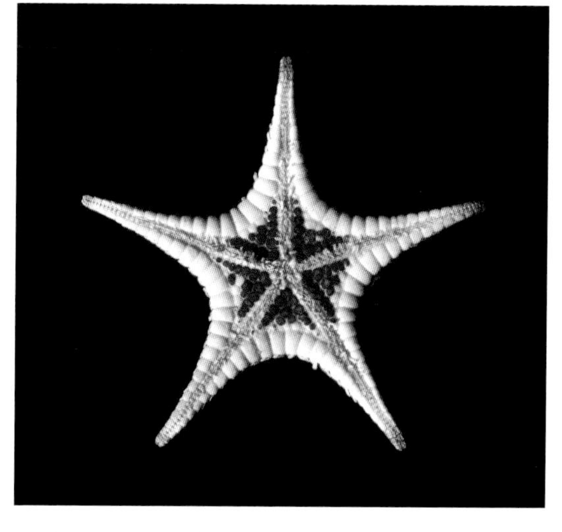

Port Hedland, Western Australia (L. Marsh)

but are flatter and each usually has a large, flat truncate spine marginally; actinal plates are nearly flat, covered by close fitting polygonal granules, slightly larger than those on the abactinal plates; adambulacrals carry a comb of 6–7 usually slender furrow spines, a pair of subequal to very unequal, large, flat subambulacral spines, the rest of the plate is granule-covered; pedicellariae are very variable in occurrence, there is usually a large erect tong-shaped, straight pedicellaria next to the subambulacral spine, few or many erect split-granule pedicellariae on actinal plates; abactinally there are often a few very small embedded bivalved pedicellariae (≤ 500 µm long) on the abactinal and superomarginal plates.

MAXIMUM SIZE

R/r = 82/33 mm (R = 2.5r).

COLOUR

Variable, often grey-brown or beige, usually with some darker plates interradially on disc; those from deep water are often red. Actinal interradial areas have a patch of dark plates.

FOOD

Stomach is filled with bottom material, sand, mud and Foraminfera (including *Marginopora)*.

HABITAT, DEPTH AND AUSTRALIAN DISTRIBUTION

On mud or muddy sand, 5–290 m. From Shark Bay and the west coast shelf of WA at 26°S around northern Australia to off the Clarence River, NSW.

TYPE LOCALITY

'China or Japan' (as *S. childreni*) and Cape York, Queensland (as *S. incei*).

FURTHER DISTRIBUTION

From east Africa and the Red Sea to Indonesia, the Philippines, north to Korea, Hainan Island and Japan.

REFERENCES

Clark, A.M., 1993: 285–286.
Clark, H.L., 1946: 97–98 (as *Stellaster incei* Gray, 1847).
Clark, A.M. and Rowe, 1971: 49; pl. 5 fig. 6 (as *Stellaster equestris* (Bruzelius, 1805)).
Mah, 2019: as *Stellaster childreni.*

REMARKS

This species has had several name changes over many years. Döderlein (1935) regarded *Stellaster equestris* (Bruzelius, 1805) as a super-species with synonyms *S. childreni* and *S. incei* among others; A.M. Clark (1993) found that Döderlein's use of the name *equestris* is invalid and believed that *S. childreni* Gray, 1840 should be revived as the valid name for this Indo-Pacific species (the classification followed here). *S. incei* is now an accepted synonym of *S. childreni* (Mah, 2019).

SMALL-SPINED SEASTAR

Stellaster inspinosus

H.L. Clark, 1916

Cockburn Sound, Western Australia (L. Marsh).

DESCRIPTION

Usually larger and more robust than *Stellaster childreni*; abactinal plates convex often with a small bare, conical tubercle at the base of the carinal series and on many other plates of the carinal series, sometimes also on some of the plates either side of the carinal; superomarginals also more convex than in *S. childreni*; **inferomarginal spines are often lacking on many plates**, hence the specific name; the actinal surface is flat; adambulacral plates carry usually 6–8 (sometimes five or nine) furrow spines and a single, broad flat, truncate subambulacral spine (rarely a thin second spine); pedicellariae are more numerous than in *S. childreni*, there are 2–3 straight or twisted erect pedicellariae between the furrow and subambulacral spines and numerous small erect split-granule pedicellariae, with clapper-shaped valves on the adambulacral plates and on the row of actinal plates next to the adambulacrals; small straight embedded bivalved pedicellariae 500 μm long on the actinal plates and some larger ones, up to 1.5 mm long, on abactinal plates; **unlike *S. childreni* there are no crystal bodies on the abactinal plates.**

MAXIMUM SIZE

R/r = 90/40 mm (R = 2.3r).

COLOUR

Pinkish brown to reddish with some actinal plates plum red. The remainder of the actinal surface cream.

FOOD

Stomach is filled with mud, from which it presumably extracts small organisms for food.

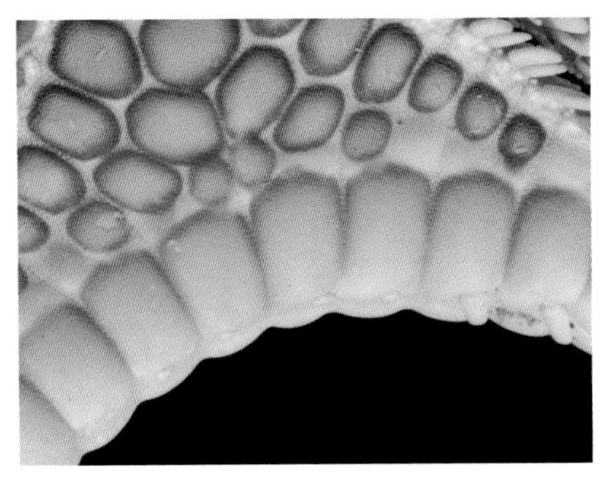

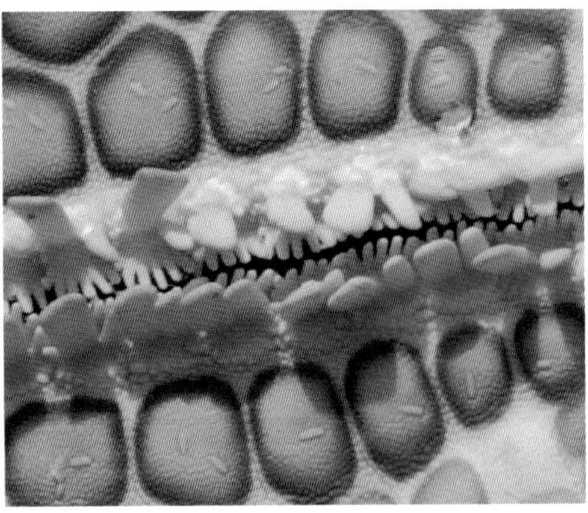

Inferomarginal [top], actinal plate and furrow characteristics [bottom], Western Australian Museum (WAMZ8092).

HABITAT, DEPTH AND AUSTRALIAN DISTRIBUTION

On silty sand or mud, 2–174 m. On the west coast shelf from Cape Naturaliste north to the Dampier Archipelago, WA, common within that range in Cockburn Sound and the Swan River estuary. Western Australian endemic.

TYPE LOCALITY

Between Cape Naturaliste and Geraldton, WA.

FURTHER DISTRIBUTION

Australian endemic.

REFERENCES

Clark, A.M., 1993: 285–286.

Clark, H.L., 1946: 97–98.

REMARKS

Stellaster inspinosus and *S. childreni* are sympatric in the Dampier Archipelago but are morphologically distinct, although in Shark Bay some may be hybrids between the two species.

PRINCELY SEASTAR
Stellaster princeps
Sladen, 1889

North West Shelf, Western Australia (L. Marsh).

North West Shelf, Western Australia (L. Marsh).

DESCRIPTION
A large, rigid seastar, unlike any other species of the genus (by far the largest species of *Stellaster*); the five arms are elongate, broad at the base tapering to an acute tip; disc large, convex, interradial arcs widely rounded; the abactinal area is covered by polygonal plates, notched for groups of papulae on disc and inner half of arms; plates in about seven rows at the base of the arm decreasing to none at the end where the last two superomarginals meet; the plates are covered with flat granules

embedded in a thin membrane; the first carinal plate on each arm bears a stout, conical pointed tubercle followed by 3–5 smaller ones at intervals along the arm; the radial disc plates also bear small pointed tubercles; some of the convex superomarginal plates bear small tubercles; **the inferomarginals each carry three large, flattened spines** visible from the upper surface; there are 7–8 short flattened furrow spines in a comb; there are two large flattened subambulacral spines and a small bivalved pedicellaria, the remainder of the plate is finely granulated; actinal plates are slightly convex, the plates in the row next to the adambulacral plates have several (3–7) small bivalved pedicellariae; small, isolated, pincer-like pedicellariae are scattered on the abactinal plates as well as a few embedded bivalved pedicellariae; crystal bodies are absent.

MAXIMUM SIZE

R/r = 200/63 mm (R = 3.2r).

COLOUR

Abactinal surface old rose with a darker pentagon on the disc and 2–3 incomplete darker arm bands. Actinal interradial plates purple to violet; rest of actinal surface cream.

FOOD

Carnivorous, preys on other seastars (*Astropecten* species) and on heart urchins, sometimes ingesting them whole.

HABITAT, DEPTH AND AUSTRALIAN DISTRIBUTION

On soft substrates, 0–210 m. From near the Montebello Islands, WA to Torres Strait, Queensland. Northern Australian endemic.

TYPE LOCALITY

Booby Island, Torres Strait, Queensland.

FURTHER DISTRIBUTION

Australian endemic.

REFERENCES

Clark, H.L., 1938: 99–101; pl. 4 figs 1–2.
Sladen, 1889: 323–325; pl. 58 figs 1–2.

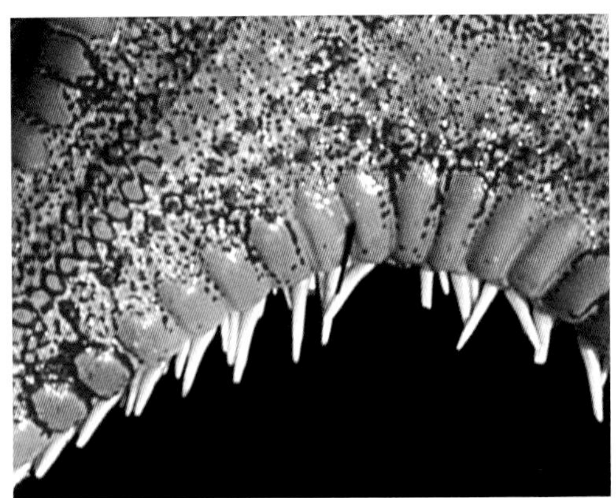

Interradial characteristics [top] and Inferomarginal spines [bottom], Swan River estuary, Western Australia (G. Edgar).

SQUAMULOSE SEASTAR
Stellaster squamulosus
(Studer, 1884)

DESCRIPTION

A very small species, distinguished from the other *Stellaster* species by having **3–6 enlarged, flattened granules that form an 'operculum' over the papular pores** on the disc and arms; the anus is closed by similar but larger granules; the disc is flat, pentagonal, the arms are elongate, pointed; the disc has polygonal plates covered with low, flat granules; 1–2 carinal plates on each arm carry a small bare tubercle; superomarginal plates convex, bare, inferomarginals each with a short, flattened, pointed or round-tipped spine; actinal plates granule-covered; a few small split-granule pedicellariae on abactinal surface and on the actinal surface near the mouth; there are 6–8 furrow spines; four enlarged granules and 3–5 smaller granules on the outer edge of the adambulacral plate.

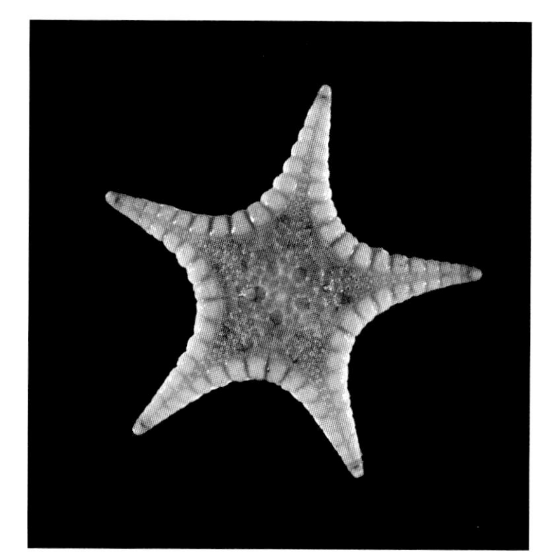

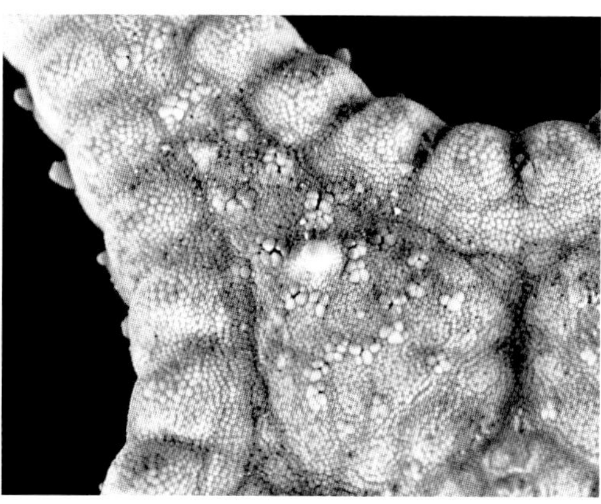

North West Shelf, Western Australia [top] (J. Marshall); abactinal papular pores [bottom], Western Australian Museum (WAMZ18036).

MAXIMUM SIZE

R/r = 25/9 mm (R = 2.5–2.7r).

COLOUR

Pale pink.

HABITAT, DEPTH AND AUSTRALIAN DISTRIBUTION

On sand or gravel with shells, coral or sponges, 28–200 m. From west of Carnarvon to off La Grange Bay, WA. Tentative record from off Gillett Cay, Swain Reefs, Queensland.

TYPE LOCALITY

Off north-western Australia.

FURTHER DISTRIBUTION

Indonesia.

REFERENCES

Studer, 1884: 33–35; pl. 4 figs 6a–c.

NOTABILIS SEASTAR

Styphlaster notabilis

H.L. Clark, 1938

Western Australian Museum (WAMZ2845).

DESCRIPTION

Large disc; arms short and stout, wide and rounded at tip; **abactinal plates polygonal or rounded, convex and well defined with a cluster of large spherical granules in the centre of each; one of the central granules may become a small tubercle**, often higher than thick; on several carinal plates the tubercles become blunt spines about 2 mm high; margins of plates and papular areas covered with polygonal granules, no pedicellariae; superomarginals convex, covered along the margins with flat-topped polygonal granules, remainder of the plate with large convex granules, largest on the upper end of the interradial plates where they are 1 mm or more in diameter; last pair of superomarginals are broadly in contact near the arm end; inferomarginals completely granule-covered, distal inferomarginals project beyond the superomarginals; there are 8–9 superomarginals and usually 11 (occasionally

10–12) inferomarginals on each side of each arm; actinal plates convex, granule-covered, similar to the inferomarginals, without pedicellariae; 3–5 subequal furrow spines, three stout subambulacral spines followed by three smaller spines and a fourth row of 3–5 small prismatic truncate spinelets; papulae are abactinal and intermarginal in the arm angle, none actinally; interbrachial septum is membranous.

MAXIMUM SIZE

R/r/br = 60/29/27 mm (R = 2.1r).

COLOUR

Lavender brown, disc and inferomarginals brown, superomarginals more lavender with white sutures between plates. Actinal surface pink to reddish and adambulacral armature creamy-white.

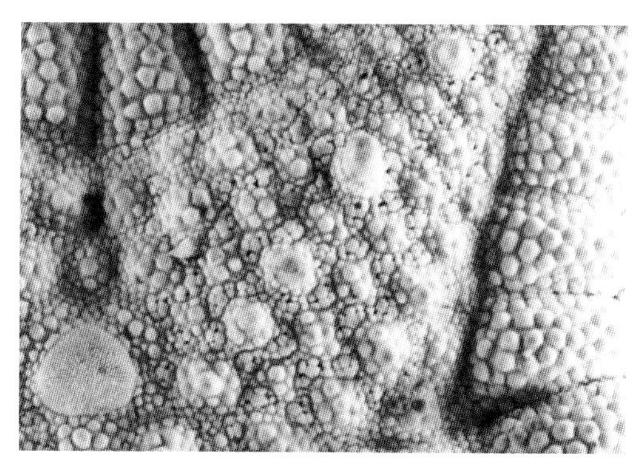

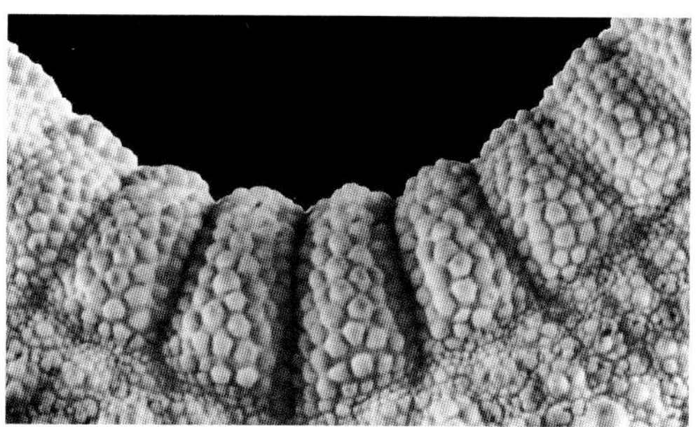

Abactinal [top] and superomarginal plate characteristics [bottom], Western Australian Museum (WAMZ2845).

HABITAT, DEPTH AND AUSTRALIAN DISTRIBUTION

On soft substrates, 3–46 m. From coastal and shelf waters; Kalbarri and the north-west coast to Troughton Island off the Kimberley coast, WA. North-western Australian endemic.

TYPE LOCALITY

Off Roebuck Bay, Broome.

FURTHER DISTRIBUTION

Australian endemic.

REFERENCES

Clark, H.L., 1938: 89; pl. 17 figs 3–4.
Clark, A.M. and Rowe, 1971: 49.

GENUS *TOSIA* Gray, 1840

Species of this genus have a pentagonal to weakly stellate body shape that is robust and thickened, and lacks spines and spinelets. The abactinal plate surfaces are bare and flat to slightly convex, never tabulate. Small granules occur around the abactimal, marginal and actinal plates. The penultimate superomarginals are enlarged, swollen, and sometimes elongate. The penultimate inferomarginals are never enlarged. The pedicellariae, if present, have short, rounded valves (split-granule form). The actinal interradial plates are covered by granules that are bare or with a prominent bald centre spot.

Type species: *Tosia australis* Gray, 1840 (Australia)

There is still some disagreement about the number of species in the genus. *Tosia magnifica* is accepted, *T. australis* is very variable with Rowe and Gates (1995) maintaining *T. nobilis* as a seperate species. Marsh (1991) considered *T. nobilis* and *T. astrologorum* as variants of *T. australis*. More recently, Naughton and O'Hara (2009) confirmed this with molecular and morphological studies and suggested that additional research into Western Australian specimens would be beneficial. *T. neossia* is distinguished from *T. australis* particularly by its brooding habit. *Tosia queenlandensis* Livingstone, 1932b was included in a new genus *Anchitosia* by Mah (2007) as it differs from *Tosia* in having enlarged terminal plates. *Tosia* is an Australian endemic genus.

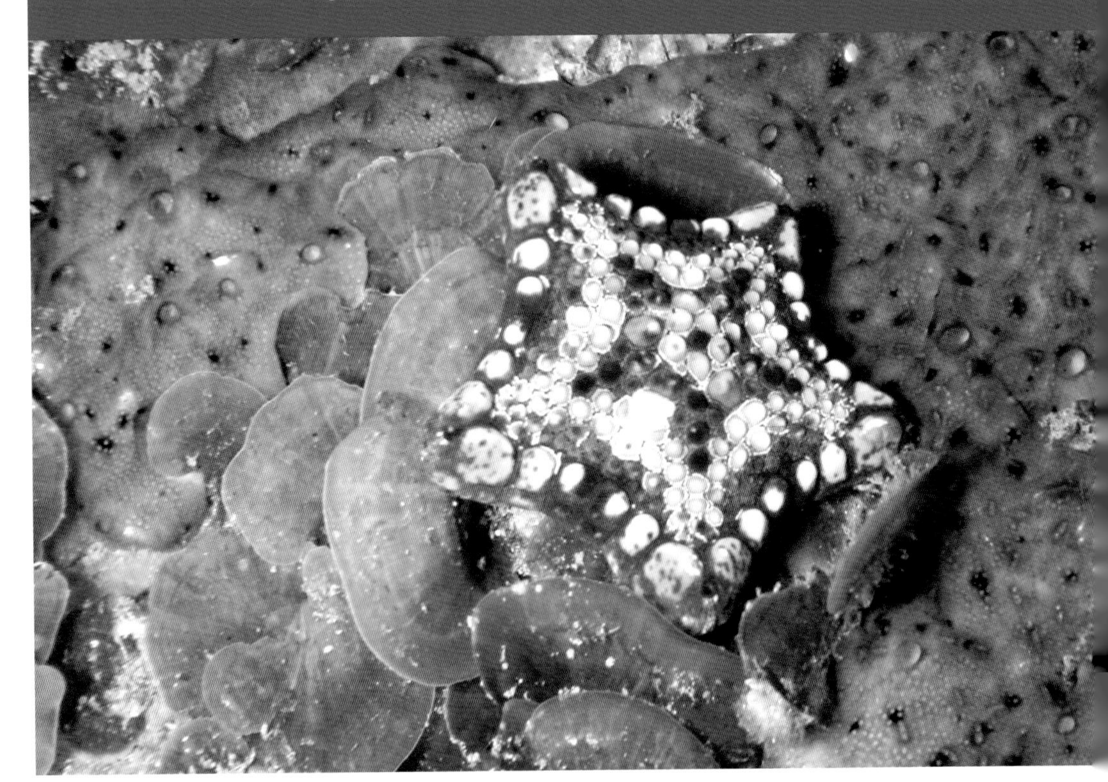

Tosia australis, Fleurieu Peninsula, South Australia (P. Southwood).

Tosia australis, Port Phillip, Victoria (E. Hallein).

KEY TO THE SPECIES OF *TOSIA* IN SHALLOW AUSTRALIAN WATERS

1	Numerous superomarginal plates (≤ 16); granules on actinal plates.	***T. magnifica*** page 282
	Fewer superomarginal plates (6–8).	2
2	No granules on oral plates.	***T. australis*** page 280
	Granules on oral plates.	***T. neossia*** page 284

SOUTHERN BISCUIT STAR
Tosia australis
Gray, 1840

South Australian Museum (SAMAK3225).

DESCRIPTION
Body pentagonal to stellate; abactinal and superomarginal plates flat to strongly convex and bare, but each is surrounded by a border of granules; **3–5 superomarginals on each side of each arm, the last or second last may be much enlarged or elongated; terminal plate very small**; last superomarginals usually meet mid-radially next to the terminal plate; adambulacral plates covered by granules, with two furrow spines and two subamublacral spines; **split-granule pedicellariae are usually present**; the extreme variability of this species has resulted in it being described under seven different species names.

MAXIMUM SIZE
R/r = 51/29 mm (R = 1.2–2r).

COLOUR
Highly variable, commonly orange, salmon pink, red, lilac, purple, sometimes with black or white markings, most often patterned in various shades of brown or cream and light orange-brown. Some of these colours may refer to *T. neossia*, which has been separated from *T. australis* (Naughton and O'Hara, 2009).

FOOD

Sponges (95% of diet), bryozoans, encrusting ascidians, molluscs and detritus.

REPRODUCTION

Gametes are released from the abactinal surface in January; eggs are buoyant and develop into ciliated swimming, non feeding, brachiolaria larvae.

HABITAT, DEPTH AND AUSTRALIAN DISTRIBUTION

Under boulders and among rocks and algae and on jetty piles in sheltered waters, 0–40 m. From Victoria, around the south coast of Australia including Tasmania to Kalbarri on the mid-west coast of WA. Southern Australian endemic.

Cockburn Sound, Western Australia [top]; Geographe Bay, Western Australia [bottom] (C. Bryce).

TYPE LOCALITY

Swan River estuary, WA.

FURTHER DISTRIBUTION

Australian endemic.

REFERENCES

Clark, A.M., 1953b: 404–406; fig. 14; pl. 45 figs 1–2; pl. 46 fig. 3.

Clark, H.L., 1938: 78–79.

Mah, 2007: 326–328; figs 8d–e, 9a–f.

Naughton and O'Hara, 2009: 348–366.

Rowe and Gates, 1995: 71 (as *Tosia australis* and *T. nobilis* (Müller and Troschel, 1843)).

Sloan, 1980: 69 (food).

REMARKS

A.M. Clark (1953b) maintained *Tosia nobilis* as a separate species from those found in Western Australia, but examination of the range of variation from within one locality, as well as from many localities, shows that only one species, *T. australis* can be maintained.

MAGNIFICENT BISCUIT STAR
Tosia magnifica
(Müller and Troschel, 1842)

Bruny Island, Tasmania (N. Coleman).

DESCRIPTION

The largest species of *Tosia*; abactinal plates flat to slightly convex, each surrounded by a row of granules; superomarginals even in size except for the distalmost, which is usually much smaller than the second last plate; **the terminal plate is small; there are up to eight superomarginal plates on each side of each arm** (up to 16 per interradius), while specimens of R = 10 mm usually have four; **some or all of the actinal plates are granule-covered**; there are 2–3 furrow spines, 2–3 subambulacrals and the remainder of the plate is granule-covered; there are split-granule pedicellariae.

MAXIMUM SIZE

R = 77 mm (R = 1.2–1.7r), generally R < 50 mm.

COLOUR

Variable, typically shades of brown, orange, red, beige and grey.

FOOD

Sponges, ascidians and algae.

HABITAT, DEPTH AND AUSTRALIAN DISTRIBUTION

On sandy or muddy bottoms and jetty piles, 0–200 m. Predominantly found in Tasmania, also recorded from Victoria and SA. Old records from WA have not been confirmed. Southern Australian endemic.

TYPE LOCALITY

Tasmania.

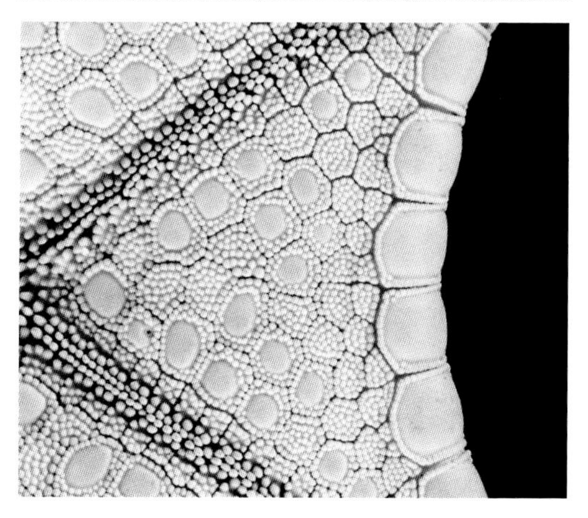

Bruny Island, Tasmania [top] (N. Coleman); actinal plate characteristics [bottom], Western Australian Museum (WAMZ12838).

FURTHER DISTRIBUTION

Australian endemic.

REFERENCES

Clark, A.M., 1953b: 408–411; pl. 45 fig. 5; pl. 46 figs 4–5.

Clark, H.L., 1946: 94 (as *Tosia aurata* Gray, 1847).

Livingstone, 1932b: 377; pl. 43 figs 3–9; pl. 44 fig. 8.

Mah, 2007: 328–329; fig. 8a–c.

REMARKS

This species is readily distinguished from *Tosia australis* by having a much larger number of marginal plates and the granule covering of the actinal plates.

NESTING BISCUIT STAR

Tosia neossia

Naughton and O'Hara, 2009

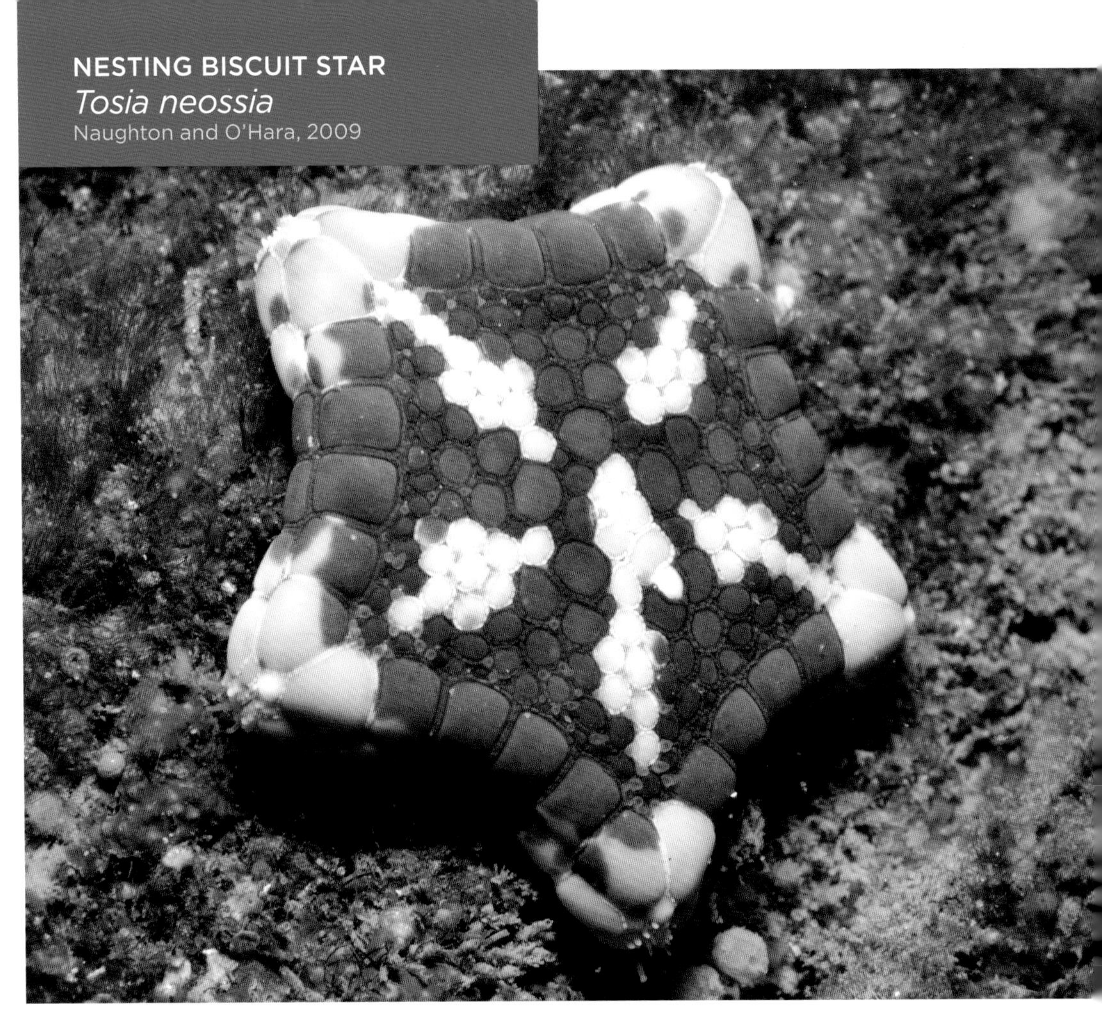

Cape Paterson, Victoria (J. Finn).

DESCRIPTION

Body form variable from pentagonal to weakly stellate with curved interradial arcs; abactinal plates flat or slightly convex; marginal plates enlarged, 6–8 per interradius, often longer than wide, flat or slightly convex; penultimate superomarginal never swollen or larger than other superomarginals; two paddle-shaped embedded pedicellariae often occur on the abactinal surface; inferomarginal plates are smooth, each surrounded by a single row of granules; **most oral plates are covered by granules**; bivalved (sometimes trivalved) paddle-shaped embedded pedicellarae are usually found interradially; two gonopores (rarely visible) per interradial area open one plate distant from the mouth; there are two furrow spines per plate followed by four to six subambulacral spinelets in three rows of two per plate.

MAXIMUM SIZE

R = 23.5 mm (R = 1.25–1.55r).

COLOUR

Consistently a pattern of two colours abactinally; reddish orange with black, brown or cream accents, black or brown with pinkish, cream or white accents, black with brown accents or brown with black accents. The accents are predominantly interradial in position. Actinal surface is predominantly cream white.

FOOD

Encrusting compound ascidians, bryozoans, sponges.

REPRODUCTION

Spawning takes place in winter (July to early September). Gametes are released from interradial pores on the actinal surface near the mouth; eggs are 500–600 μm in diameter, they are deposited in hollows and crevices of the substrate and develop into non feeding, non swimming, non ciliated lecithotrophic brachiolariae that develop a tripod form and adhere to the substrate by a central adhesive disc; metamorphosis begins about one week after fertilisation and is complete by about 30 days.

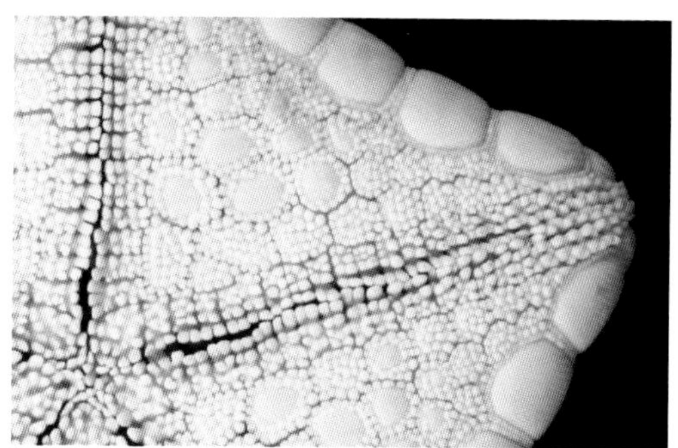

Cape Paterson, Victoria [top] (J. Finn); actinal plate characteristics [bottom], South Australian Museum (SAMAK3261).

HABITAT, DEPTH AND AUSTRALIAN DISTRIBUTION

On rocks and jetty piles in sheltered waters. From Westernport Bay, Victoria (38°41'S, 145°37'E), the eastern and southern coasts of Tasmania (43°19'S, 146°32'E), to Port Macdonnell, SA (38°03'S, 140°02'E). South-eastern Australian endemic.

TYPE LOCALITY

Apollo Bay, Victoria.

FURTHER DISTRIBUTION

Australian endemic.

REFERENCES

Coleman, 2007: 3 (as *Tosia australis* with brooded juveniles).

Naughton and O'Hara, 2009: 348–366.

REMARKS

Tosia neossia differs from *T. magnifica* by its much smaller size and fewer superomarginal plates (6–8 in *T. neossia*, up to 16 in *T. magnifica*). It differs from *T. australis* by having actinal gonopores (abactinal in *T. australis*), granulated actinal plates and different colour patterns.

MITHRODIIDAE

A small family comprising only two genera is found in the Indo-Pacific, Central American region and western Atlantic. It is distinguished from other families of Valvatida by the reticulate **skeleton bearing spines or tubercles. The whole surface, including spines, is overlaid by a thick skin bearing granules, scales or spinelets**. The pedicellariae are erect, slender, and multivalved.

Both genera are represented in Australian seas, *Mithrodia* by one species and *Thromidia* by two. In the Indo-Pacific region three species of *Mithrodia* and four of *Thromidia* are recognised.

Mithrodia clavigera, Great Barrier Reef, Queensland (A. Hoggett).

GENUS *MITHRODIA* Gray, 1840

Species of this genus typically have five cylindrical arms that are elongate, slender and spiny. The skeleton is open or compact and reticulate, with plates bearing tubercles. The seastar surface is covered by a thick skin with scale or thorn-like granules. Tubercles or spines are absent from the papular areas. Large, stout spines occur actinally and abactinally. The marginal plates are inconspicuous but their position is indicated by two of the rows of stout spines. Actinal intermediate areas are narrow with only one row of plates. Multivalved pedicellariae are present actinally and abactinally.

Type species: *Mithrodia clavigera* (Lamarck, 1816) (Indo-Pacific)

Only the type species (page 287) is found in shallow Australian waters.

NAIL-ARMED SEASTAR
Mithrodia clavigera
(Lamarck, 1816)

DESCRIPTION
Small disc; slender arms, usually five, cylindrical for most of length, tapering to a fairly acute tip; skeleton reticulate with plates and connecting ossicles bearing tubercles or blunt spines, **the whole body and spines covered by a thick skin with scales or thorn-like granules**; some of the radial, superomarginal and inferomarginal plates, and a row next to the furrow, have elongate cylindrical blunt-ended scale-covered spines; furrow spines slender, cylindrical, 10–12 per plate in webbed fans; subambulacral spines one per plate, scale-covered except on the side next to the furrow.

MAXIMUM SIZE
R/r = 185/15 mm (R = 12.3r).

COLOUR
Cream, fawn or grey, lightly or heavily blotched, sometimes banded with dark brown or dark grey. Sometimes arm ends are nearly black.

REPRODUCTION
The larvae are planktotrophic.

HABITAT, DEPTH AND AUSTRALIAN DISTRIBUTION
Usually on intertidal coral reef flats, occasionally to 106 m. It is rare in north-western Australia on Ningaloo Reef and on the offshore reefs, Rowley Shoals and Scott Reef, WA, and Ashmore Reef, Timor Sea, and has been dredged from the North West Shelf. Also found on the Great Barrier Reef south to the Capricorn Group and Lord Howe Island, Tasman Sea.

TYPE LOCALITY
Not recorded.

FURTHER DISTRIBUTION
From east Africa and the Red Sea, north to Japan, through South-East Asia to the west coast of Central America, the western Caribbean and east of Brazil. Not found in Hawaii.

REFERENCES
Pope and Rowe, 1977: 213–215.

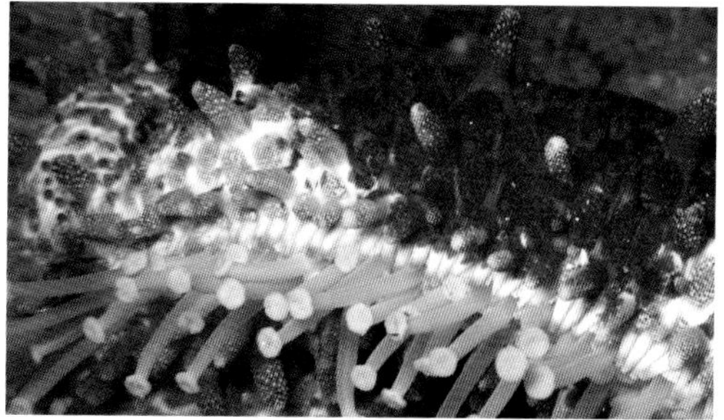

Okinawa Islands, Japan [top] (J. Marshall); Réunion Island [bottom] (P. Bourjon).

GENUS *THROMIDIA* Pope and Rowe, 1977

Species of this genus are large, thick and five-armed. Arms with more or less parallel sides with blunt, rounded tips. The skeleton is reticulate, and the skeletal plates have small tubercles. The seastar surface is covered by a thick skin with scale or thorn-like granules. The larger, stout spines are restricted to a subambulacral row and one (rarely two) actinal rows. Tubercles occur in the papular areas. Marginal plates are inconspicuous and without spines. Actinal intermediate areas are wide with several rows of plates. Multivalved pedicellariae are present only actinally, and usually restricted to areas adjacent to, and between, the spines of the first actinal and subambulacral rows.

Type species: *Thromidia catalai* Pope and Rowe, 1977 (Central Indo-Pacific)

Two species are found in Australian waters in depths of less than 30 m; *T. brycei* (page 289) and *T. catalai* (page 290).

Thromidia brycei, Dirk Hartog Island, Western Australia (C. Bryce).

BRYCE'S SEASTAR
Thromidia brycei
Marsh, 2009

DESCRIPTION
A large, stout seastar with a fairly small disc and five slightly tapering, thick arms, slightly constricted at the base; **the body is covered by thick skin over a reticulate skeleton bearing granule-covered tubercles 1–2.5 mm in height; the apical granules are usually rounded, those on the sides pointed**; minute spinelets between the tubercles are usually conical with sharp points; **the tubercles continue at the same size and density to the arm tips**; papular areas between the skeletal meshes are small, triangular to irregular in shape, rarely more than 5 mm in diameter with 12–30 pores; furrow spines are in a webbed fan of 10–12 spines; there is one subambulacral spine per plate, bare on its inner side; pedicellariae of 4–5 slender upright valves are found actinally and abactinally.

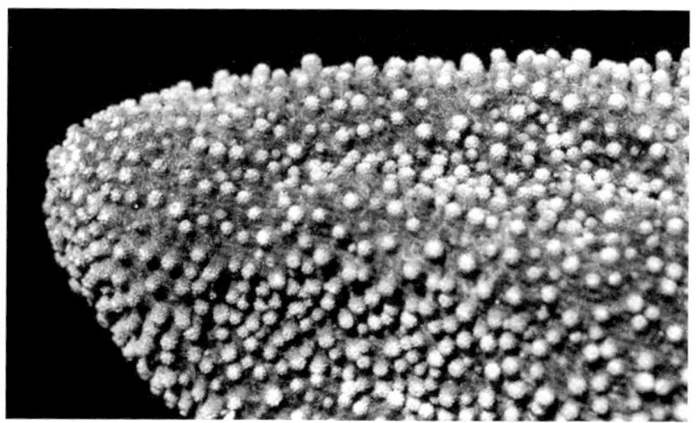

Houtman Abrolhos, Western Australia [top] (C. Bryce); tubercle characteristics [bottom], holotype, Western Australian Museum (WAMZ1716).

MAXIMUM SIZE
R/r = 280/39 mm (R = 6–8r).

COLOUR
Pinkish cream to light brown, heavily blotched with brown on disc and arms, sometimes all brown, arm tips not distinctively coloured.

HABITAT, DEPTH AND AUSTRALIAN DISTRIBUTION
On substrates of rock, algae and sponges, sometimes sand or mud, 17–149 m. From the Houtman Abrolhos islands to north of Eighty Mile Beach, WA. North-western Australian endemic.

FURTHER DISTRIBUTION
Australian endemic.

REFERENCES
Marsh, 2009: 145–151; figs 1–3.

CATALA'S SEASTAR
Thromidia catalai
Pope and Rowe, 1977

Maluku Islands, Indonesia (L. Marsh)

DESCRIPTION

A very large, stout species with a fairly small disc and five stout arms, tapering slightly to a blunt end; **the skeleton forms an open meshwork enclosing large papular areas (up to 10 mm across)** in which small tubercles are present; **numerous small tubercles (~1.5 mm high by 1 mm wide) are crowded over the whole surface, on plates and in the papular areas, but are widely spaced near the arm ends**; the surface, including the tubercles, is covered by a thick skin with scale or thorn-like granules; the skin between tubercles is closely covered by very small pointed granules; **five furrow spines in a webbed fan backed by a single, stout subambulacral spine** (8 mm high by 2 mm wide) covered (except on part of the face next to the furrow spines) with scales, which are sometimes on the furrow spines as well; there are slightly smaller spines on the first row of actinal plates, opposite every second subambulacral spine; pedicellariae of usually five slender, erect valves arising from a calcareous ring are found outside the subambulacral spines, and sometimes on the actinal and abactinal surface.

MAXIMUM SIZE

R/r = 350/78 mm (R = 4–5r).

COLOUR

Generally pink with cinnamon brown arm ends.

FOOD

Encrusting organisms including sponges.

HABITAT, DEPTH AND AUSTRALIAN DISTRIBUTION

A rare species, found on sand or rubble bottom in the outer part of lagoons and on inner and outer reef slopes, 10–105 m. Only known from the Capricorn Group, Great Barrier Reef, Queensland.

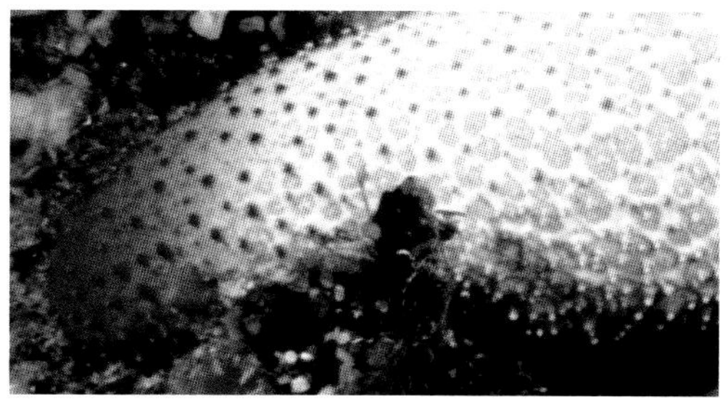

Solomon Islands, tubercle characteristics [bottom] (C. Newbert).

TYPE LOCALITY

Tabu Reef, off Amedée Island, New Caledonia.

FURTHER DISTRIBUTION

Indonesia, Papua New Guinea, Solomon Islands, New Caledonia, Hawaii, Guam and the Ogasawara Islands, Japan.

REFERENCES

Pope and Rowe, 1977: 203–207; 211 (key); figs 1–6, 9.

REMARKS

A specimen of *Thromidia catalai* from New Caledonia was listed in the Guiness Book of Animal Facts and Feats, 3rd ed., (Wood, 1982) as the heaviest seastar on record at 5.9 kg. Related species are found in the Indian Ocean and South Africa.

OPHIDIASTERIDAE

Ophidiasterids are distinguished from other members of the order Valvatida by the following characters. They have a small disc and five (occasionally more) arms that are usually long and slender, and often more or less cylindrical. The body is normally covered by skin bearing a few or many granules, although some species (e.g. *Leiaster*) lack granules. The marginal plates are usually small. The skeleton is tessellate, with the plates often arranged in longitudinal series. Small superambulacral plates are generally present. The pedicellariae, when present, are small, excavate, or lacking, and often of sugar-tongs form. Papulae are sometimes present on the actinal surface as well as the abactinal.

The family occurs in tropical and subtropical shelf and shallow waters. There are 25 known genera of which 14 occur in Australian waters, 10 in depths of less than 30 m. Forty species are recorded from Australia, of these 32 are found in depths of less than 30 m.

KEY TO THE GENERA OF OPHIDIASTERIDAE IN SHALLOW AUSTRALIAN WATERS

1	No granules on body surface, which is covered by thick smooth skin except for the furrow and subambulacral spines; arms long and slender, cylindrical to slightly tapering.	***Leiaster*** page 316
	Granules, spines or tubercles partly or wholly cover the skin over the skeletal plates.	**2**
2	Skeletal plates completely covered by granules (with or without spines or tubercles), or granules may be restricted to a band around the base of plates.	**3**
	Armament restricted to a cluster of coarse granules, tubercles or a single spine in the centre of each abactinal plate, the remaining surface is bare skin.	**4**
3	Skeletal plates convex, bare except for fine granules around the base of each plate; pores present on abactinal and actinal surface; $R \leq 23$ mm.	***Bunaster*** page 294
	Skeletal plates covered by granules or tubercles, centre of some plates bare in some genera.	**5**

Couplet	Description	Go to
4	Body covered by bare skin with seven rows of plates with a single spine in the centre of each plate; arms cylindrical.	***Cistina*** page 294
	Body skin covered with a cluster of coarse granules in the centre of each plate; in dry specimens crystal bodies on the plates may give a bumpy texture.	***Dactylosaster*** page 294
5	Arms more or less cylindrical, tapering in some genera.	**6**
	Arms flat actinally, rounded above; actinal plates in three rows on disc; during growth papulae rows increase from six to eight, 10 in large adults.	***Hacelia*** page 314
6	Skeletal plates in distinct rows.	**7**
	Skeletal plates irregular not in rows.	**8**
7	Papular pore areas in six rows.	***Tamaria*** page 351
	Papular pore areas always in eight rows.	***Ophidiaster*** page 339
8	Abactinal plates irregular in arrangement; papular pore areas not in rows; marginal plates not distinct.	***Linckia*** page 322
	Abactinal plates irregular in arrangement except at base of arms where they may be in rows; marginal plates prominent.	**9**

Gomophia sphenisci, Cassini Island, Western Australia (C. Bryce).

9	Superomarginal plates alternating large and small; intermarginal plates present in some species.	***Gomophia*** page 302
	Superomarginal plates decreasing regularly in size distally, not alternating large and small; no intermarginal plates.	***Nardoa*** page 331

GENUS *BUNASTER* Döderlein, 1896

Species of this genus have a small disc and elongate, more or less cylindrical arms, and are of small size (R ≤ 23 mm). **The abactinal and marginal plates are bare and convex, with small granules between the plates.** The marginal plates are prominent and wider than long when viewed from above. They are set at an angle to the direction of the arm. **Intermarginal plates are present.** Papulae are in 6–10 rows, and small bivalved pedicellariae are usually present, while excavate pedicellariae are rare. The genus is distributed from the South China Sea to Australia, from the shore to depths of ~200 m. The three species in the genus have widely separated distributions.

Type species: *Bunaster ritteri* Döderlein, 1896 (Indonesia, Australia)

Three species occur in shallow Australian waters.

GENUS *CISTINA* Gray, 1840

Type species: *Cistina columbiae* Gray, 1840 (Indo-Pacific)

Monotypic genus. Diagnosis as for the species (page 298).

GENUS *DACTYLOSASTER* Gray, 1840

Type species: *Dactylosaster cylindricus* (Lamarck, 1816) (Indo-Pacific)

Only the type species (page 300) occurs in shallow Australian waters.

KEY TO THE SPECIES OF *BUNASTER* IN SHALLOW AUSTRALIAN WATERS

1	Papulae in 10 rows, 1–4 pores per papular area; actinal plates in two series; South China Sea to Timor Sea.	***B. ritteri*** page 295
	Papulae in 6–8 rows, singular.	**2**
2	Papulae in six rows; actinal plates in a single series; Great Barrier Reef, Queensland, Papua New Guinea.	***B. uniserialis*** page 296
	Papulae in 6–8 rows; actinal plates in two (occasionally 3–4) series; southern Western Australia.	***B. variegatus*** page 297

RITTER'S SEASTAR
Bunaster ritteri
Döderlein, 1896

DESCRIPTION
The disc is convex and the arms taper to a blunt tip and they are round in section; the convex plates are bare, resembling miniature irregular boulders set in coarse pebbly mortar; the superomarginals are set at an angle to the direction of the arm; there are 1–2 rows of intermarginal plates; **the papulae are in 10 rows** with 1–4 pores per area; **actinal plates are in two rows**; small bivalved pedicellariae are common, spatulate tong-shaped pedicellariae are rare.

MAXIMUM SIZE
R/r = 23/6 mm (R = 3.6–4.7r).

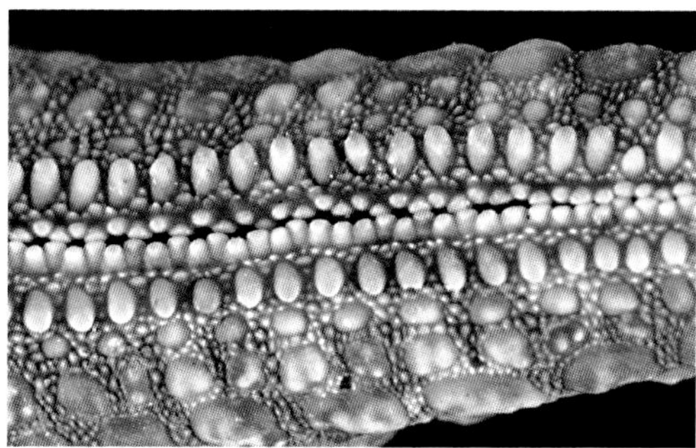

Loloata Island, Papua New Guinea [top] (L. Marsh); actinal plate characteristics [bottom], Western Australian Museum (WAMZ15721).

COLOUR
Mottled shades of greenish grey.

HABITAT, DEPTH AND AUSTRALIAN DISTRIBUTION
On or under boulders on coral reefs and seagrass flats, 0–54 m. In Australia it is only known from Ashmore and Cartier Reefs, Timor Sea.

TYPE LOCALITY
Ambon, Indonesia.

FURTHER DISTRIBUTION
Indonesia, Papua New Guinea, the Philippines and South China Sea.

REFERENCES
Fisher, 1919: 398–399; pl. 95 figs 8, 8a; pl. 124 fig. 4. (as *Bunaster lithodes*).

Marsh, 1991: 422–425; figs 1a–b, 2a–b, 3a–b, 4a–b.

REMARKS
Marsh (1991) synonymised *Bunaster lithodes* Fisher from the Philippines with *B. ritteri* Döderlein and this decision is upheld here. Both species are considered valid on the World Register of Marine Species (Mah, 2019).

CHAIN-LINK SEASTAR
Bunaster uniserialis
H.L. Clark, 1921

DESCRIPTION
The abactinal surface is flat, the arms are slender and quadrangular in section; abactinal plates are separated by rounded well spaced granules; **papulae single, in six rows; actinal plates in a single series**; enlarged granules at the base of subambulacral spines in a continuous row, no granules between subambulacral and furrow spines.

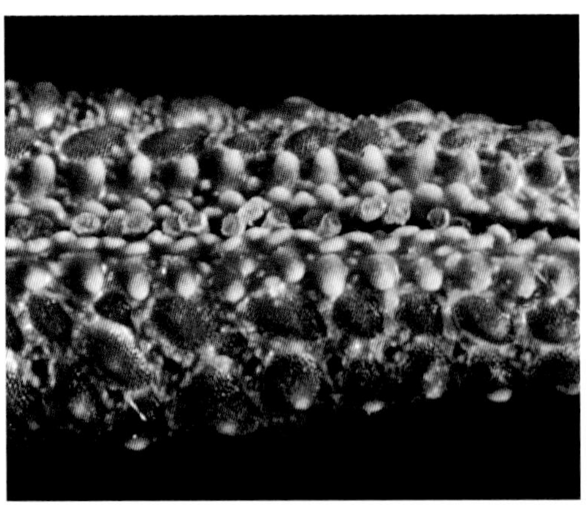

Lord Howe Island, Tasman Sea [top] (N. Coleman); actinal plate characteristics [bottom], Western Australian Museum (WAMZ18416).

MAXIMUM SIZE
R/r = 11.3/2.7 mm (R = 3.4–5.2r).

COLOUR
Often variegated maroon, brown and bluish white. Mottled mauve animals have been found under encrusting purple ascidians.

HABITAT, DEPTH AND AUSTRALIAN DISTRIBUTION
A coral reef species, found cryptically on reef flats, on seaward reef crests and outer reef slopes, 0–25 m. The Great Barrier Reef, Queensland, from Torres Strait to the Capricorn Group (23°29'S), Osprey Reef, Coral Sea and Lord Howe Island, Tasman Sea.

TYPE LOCALITY
Murray Islands, Torres Strait, Queensland.

FURTHER DISTRIBUTION
Madang, Papua New Guinea.

REFERENCES
Marsh, 1991: 425–429; figs 1c, 2c, 3b, 4c.

VARIEGATED SEASTAR
Bunaster variegatus
H.L. Clark, 1938

DESCRIPTION
The disc is moderately convex and the arms taper to a blunt tip; the convex, bare abactinal plates are separated by closely spaced granules; the carinal and marginal plates are angled to the direction of the arms; intermarginal plates are present; **papulae are single, in 6–8 rows depending on the size** of the animal and **actinal plates are in two or three rows**; furrow spines are paired with one subambulacral spine per plate; small, white bivalved pedicellariae are sometimes present but no tong-shaped pedicellariae have been found.

MAXIMUM SIZE
R/r = 21/5 mm (R = 2.9–4.7r).

COLOUR
Variable, for example pink disc, arms cream with dark brown or pink bands or blotches. Commonly shades of old rose, pure white, tan or dark brown.

HABITAT, DEPTH AND AUSTRALIAN DISTRIBUTION
Cryptic, on the underside of boulders or coral slabs or in crevices of rock or coral, 0–190 m. From west of Cape Leeuwin (34°25'S) to outside South Passage, Shark Bay (26°S), WA. South-western Australian endemic.

TYPE LOCALITY
Bunker Bay, Cape Naturaliste, WA.

FURTHER DISTRIBUTION
Australian endemic.

REFERENCES
Marsh, 1991: 429–432; figs 1d, 2d, 4d.

Houtman Abrolhos, Western Australia [left] (C. Bryce); carinal and marginal plate characteristics [above], Western Australian Museum (WAMZ18404).

COLUMBIAN SEASTAR
Cistina columbiae
Gray, 1840

Ningaloo Reef, Western Australia (L. Marsh).

DESCRIPTION
Disc small; five cylindrical arms with blunt tips; **body covered by fairly thick skin, without granules but with seven rows of short, stout spines**, one on each skeletal plate; plates linked to the next row by connecting ossicles, leaving papular areas (eight rows) between; furrow spines slender, paired, joined by webbing, subambulacrals relatively large, tapering, one on each adambulacral plate near the base of the arms, on alternate plates distally; no pedicellariae.

MAXIMUM SIZE
R/r = 85/9 mm (R = ~9r).

Ari Atoll, Maldives (F. Ducarme).

COLOUR
Pinkish red or orange-brown, sometimes with darker blotches.

FOOD
Probably feeds on substrate biofilm.

REPRODUCTION
Spawned after removal from the reef in Queensland during August (Keesing, personal communication).

HABITAT, DEPTH AND AUSTRALIAN DISTRIBUTION
A cryptic coral reef species, found under rocks usually on reef slopes, 1–44 m. Ningaloo Reef, Scott Reef, Mermaid Reef, WA, Hibernia and Ashmore Reefs, Timor Sea and the Great Barrier Reef from Ribbon Reefs to Swain Reefs, Queensland.

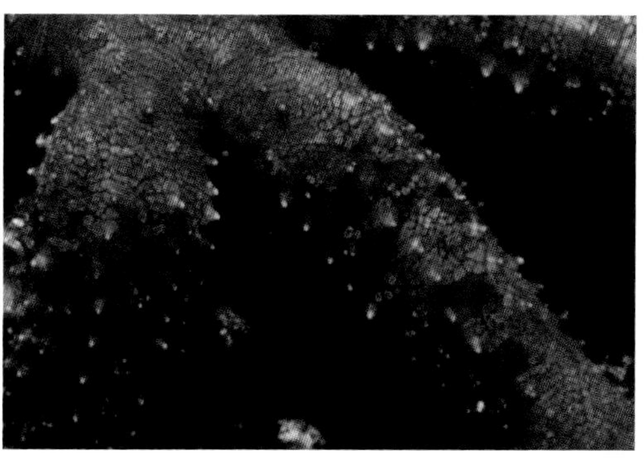

Hibernia Reef, Timor Sea [top]; abactinal arm plate characteristics [bottom] (C. Bryce).

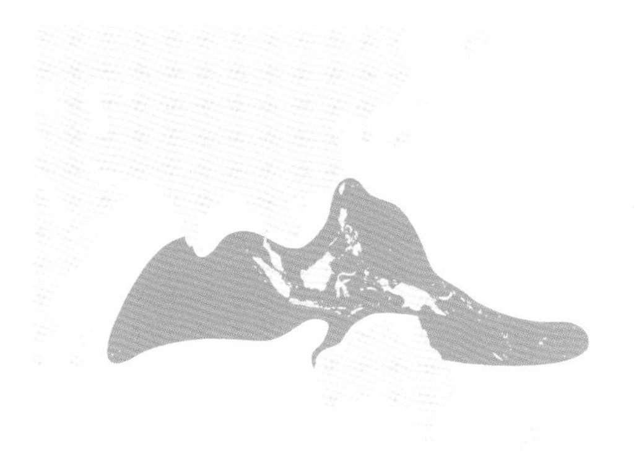

TYPE LOCALITY
West coast of Colombia (may be erroneous).

FURTHER DISTRIBUTION
Widespread through the Indo-West Pacific but is apparently rare. From Mauritius, Lakshadweep Islands, Cocos (Keeling) Islands, the Moluccas (Indonesia) to Guam, Taiwan, New Caledonia and Niue Island (East of Tonga). The records from Ashmore Reef and the Great Barrier Reef, Cocos (Keeling) Islands, the Moluccas, Taiwan and Niue Island are based on specimens in the Western Australian Museum.

REFERENCES
Blake, 1978: 234–241; pl. 1 (taxonomic position).
Clark and Rowe, 1971: 72.
Yamaguchi, 1975: 20–21 (food).

ROUND-ARMED SEASTAR
Dactylosaster cylindricus
(Lamarck, 1816)

DESCRIPTION
Disc small; five cylindrical arms with blunt tips; **body covered by smooth skin with 5–15 coarse, sometimes flattened granules in the centre of each plate**; in dried specimens, crystal bodies show through the skin giving it a slightly rough appearance; papulae are in groups of 6–12 in eight rows along the arms; furrow spines are short, cylindrical, in unequal pairs, a small flat granule separates the spines; there is a single, large, cylindrical, slightly curved subambulacral spine with a round or pointed end, one per adambulacral plate basally, usually on every second plate distally.

MAXIMUM SIZE
R/r = 92/9 mm (R = ~10r).

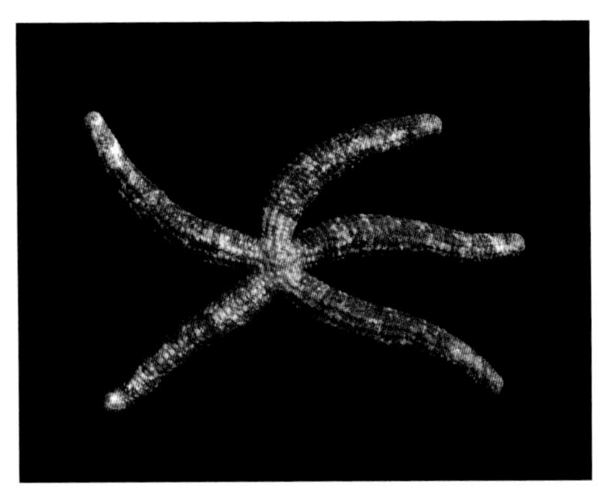

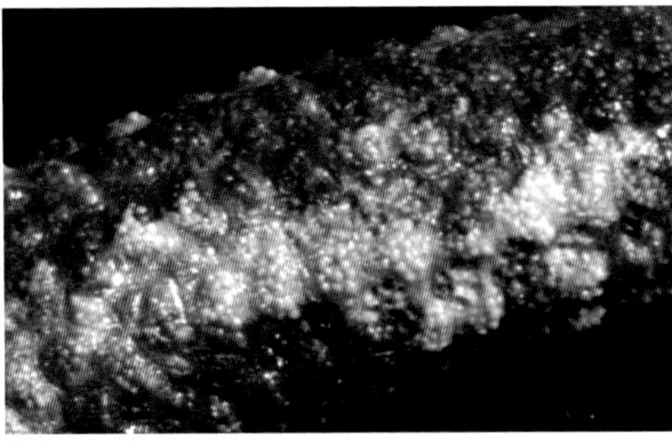

Cocos (Keeling) Islands [top], body plate characteristics [bottom] (L. Marsh).

COLOUR
Cream to beige mottled with red, sometimes uniform dark red.

FOOD
Probably feeds on substrate biofilm of algae and microorganisms.

HABITAT, DEPTH AND AUSTRALIAN DISTRIBUTION
Under boulders or dead coral slabs, occasionally exposed on coral reef flats, and on reef slopes, 0–6 m. In WA it has only been found on Ningaloo Reef and the Rowley Shoals. It is also found on the Great Barrier Reef from Torres Strait to the Capricorn Group, Queensland, and at the Cocos (Keeling) Islands and Middleton Reef, Tasman Sea.

TYPE LOCALITY
'Australian Seas' (may be erroneous). Probably Mauritius.

FURTHER DISTRIBUTION
East Africa, Red Sea through South-East Asia to Hawaii, Clipperton Island, Eastern Pacific (10°N, 110°W) and Henderson Island (south-eastern Polynesia).

REFERENCES
Clark, A.M. and Rowe, 1971: 59; pl. 8 fig. 5.
Maluf, 1988: 120.
Marsh and Marshall, 1983: 675.

Great Barrier Reef, Queensland (J. Keesing).

Cocos (Keeling) Islands (C. Bryce).

GENUS *GOMOPHIA* Gray 1840

Species of this genus have long, tapering arms, more or less round in section. **Superomarginal plates are of two sizes, alternating more or less regularly** along the arm. The larger superomarginals are as large, or larger, than the inferomarginals and may be flat to convex, while the smaller superomarginals are smaller than the inferomarginals and are flat or nearly so. **Intermarginal plates may be present.** The abactinal plates are not in regular rows except at the base of the arm, some are enlarged and moderately convex to hemispherical, cylindrical or conical. Granules are rounded or polygonal, closely or widely spaced, covering the whole surface. Granules on the apex of the larger plates are very prominent and pointed in some species. Papulae are in groups on the abactinal surface and are present intermarginally, but are usually absent on the actinal surface. The adambulacral armature comprises a furrow row and one or two subambulacral rows. There are no pedicellariae.

Type species: *Gomophia egyptiaca* Gray, 1840 (Indo-West Pacific, Red Sea)
There are a number of species in this genus, six are found in Australian seas in depths of less than 30 m and are discussed here. Many of the species were previously included in the genus *Nardoa* (see Rowe and Gates 1995).

Gray (1840) described *Gomophia* as having large rounded tubercles on the abactinal surface of the disc and a series of large conical tubercular spines on the arms. Rowe (in Rowe and Gates, 1995) expanded the diagnosis to note that the superomarginal plates alternate large then small as they decrease in size along the arm, whereas in *Nardoa* the plates decrease in size regularly along the arms. Rowe (in Rowe and Gates, 1995) observed that not only species with conical tubercles but some with rounded tubercles also had superomarginal plates alternating large and small along the arms, and transferred three species of *Nardoa* (*N. mamillifera*, *N. rosea* and *N. sphenisci*) to *Gomophia* despite the lack of conical tubercles on the arms. Mah (2019) lists these three species under *Nardoa* with *Gomophia* as an alternative representation. In this work, Rowe and Gates (1995) are followed pending genetic studies to clarify the relationship of the two genera.

KEY TO THE SPECIES OF *GOMOPHIA* IN SHALLOW AUSTRALIAN WATERS

1	Intermarginal plates present.	**2**
	No intermarginal plates.	**4**
2	Large or small conical tubercles on disc and arms.	**3**
	No conical tubercles; abactinal plates convex.	***G. sphenisci*** page 310

Gomophia watsoni, Great Barrier Reef, Queensland (A. Hoggett).

3	Large conical tubercles, pointed but without a single pointed apical granule.	***G. egeriae*** page 304
	Small conical tubercles, each tipped by a single pointed granule.	***G. watsoni*** page 311
4	Abactinal plates form distinct hemispherical tubercles 2–3 mm high.	**5**
	Abactinal plates convex but not forming hemispherical tubercles, not more than 2 mm high.	***G. rosea*** page 308
5	Distal abactinal plates elongate; granules on tubercles rounded, 2–3 per mm and 7–8 per mm between tubercles.	***G. gomophia*** page 305
	Distal abactinal plates not elongate; granules on tubercles, 3–4 per mm and 6–7 per mm between tubercles.	***G. mamillifera*** page 307

EGERIA SEASTAR
Gomophia egeriae
A.M. Clark, 1967

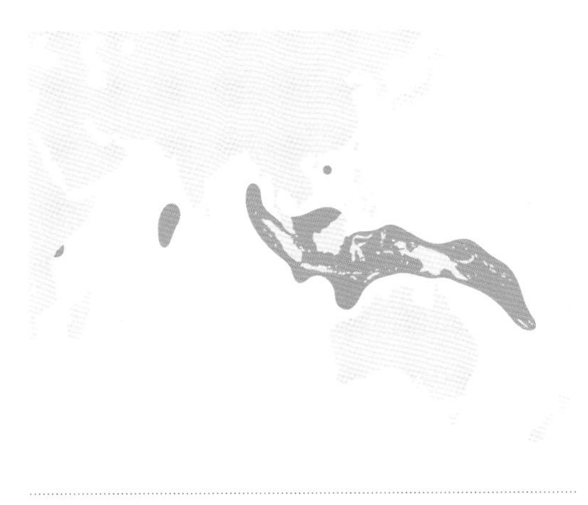

DESCRIPTION
Small disc; arms round in section, tapering to an acute tip; **large conical tubercles (3 mm high) with a rounded upper surface (not tipped with a single pointed granule)**, 20–50 tubercles on one arm; tuberculate superomarginals alternate with 1–4 small plates; **few intermarginal plates** often confined to the arm angle; actinal plates in a single row; furrow spines prismatic to flat, in slightly oblique fans of 4–5 spines, subambulacral spines shorter, more rounded, in fans of 3–4 with a few enlarged granules towards the outer part of the plate.

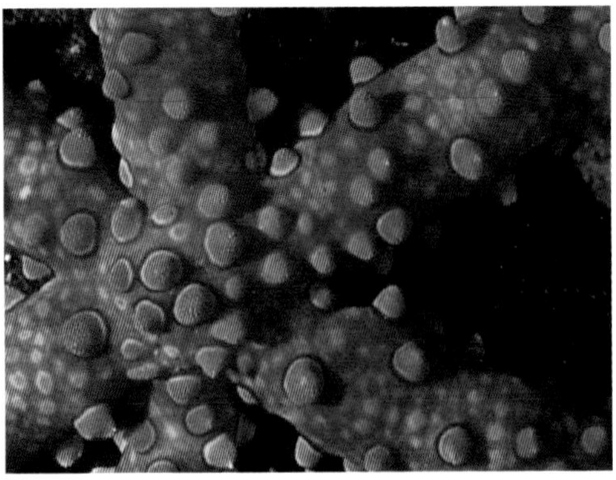

Bali, Indonesia [top], tubercle characteristics, [bottom] (C. Bryce).

MAXIMUM SIZE
R/r = 90/11 mm (R = 6–9.9r, 6.25–8.6br).

COLOUR
Usually a dusky pink with 4–6 irregular bands of orange-red on the arms. Tubercles and arm ends orange-red, ringed with dark wine red at base.

FOOD
Algae and detritus.

HABITAT, DEPTH AND AUSTRALIAN DISTRIBUTION
Outer reef slopes, 10–84 m. In Australia it is only known from Mermaid Reef, Rowley Shoals, WA, Christmas Island (Indian Ocean) and Torres Strait.

TYPE LOCALITY
Macclesfield Bank, South China Sea.

FURTHER DISTRIBUTION
Kenya, Andaman Islands, South China Sea, Moluccas, Indonesia, Papua New Guinea, Solomon Islands and New Caledonia.

REFERENCES
Clark, A.M. 1967: 169–171.

Coleman, 2007: 35.

STUDDED SEASTAR
Gomophia gomophia
(Perrier, 1875)

Ashmore Reef, Timor Sea (C. Bryce).

DESCRIPTION

Small disc; arms round in section, long, tapering to acute tips; **scattered hemispherical tubercles (3 mm wide by 2 mm high) on disc and arms**, decreasing distally in size and convexity; distal arm plates elongate; superomarginal plates alternate in size, **no intermarginal plates**; actinal plates in three rows at base of arm, one distally; 4–5 furrow spines, 3–4 subambulacral spines followed by 2–3 enlarged granules; **granules on abactinal tubercles are large, rounded and not crowded (2.8 per mm); granules are**

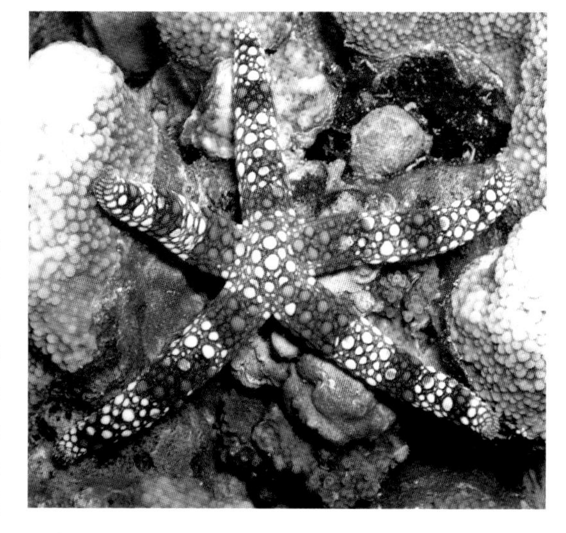

Ashmore Reef, Timor Sea (C. Bryce).

very fine (7.7 per mm) between tubercles where they are also well spaced; papulae are up to 24 per papular area with a few on the actinal surface.

MAXIMUM SIZE

R/r = 135/17 mm (R = 8r, 8br).

COLOUR

Red to red-brown with 3–7 slightly darker or lighter coloured, or cream bands on the arms. Tubercles on the disc may be cream.

HABITAT, DEPTH AND AUSTRALIAN DISTRIBUTION

On rubbly to sandy sheltered coral reef slopes, and on lagoon floors, 1–40 m. Ashmore Reef, Timor Sea.

TYPE LOCALITY

New Caledonia.

FURTHER DISTRIBUTION

Moluccas, Indonesia, the Philippines, Xisha Islands (South China Sea), Ryukyu Islands, Solomon Islands to New Caledonia.

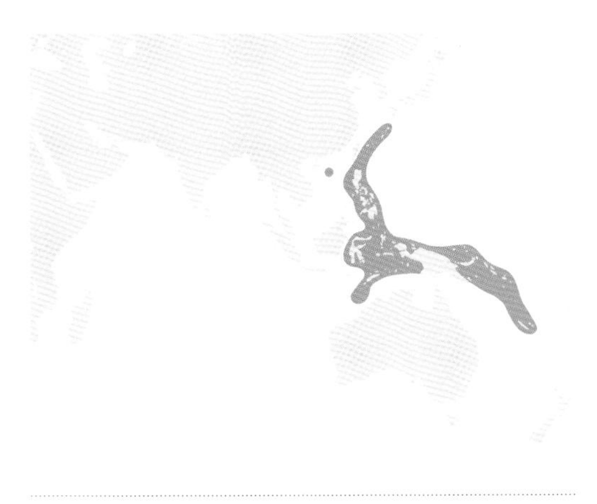

REFERENCES

Clark, A.M., 1967: 172–173; pl. 2 figs 1–4 (as *Nardoa gomophia*).
Clark, H.L., 1921: 52 (as *Nardoa novaecaledoniae*).
Clark, A.M. and Rowe, 1971: 63 (as *Nardoa gomophia*).
Rowe and Gates, 1995: 83 (transferred to genus *Gomophia*).

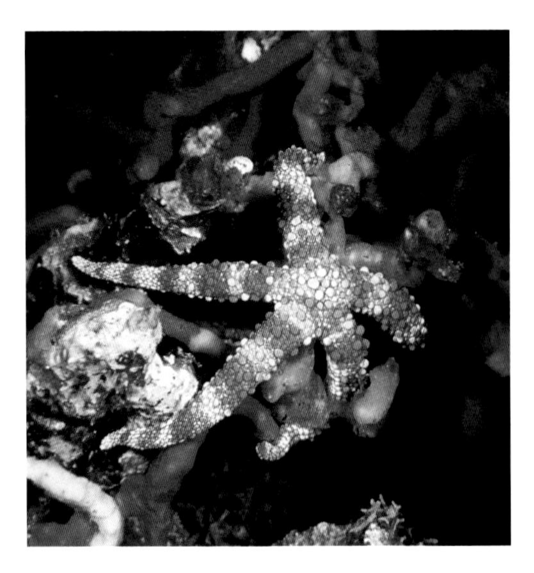

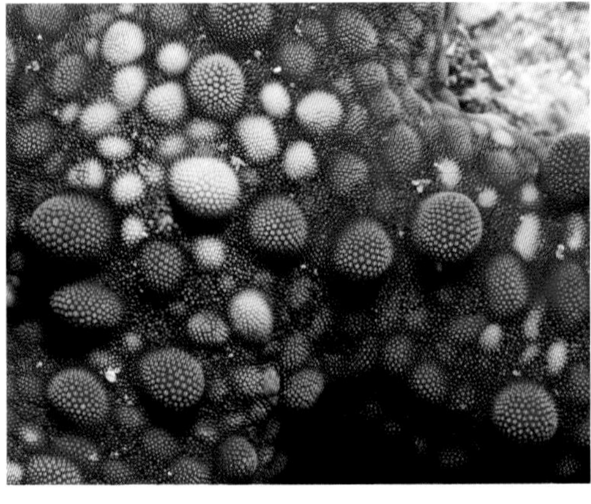

Rabaul, New Britain (R. Steene) [left]; tubercle characteristics [above], Raja Ampat, Indonesia (S. Morrison).

NIPPLED SEASTAR
Gomophia mamillifera
(Livingstone, 1930)

DESCRIPTION

Small disc; cylindrical stout arms slightly tapering to a blunt tip; **numerous hemispherical tubercles on the disc and arms** extending to the arm ends; **tubercles up to 4 mm diameter, 2–3 mm high, distal arm plates not elongate**; large convex superomarginals alternate irregularly with small plates; **no intermarginal plates**; on the abactinal surface granules are well spaced, rounded or sometimes bluntly pointed, **on the apex of tubercles there are 3–4 per mm, around the base of tubercles 6–7 per mm and 4–5 per mm in papular areas**; actinal plates in 2–3 rows in the arm angle, one distally; 3–4 (sometimes five) furrow spines, 2–4 subambulacral spines and 0–4 enlarged granules; abactinal papular areas have 10–14 pores (up to 20) and there are a few on the actinal surface.

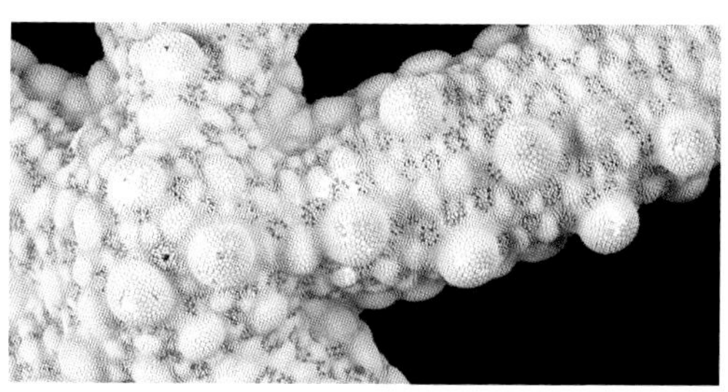

Great Barrier Reef, Queensland [top]; tubercle characteristics [bottom] (L. Marsh).

MAXIMUM SIZE

R/r = 100/14 mm (R = ~6.6r, ~5.9br).

COLOUR

Dark pinkish red or brown with 2–3 paler bands on the arms.

HABITAT, DEPTH AND AUSTRALIAN DISTRIBUTION

On coral reefs and reef flats, 0–16 m. Murray Islands south to Lodestone Reef, Great Barrier Reef, Queensland.

TYPE LOCALITY

Murray Islands, Torres Strait, Queensland.

FURTHER DISTRIBUTION

The Loyalty Islands (New Caledonia), Fiji and Taiwan.

REFERENCES

Clark, A.M., 1967: 179–180 (as *Nardoa mamillifera*).

Clark, H.L., 1946: 116 (as *Nardoa mamillifera* Livingstone, 1930).

Clark, A.M. and Rowe, 1971: 63 (as *Nardoa mamillifera*).

Livingstone, 1930: 20–22; pl. 7 figs 1–5 (as *Nardoa mamillifera*).

Rowe and Gates, 1995: 83–84 (transferred to genus *Gomophia*).

ROSE SEASTAR

Gomophia rosea

(H.L. Clark, 1921)

Australian Museum (AMJ6808).

DESCRIPTION

Small disc; slender, slightly tapering arms with **large and small moderately convex plates up to 3.2 mm wide and 1.6 mm high** continuing to the arm ends; abactinal skeleton more compact than in *G. gomophia*; **arm plates do not become elongate distally; granules on the larger, convex abactinal plates are rounded to polygonal**, slightly spaced and very convex, of unequal size giving the surface a rough appearance, **3–4 granules per mm; between the convex plates** granules are rounded, **convex, well spaced (5–6 per mm)**; superomarginal plates alternate in size, there are **no intermarginal plates**; papular areas have 10–15 pores per area abactinally, there are papulae between the marginal plates and below the inferomarginals; there are two rows of actinal plates in the arm angle, one distally; 4–5 furrow spines, 2–3 subambulacral spines and 2–3 enlarged granules; actinal granules have a rough appearance.

MAXIMUM SIZE

R/r = 95/10 mm (R = ~7.7r, ~6.9br).

COLOUR

Dull green, brown to red, with the convex plates cream to red and the furrow rose red.

HABITAT, DEPTH AND AUSTRALIAN DISTRIBUTION

Cryptic, on coral reef flats and slopes, 0–10 m. From Murray Islands and the northern Great Barrier Reef, Queensland. Queensland endemic.

TYPE LOCALITY

Murray Islands, Torres Strait, Queensland.

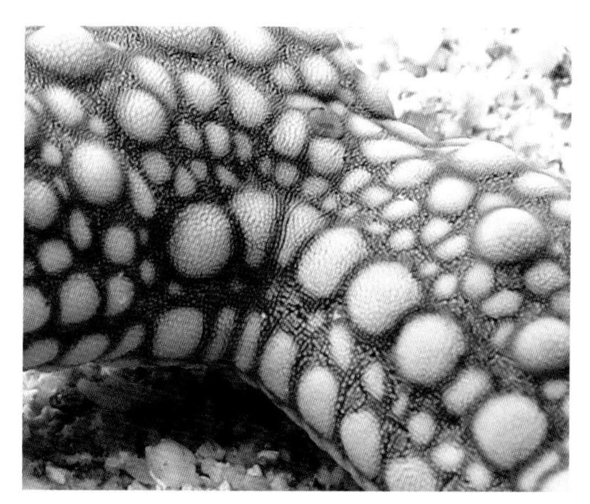

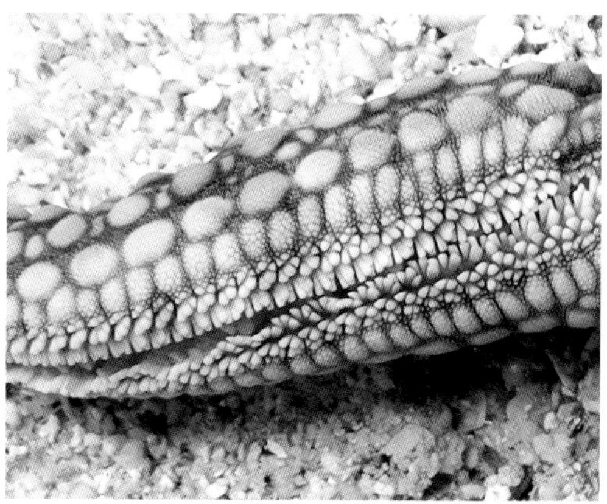

Abactinal [top], actinal plate and furrow characteristics [bottom], Great Barrier Reef, Queensland (A. Hoggett).

FURTHER DISTRIBUTION

Australian endemic.

REFERENCES

Clark, H.L., 1921: 29, 53; figs 1–2; pl. 10 fig 1 (as *Nardoa rosea*).

Clark, H.L., 1946: 115 (as *Nardoa rosea*).

Clark, A.M. and Rowe, 1971: 63 (as *Nardoa rosea*).

Rowe and Gates, 1995: 84 (transferred to genus *Gomophia*).

REMARKS

Specimens photographed in Papua New Guinea appear to be *Gomophia rosea*, while a specimen from the southern Great Barrier Reef is an undescribed species of *Gomophia*.

WEDGED SEASTAR
Gomophia sphenisci
(A.M. Clark, 1967)

DESCRIPTION

Small disc, arms round in section, tapering to an acute tip; **most abactinal plates convex (2.4–3.5 mm across) but not forming distinct tubercles**; superomarginal plates alternate in size often irregularly; intermarginal plates few or many; **granules on abactinal plates** convex, polygonal, closely spaced, **larger on the apex of the convex plates (3–6 per mm) than between them where there are 7–9 per mm**; actinal plates in one row with 1–3 plates of a second row in the arm angle; furrow spines 4–5, stout, blunt, angular in slightly overlapping fans, three shorter, stouter subambulacral spines with slightly enlarged granules behind them.

MAXIMUM SIZE

R/r = 120/13 mm (R = 4.5r).

COLOUR

Almost white, or pale grey, sometimes brown, with rust coloured arm tips.

HABITAT, DEPTH AND AUSTRALIAN DISTRIBUTION

On sand or shell bottom, 22–36 m. From the Dampier Archipelago, WA to Parry Shoal, NT, apparently rare. Northern Australian endemic.

TYPE LOCALITY

Holothuria Bank, WA.

FURTHER DISTRIBUTION

Australian endemic.

REFERENCES

Clark, A.M., 1967: 173–175; pl. 3 figs 1–3 (as *Nardoa sphenisci*).

Clark, A.M. and Rowe, 1971: 63 (as *Nardoa sphenisci*).

Rowe and Gates, 1995: 84 (transferred to genus *Gomophia*).

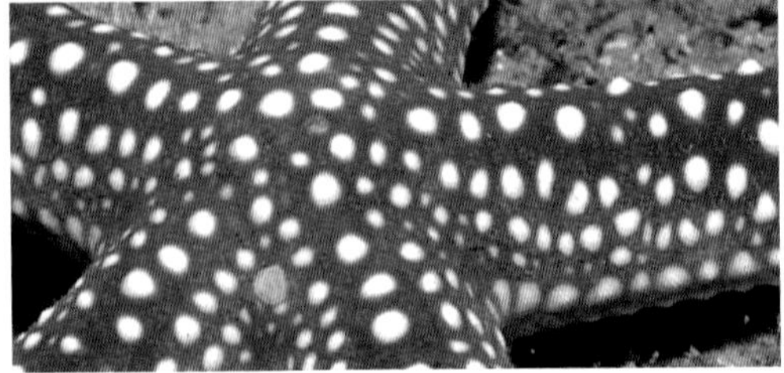

Dampier Archipelago, Western Australia [left] (S. Morrison); arm tip [top] and tubercle characteristics [bottom], Cassini Island, Western Australia (C. Bryce).

WATSON'S SEASTAR

Gomophia watsoni

(Livingstone, 1936)

Great Barrier Reef, Queensland (A. Hoggett).

DESCRIPTION

Disc small; arms round in section, tapering to an acute tip, with **scattered small conical tubercles, 20–45 per arm, about 1.5 mm high, each tipped with an enlarged pointed granule**; superomarginal plates forming a distinct series bearing tubercles alternating with smaller flat plates; intermarginal plates in 1–2 rows separating the two marginal series; furrow spines flat to wedge-shaped, in an oblique fan of 3–4, subambulacral spines shorter, more rounded, usually three; actinal plates in a single row except in the arm angle where there may be 1–2 plates in a second row.

MAXIMUM SIZE

R/r = 72/11 mm (R = 3.4–8.4r, 3.1–7.3br).

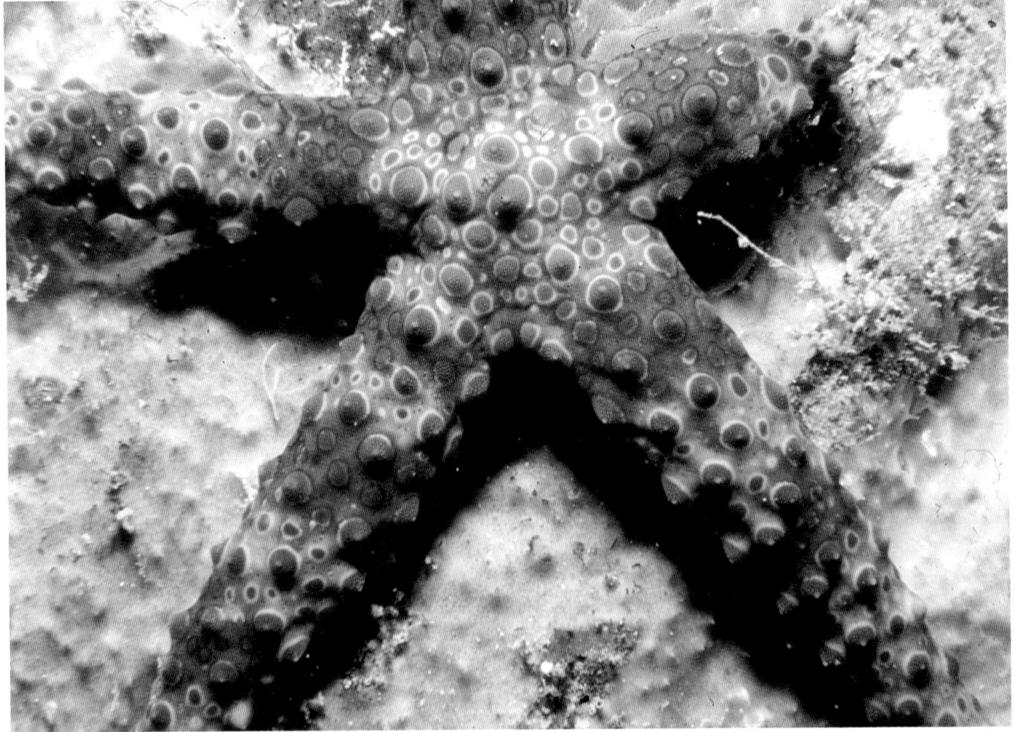

Cartier Island, Timor Sea [top] (C. Bryce); Heron Island, Queensland [bottom] (N. Coleman).

COLOUR

Variable, often grey-brown, brown-purple, red-brown, cream, arms sometimes banded. Tubercles often blue, grey or violet, ringed at the base with white or dark brown.

FOOD

Observed feeding on a small sponge at night.

REPRODUCTION

May also be similar to *Gomophia egyptiaca*, which has lecithotrophic development.

HABITAT, DEPTH AND AUSTRALIAN DISTRIBUTION

Cryptic, on coral reef flats and slopes, 0–35 m. Christmas Island, Cartier Island, Timor Sea and Lizard Island, to Capricorn Group, Great Barrier Reef, Queensland.

TYPE LOCALITY

Bushy Island off Mackay, Queensland.

FURTHER DISTRIBUTION

Rodrigues Island (east of Mauritius), East Malaysia, Indonesia, Papua New Guinea, Solomon Islands, New Caledonia, Palau.

REFERENCES

Clark, A.M., 1967: 176–177 (transferred to genus *Gomophia*).

Clark, H.L., 1946: 121 (as *Ophidiaster watsoni*).

Clark, A.M. and Rowe, 1971: 64 (as *Gomophia egyptiaca*).

Coleman, 2007: 37 (food).

Livingstone, 1936: 386; pl. 28 figs 1, 3, 5, 7 (as *Ophidiaster watsoni*).

Rowe and Gates, 1995: 84.

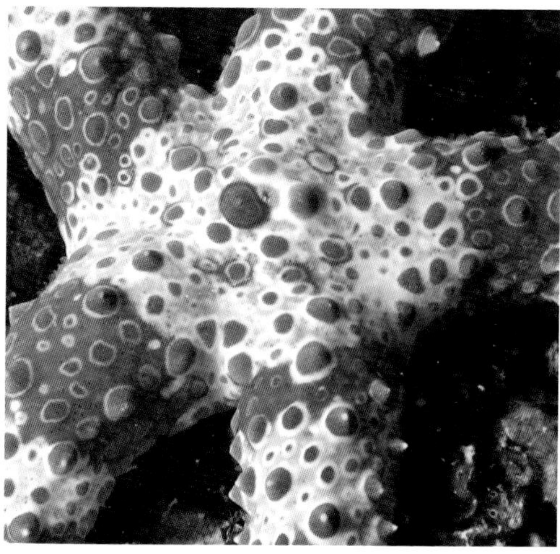

Actinal plate [above] and tubercle characteristics [right], Great Barrier Reef, Queensland (A. Hoggett).

GENUS *HACELIA* Gray 1840

Species of this genus have arms that are tapering, flattened actinally and rounded above. The body covering is granulose and the abactinal plates are in regular longitudinal series. The papulae are in 10 rows in adults on the actinal as well as on the abactinal surface. The actinal plates are in three rows. Juvenile specimens may have as few as six rows of papulae but are the same as adults in other characteristics.

Type species: *Hacelia attenuata* Gray, 1840 (Mediterranean Sea)

Only one species, *H. heliosticha* (page 315), is found in shallow Australian waters. A second Australian species, *H. tyloplax* (H.L. Clark, 1914), is found below 73 m off the west coast of WA. *Ophidiaster helicostichus* was transferred to *Hacelia* by H.L. Clark (1909b), followed by H.L. Clark (1938, 1946) and by Rowe and Gates (1995) because of the 10 rows of pores in very large specimens. This classification is followed here. A.M. Clark (1967) referred it back to *Ophidiaster* as most specimens in the Natural History Museum in London had eight rows of papulae. She also noted that small specimens with only six rows of papulae could be confused with *Tamaria*. Mah (2019) notes *Ophidiaster helicostichus* as an alternate representation for this species.

Hacelia helicosticha, Dampier Archipelago, Western Australia (C. Bryce).

COILED SEASTAR
Hacelia helicosticha
(Sladen, 1889)

DESCRIPTION
Disc small, convex; five arms taper evenly to a narrow tip; convex abactinal plates and connecting ossicles covered with evenly sized, close fitting, flat or convex polygonal granules (four per mm) contrasting with very small, sometimes erect, pointed uneven-sized granules of the papular areas (7–8 per mm); **papulae in 10 rows in large specimens, generally eight rows except in very small specimens, which have six**; number of papulae is size dependent, usually 10–20 pores per papular area, up to 35 per area in very large specimens, papular areas are sunken; **actinal plates are in 3–4 rows at the base of the arm** and covered in coarse convex rounded, slightly spaced granules, more uneven than on the abactinal plates where there are 2.5–3 per mm; furrow spines in unequal pairs, truncate to round-ended, flat on the inner side, convex on the outer side, no granules between them; adambulacral spines almond-shaped proximally, conical distally; sugar-tongs pedicellariae with 2–5 teeth fit into a fairly thin walled alveolus usually without notches for the teeth, the side walls of the alveolus are sometimes very swollen; pedicellariae are on the plates, not in papular areas on the abactinal and actinal surfaces.

MAXIMUM SIZE
R/r = 165/17 mm (R = 6.1–10.9r).

COLOUR
Variable, for example cream with irregular purple bands on the arms; cream with dark purple papular areas and purple arm tips; or brown or grey-brown with dark brown papular areas. Actinal surface light coloured.

HABITAT, DEPTH AND AUSTRALIAN DISTRIBUTION
Muddy sand and mud, 8–173 m. Found from Exmouth Gulf, the North West Shelf and Kimberley coast, WA to Torres Strait and south to Port Curtis (23°42'S) on the Queensland coast.

TYPE LOCALITY
Booby Island, Torres Strait.

FURTHER DISTRIBUTION
West Pacific Ocean.

REFERENCES
Clark, A.M., 1967: 195–196 (as *Ophidiaster helicostichus*).
Clark, H.L., 1909b: 111.
Clark, H.L., 1946: 122–123.
Clark, A.M. and Rowe, 1971: 60 (as *Ophidiaster helicostichus*).
Marsh, 1976: 291 (transferred to genus *Hacelia*).
Rowe and Gates, 1995: 84.

GENUS *LEIASTER* Peters, 1852

Seastars of this genus have a small disc and long, more or less cylindrical arms, covered by smooth skin, and entirely without granules. The papular areas are in eight series, and crystal bodies are present in some species.

Type species: *Leiaster coriaceus* Peters, 1852 (Indo-Pacific)

Three species are recorded in Australian waters from depths of less than 30 m. One undescribed species is known from Great Barrier Reef and the West Pacific.

Leiaster coriaceus, Dirk Hartog Island, Western Australia (C. Bryce).

KEY TO THE SPECIES OF *LEIASTER* IN SHALLOW AUSTRALIAN WATERS

1	Crystal bodies on abactinal and marginal plates; furrow spines paired not grooved.	***L. leachii*** page 319
	Without crystal bodies on any plates.	**2**
2	Furrow spines flat, in groups of three with a deep longitudinal groove (visible with skin removed).	***L. coriaceus*** page 317
	Furrow spines cylindrical, in pairs, not grooved (visible with skin removed).	***L. glaber*** page 318

LEATHERY SEASTAR
Leiaster coriaceus
Peters, 1852

DESCRIPTION

The body is covered by thick skin not completely obscuring the outlines of the plates; the abactinal and marginal plates are more or less triangular, not four-lobed as in *L. leachii*; there are no crystal bodies on any plates; papular areas are poorly defined sometimes extending onto the skin covering the skeletal plates; **furrow spines are typically in groups of three, slightly widened at the tips and with a deep longitudinal groove** that is not evident without removing the overlying skin; pedicellariae are variable in occurrence, most abundant on the actinal surface.

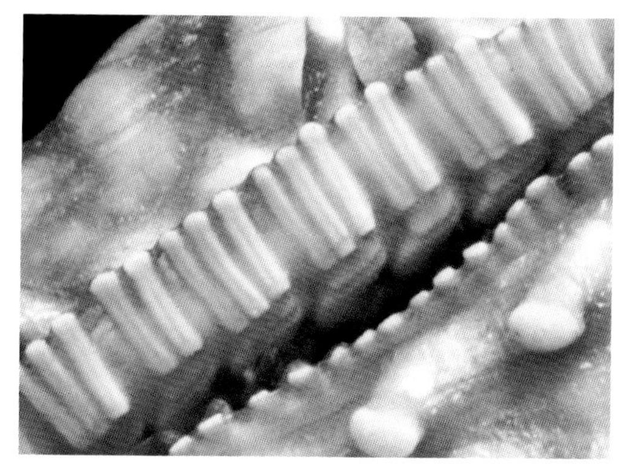

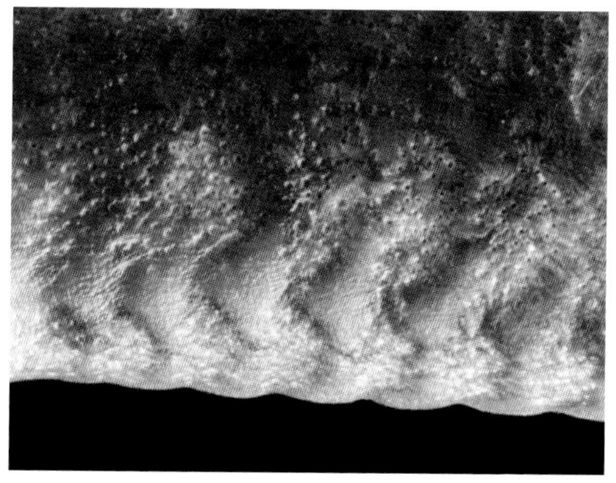

Furrow [top] and abactinal plate characteristics [bottom], Western Australian Museum (WAMZ2980).

MAXIMUM SIZE

R/r/br = 135/15/14 mm (R = 9r, 9.6br).

COLOUR

Cream or yellow with reddish brown blotches.

HABITAT, DEPTH AND AUSTRALIAN DISTRIBUTION

Found on a mixed sand and rock substrate, 10–102 m; uncommon. Recorded from off Beagle Island (29°48'S) to Point Cloates, WA and on the Great Barrier Reef, and Moreton Bay, Queensland.

TYPE LOCALITY

Querimba Island, Mozambique.

FURTHER DISTRIBUTION

From the east coast of Africa and Madagascar to the Red Sea, the Philippines and New Caledonia; it is also found in the eastern Pacific from Lower California and Panama.

REFERENCES

Clark, A.M. and Rowe, 1971: 57.

Guille et al., 1986: 136–137.

Jangoux, 1980: 96–99; fig. 1b–c, pl. 7 figs 1–4.

SMOOTH-SKINNED SEASTAR
Leiaster glaber
Peters, 1852

DESCRIPTION
Five subcylindrical arms, slightly constricted at the base, abruptly tapering to a blunt point; abactinal surface of disc slightly convex; test covered by a thick, tough, smooth skin, plates visible after drying; **no crystal bodies**, few or no pedicellariae; papular areas on arms in eight series, about 15 pores to an area abactinally, 6–10 actinally; 2–3 granules embedded in skin in papular areas; abactinal plates four sided with rounded corners and slightly excavated sides, longitudinally imbricated and joined laterally by transverse small ossicles; **adambulacral plates each carry three furrow spines that are short, equal, flattened with rounded tips, not grooved on inner side**; subambulacral spines are on every second plate; they are upright, robust, slightly compressed.

Western Australian Museum [top] (WAMZ269401); papular area characteristics [bottom], Western Australian Museum (WAMZ23363).

MAXIMUM SIZE
R/r = ~95/7 mm.

COLOUR
Variable, usually deep maroon, magenta at arm tips with yellow tube feet.

HABITAT, DEPTH AND AUSTRALIAN DISTRIBUTION
On coral rubble or sand, 20–124 m. Off Ningaloo Reef and the North West Shelf, WA.

TYPE LOCALITY
Mozambique.

FURTHER DISTRIBUTION
Andaman Islands, Taiwan, the Philippines, Hawaii, Pacific coast of America.

REFERENCES
Döderlein, 1926: 17.

Fisher, 1906: 1083; pl. 30 figs 1, 1a; pl. 31 fig. 3 (as *Leiaster callipeplus*).

Jangoux, 1980: 100.

LEACH'S SEASTAR

Leiaster leachii

(Gray, 1840)

Ningaloo Reef, Western Australia (C. Bryce).

Dirk Hartog Island, Western Australia (C. Bryce).

DESCRIPTION

Small disc; five long cylindrical arms, the whole covered with thin to thick smooth skin through which **the outline of the plates is visible in small dried specimens, but in large animals the skin obscures the plates except sometimes for a thinner area over the centre of the plates, through which a circular area of crystal bodies is visible**; papular areas are in eight rows, the number of pores is size dependent, from six per area in a specimen of R = 35 mm to over 100 in very large specimens where R > 200 mm; furrow spines are paired (occasionally three

Leiaster leachii, Lord Howe Island, Tasman Sea (J. Turnbull).

per plate), elongate, sometimes appearing grooved (due to skin covering) on their inner face; subambulacral spines are flat, truncate and very broad-ended in large specimens; pedicellariae are tongs-like, each fitting into a narrow alveolus, they may be abundant or absent.

MAXIMUM SIZE

R/r = 400/21 mm (R = 7–19r)

COLOUR

Variable, often uniform deep wine red, or cream to pink blotched with red, tan and orange with the madreporite, arm tips, and sometimes patches on the arms, magenta.

FOOD

Scavenges, has been taken in rock lobster pots, attracted to the bait.

REPRODUCTION

In Japan, spawning occurs in August. Eggs are small (140 µm) and develop into plankton feeding bipinnaria larvae that swim for at least three weeks.

HABITAT, DEPTH AND AUSTRALIAN DISTRIBUTION

On rock, sand and rubble, or on coral reefs where the long flexible arms twine around coral branches or infiltrate reef crevices, emerging at night, 1–183 m; uncommon. From the continental shelf off Two Rocks (31°30'S), the Houtman Abrolhos, Shark Bay and North West Shelf, WA, and the Great Barrier Reef, Queensland south to Stanley Reef (19°20'S) and Norfolk and Lord Howe Islands, Tasman Sea.

TYPE LOCALITY

Mauritius.

FURTHER DISTRIBUTION

It is widely distributed in the tropical Indo-West Pacific, from east Africa, Madagascar and the Red Sea to Hawaii, and north to Japan.

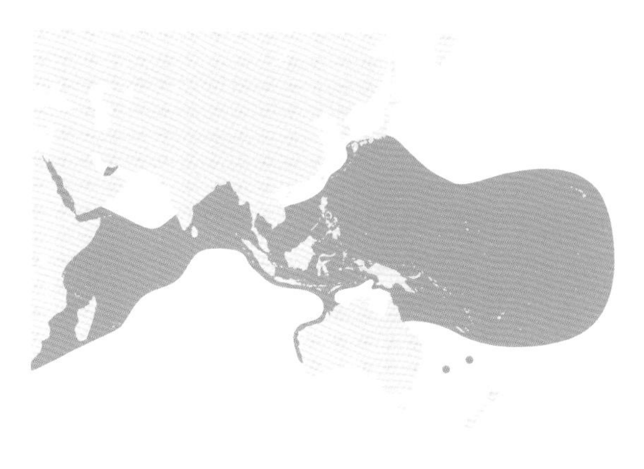

REFERENCES

Clark, H.L., 1946: 119–120
(as *Leiaster leachii* and *L. speciosus*).

Jangoux, 1980: 90–96; fig. 1a; pl. 5 figs 1–6.

Komatsu, 1973: 55–58; pl. 1 (reproduction).

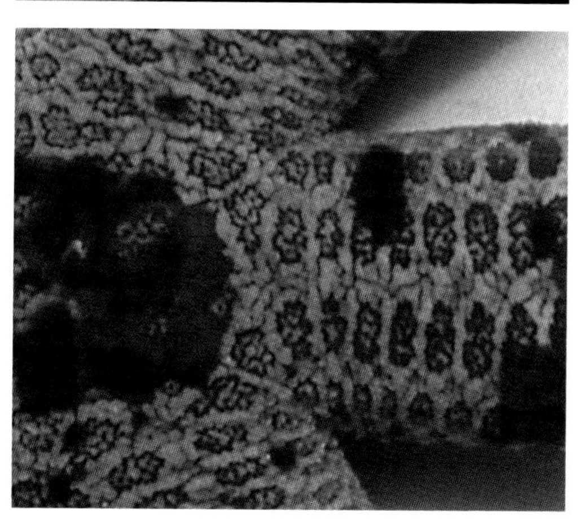

Abactinal plate characteristics [top], Western Australian Museum (WAMZ15882); papular area characteristics [bottom], Houtman Abrolhos, Western Australia (B. Wilson).

REMARKS

Jangoux (1980) recognised five species of *Leiaster* in the Indo-Pacific, two with crystal bodies on the plates and three without. However, the holotypes of the two nominal species with crystal bodies, *L. leachii* and *L. speciosus,* cannot be separated by any combination of characters when a large series of specimens is examined, and a single species, *L. leachii*, is recognised here. Both species are considered valid on the World Register of Marine Species (Mah, 2019).

A few authors refer to some seastars found on the Great Barrier Reef and in the West Pacific as *L. speciosus*, but these specimens do not conform to the holotype of that specie; they belong to an undescribed species.

Two described species (*L. leachii* and *L. coriaceus*) are so far recorded from less than 30 m depth in Australian waters, while *L. glaber*, although found at 9 m depth in the Bay of Bengal and included in this book, has only been found in deeper water in Australia.

GENUS *LINCKIA* Nardo, 1834

Species of this genus have a small disc and long cylindrical arms. The **abactinal plates are irregularly arranged and do not occur in rows**. The papular areas are irregularly scattered except between the marginal plates, there are no papulae on the actinal surface, and the whole body is granule-covered. There are no pedicellariae, and the adambulacral armature appears granuliform in two species.

Type species: *Linckia laevigata* (Linnaeus, 1758) (Indo-Pacific)
The three Indo-West Pacific species all occur on coral reefs in Australia.

Linckia laevigata, Ningaloo Reef, Western Australia (L. Marsh).

KEY TO THE SPECIES OF *LINCKIA* IN SHALLOW AUSTRALIAN WATERS

1	Granulation of actinal surface extending into ambulacral grooves so that a vertical series of minute granules separates the furrow spines from each other.	**2**
	Granulation of actinal surface not extending into ambulacral grooves; no granules between furrow spines or between furrow spines and subambulacrals; R ≤ 250 mm.	***L. guildingi*** page 324
2	Arms usually five, relatively short and stout; madreporite single; colour typically blue or blue-green (sometimes actinal surface is orange) or orange; R ≤ 205 mm.	***L. laevigata*** page 326
	Arms slender and tapering, often irregular in length and number, autotomous; usually two madreporites; colour often spotted, arm tips usually blue-grey; R ≤ 117 mm.	***L. multifora*** page 328

Linckia multifora, Scott Reef, Western Australia (L. Marsh).

GUILDING'S SEASTAR
Linckia guildingi
Gray, 1840

North Malé Atoll, Maldives (A. Machet).

DESCRIPTION

Disc small; 5–6 very long cylindrical arms tapering near the ends to a blunt tip; **usually two madreporites; autotomous when young**; abactinal plates and papular areas irregularly arranged, marginal and actinal plates in rows, no papulae on the actinal surface; **furrow spines round tipped, in unequal pairs, no granules between spines**; few or no granules between furrow and subambulacral spines, which are round and larger than but scarcely projecting above the

Juvenile, Houtman Abrolhos, Western Australi (C. Bryce).

granules of the actinal plates; usually a second row of slightly smaller subambulacral spines, lie close to the first row; no pedicellariae.

MAXIMUM SIZE

R/r = 250/~23 mm (R = 7–14r).

COLOUR

Yellow to brown, greenish; rarely blue. Smaller animals (R < 30 mm) are commonly mottled with shades of brown or purple.

FOOD

Scavenges but also probably feeds on substrate biofilm of microorganisms.

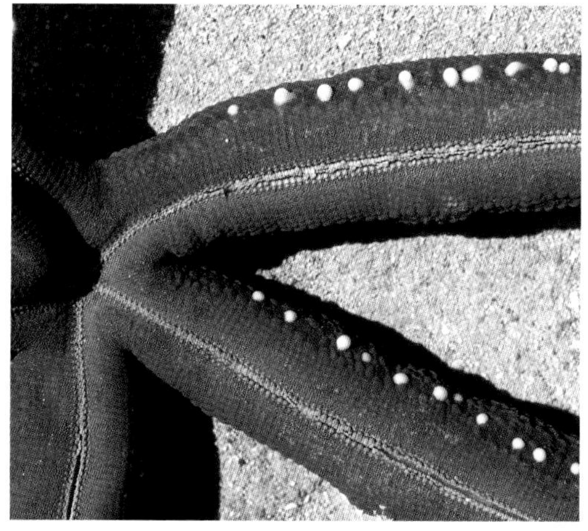

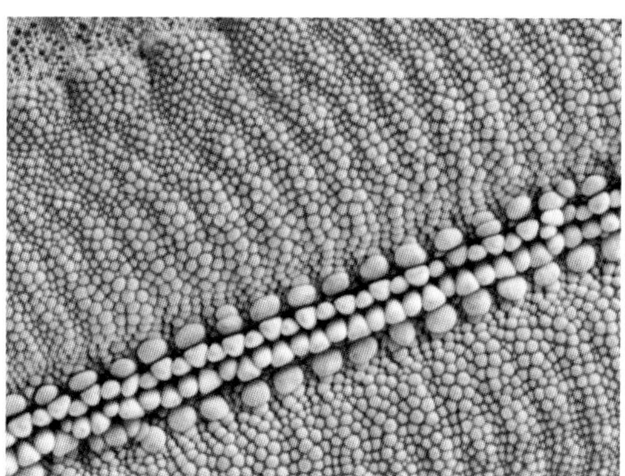

Male spawning [top], Shark Bay, Western Australia (L. Marsh); furrow characteristics [bottom], Western Australian Museum (WAMZ19667).

REPRODUCTION

Reproduces asexually by autotomy, six armed specimens and comets are sometimes found in young seastars. A male specimen spawned after capture in early April at Shark Bay in WA. Gonopores open through every second or third papular area along the arms.

HABITAT, DEPTH AND AUSTRALIAN DISTRIBUTION

On coral reefs, young usually under boulders and adults in the open, 0–114 m. From the Houtman Abrolhos (~28°S), WA around northern Australia and the Great Barrier Reef south to the Bunker Group (23°48'S), Queensland.

TYPE LOCALITY

St Vincent, West Indies.

FURTHER DISTRIBUTION

Circumtropical species, found from east Africa to Hawaii and from the West Indies to Brazil in the Atlantic.

REFERENCES

Clark, H.L., 1946: 117–118.

Clark, A.M. and Rowe, 1971: 61; fig. 14b; pl. 8 fig. 7.

Ely, 1942: 18–19; pl. 1.

BLUE SEASTAR
Linckia laevigata
(Linnaeus, 1758)

DESCRIPTION
Disc small; five arms of moderate length, fairly stout, cylindrical to end or very slightly tapering; **arms are rarely of unequal length, not autotomous; the madreporite is single**; abactinal plates and papular areas are irregularly arranged but there is sometimes a median strip on the arms without pore areas; the marginal and actinal plates are in distinct rows, there are no pedicellariae; **furrow spines are in unequal pairs, separated from each other and from the next pair by a vertical row of granules; there is a granule-covered strip between the furrow and subambulacral spines**; juveniles can be distinguished from *L. multifora* of the same size by their blunt-ended cylindrical even length arms, single madreporite and lack of spots in the colour pattern.

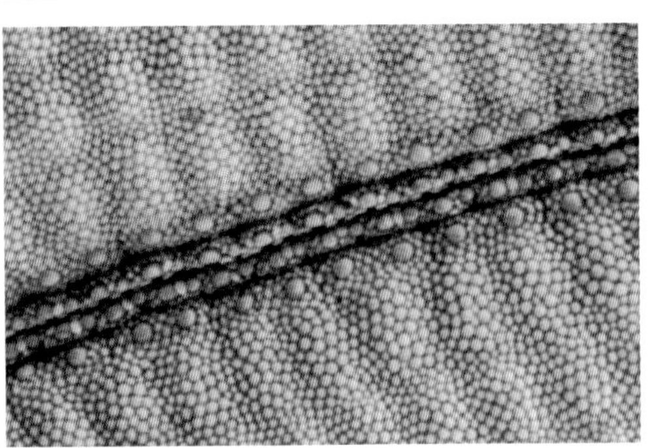

Parasitic gastropod *Thyca crystallina* [top], Komodo Island, Indonesia (S. Morrison); furrow characteristics [bottom], Western Australian Museum (WAMZ15892).

MAXIMUM SIZE
R/r = 205/25 mm (R = 6.2–8.2r).

COLOUR
Usually bright blue, sometimes greenish blue with actinal surface orange, seastars from deep water may be completely orange. Juveniles are dark coloured, purple brown, faintly mottled.

FOOD
The substrate biofilm of microorganisms and detritus, often coralline algae but no feeding scars are left.

REPRODUCTION
Reproduction takes place year round near the equator, elsewhere it peaks in early summer and continues through summer. In WA spawning was observed synchronously with mass spawning of corals in March. The eggs are small (150 µm) and grow into plankton-feeding bipinnaria larvae that go through a brachiolaria stage before metamorphosing at 3–4 weeks from fertilisation. After two years, the juveniles (R = ~50 mm) change from the juvenile to the adult colour and emerge from the reef.

Great Barrier Reef, Queensland (A. Hoggett).

HABITAT, DEPTH AND AUSTRALIAN DISTRIBUTION

Adults live in the open on coral reef flats and slopes while juveniles are rare and cryptic, 0–60 m. From Ningaloo Reef, inner North West Shelf, nearshore and offshore reefs off north-western Australia to the southern end of the Great Barrier Reef (23°48'S), Queensland.

TYPE LOCALITY

'Mediterranean Sea' (erroneous) and 'Indian Ocean'.

FURTHER DISTRIBUTION

This species is widely distributed through the tropical Indo-West Pacific from east Africa to Tahiti and Pitcairn Island in south-eastern Polynesia, north to Taiwan.

REFERENCES

Clark, H.L., 1946: 117.
Clark, A.M. and Rowe, 1971: 62.
Ely, 1942: 19–20; pls 2–3.
Yamaguchi, 1973: 377; fig. 1.
Yamaguchi, 1977a: 13–30 (development and growth).

DALMATIAN SEASTAR
Linckia multifora
(Lamarck, 1816)

Dampier Archipelago, Western Australia (C. Bryce).

DESCRIPTION

The disc is small with **slender, slightly tapering cylindrical arms; the species is autotomous so comet forms and animals with regenerating arms of different lengths are common,** and four and six armed specimens are often found; nearly all have two, sometimes three madreporites; abactinal plates and papular areas are irregularly arranged but the marginal and actinal plates are in rows; **furrow spines are in unequal pairs separated from each other by a row of granules**; a granule-covered strip separates

Dampier Archipelago, Western Australia (C. Bryce).

the furrow spines from the more or less globular subambulacral spines; there are no pedicellariae.

MAXIMUM SIZE

R/r = 117/9 mm (R = 5–13r),
R usually < 60 mm.

COLOUR

Variable, usually cream or grey to orange-pink with red and/or brown spots, sometimes blue-green with darker spots, arm tips usually mauve or grey-blue.

FOOD

The substrate biofilm of microorganisms and detritus.

REPRODUCTION

Sexual reproduction takes place in summer. Plankton-feeding bipinnaria larvae develop and go through a brachiolaria stage before metamorphosing after about four weeks. Asexual reproduction is the predominant method of increasing populations, particularly on reefs in isolated locations, throughout the year. In asexual reproduction the seastar deliberately stretches an arm until it breaks, in a process taking a few minutes, but it is 3–4 weeks before the cast off arm develops a mouth and another 10 days before new arms are apparent (a stage called a 'comet'). A new arm is regenerated by the parent seastar; in any population the majority of animals are in some stage of regeneration; comets take up to a year to reach a size where autotomy can take place again, the longest arm is always the one autotomised.

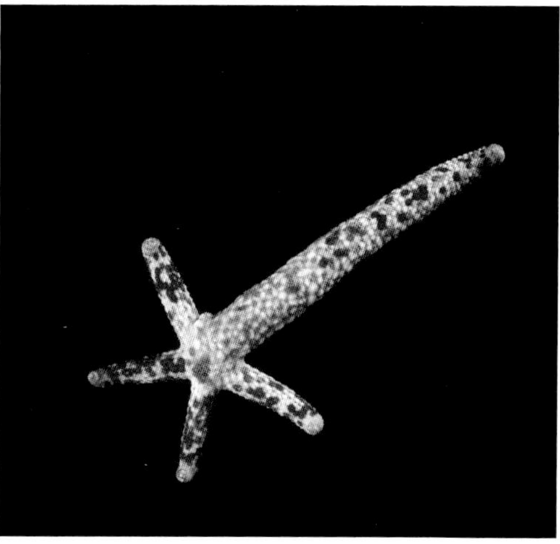

Spawning posture [top], Scott Reef, Western Australia; comet development [bottom], Dampier Archipelago, Western Australia (C. Bryce).

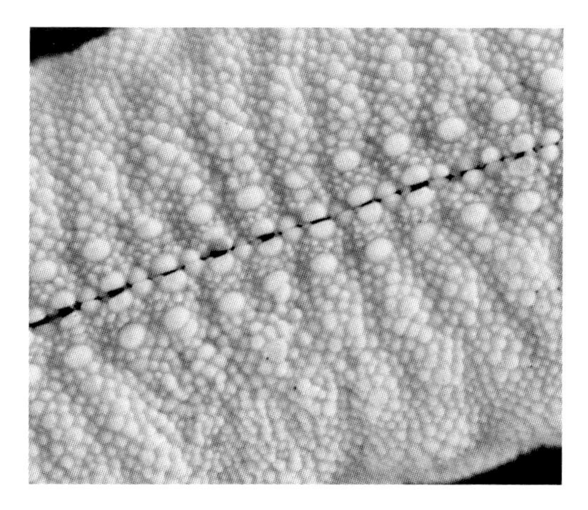

Furrow characteristics, Western Australian Museum (WAMZ19849).

HABITAT, DEPTH AND AUSTRALIAN DISTRIBUTION

On living or dead coral or rock on coral reef flats and slopes, also among seagrass on lagoonal sand flats, 0–100 m; common. From the Houtman Abrolhos, on north west coral reefs and the inner continental shelf, WA, around northern Australia to Heron Island, Great Barrier Reef, Queensland, Middleton Reef, Tasman Sea.

TYPE LOCALITY

'European seas' (erroneous).

FURTHER DISTRIBUTION

It is widespread through the tropical Indo Pacific from east Africa and the Red Sea north to Taiwan, east to Hawaii and Costa Rica.

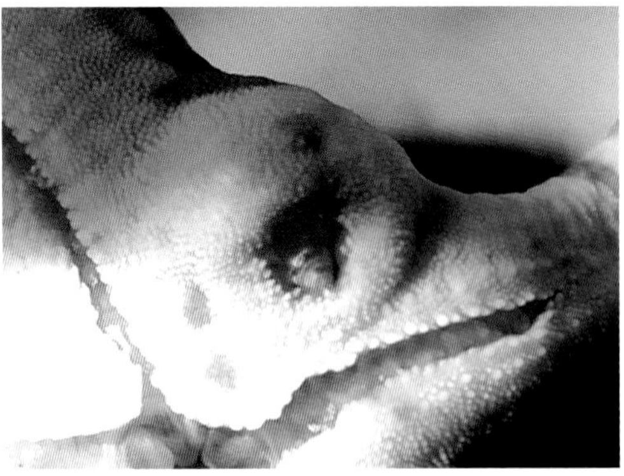

Scott Reef, Western Australia [top] (C. Bryce); parasitic gastropod *Stilifer linckiae* [bottom], Réunion Island (P. Bourjon).

REFERENCES

Crawford and Crawford, 2007: 371–387; pl. 1.

Edmondson, 1935: 1–20; fig. 1a–c, g, j (autotomy).

Ely, 1942: 19–20; pls 2–3.

Mortensen, 1938: 35–37; pl. 4 figs 3–4 (development).

Rideout, R.S., 1978: 287–295 (asexual reproduction).

Yamaguchi 1975: 20–21 (food).

REMARKS

Linckia multifora is often host to the gall-forming parasitic gastropod mollusc *Stilifer linckiae.*

GENUS *NARDOA* Gray, 1840

Species of this genus have **marginal plates that decrease regularly in size distally and do not alternate in size**. The superomarginal and inferomarginal plates are similar in size and flat, and there are **no intermarginal plates**. The abactinal plates are not in regular rows except at the base of the arms and tubercular plates are sometimes present. Granules are rounded to polygonal and form an even cover while **papulae occur abactinally, intermarginally and actinally** and are rarely reduced actinally. There is adambulacral armature of furrow spines, subambulacral spines and a row of enlarged granules. Granuliform pedicellariae are sometimes present.

Type species: *Nardoa variolata* (Bruzelius, 1805) (western Indian Ocean)

Besides the type species, five species of *Nardoa* are recognised of which three are found in Australia; none are endemic. *Nardoa gomophia* (Perrier, 1875), *N. mamillifera* Livingstone, 1930 and *Nardoa rosea* H.L. Clark, 1921 are currently listed in *Nardoa* where they have been transferred by Mah (2019). However, they are also considered to be alternate representations in *Gomophia* (Mah, 2019). *Nardoa sphenisci* A.M. Clark was referred to *Gomophia* (Rowe and Gates, 1995). The genus is found from the western Indian Ocean to the West Pacific.

Nardoa galatheae, Shark Bay, Western Australia (C. Bryce).

KEY TO THE SPECIES OF *NARDOA* IN SHALLOW AUSTRALIAN WATERS

1	Abactinal plates uniform in size along the arm.	***N. tuberculata*** page 336
	Abactinal plates decreasing in size along the arm.	**2**
2	Abactinal plates at base of arm not in regular rows.	***N. novacaledoniae*** page 334
	Abactinal plates at base of arm in regular rows.	***N. galatheae*** page 333

Nardoa tuberculata f. *pauciforis*, Great Barrier Reef, Queensland (R. McMinds).

SEA NYMPH SEASTAR
Nardoa galatheae
(Lütken, 1864)

DESCRIPTION

Disc small, convex; arms long, tapering to a narrow tip; abactinal plates at base of arms up to 4.1 mm diameter, irregular to round in shape, flat to convex, not in regular rows except at the base of the arms; **plates decrease in size distally becoming narrow and elongate**; there are 7–11 (usually nine) abactinal plates across the base of the arm between the superomarginal series; **granules on the abactinal plates usually 3–3.5 per mm, while in papular areas and between the primary plates there are usually six granules per linear mm**; papulae 7–22 per area, they occur abactinally, intermarginally and actinally; usually three rows of actinal plates in the arm angle; usually four furrow spines, 3–4 subambulacral spines, and 3–4 enlarged granules merging into the actinal granulation.

Coral Bay, Western Australia [top], abactinal plate characteristics [bottom] (S. Morrison).

MAXIMUM SIZE

R/r = 144/16 mm (R = 9r, 8br).

COLOUR

Dark brown, appearing velvety dark olive green to brown under water.

REPRODUCTION

In north-western Australia, *Nardoa galatheae* was observed spawning synchronously with mass spawning of corals during March.

HABITAT, DEPTH AND AUSTRALIAN DISTRIBUTION

In the open on tropical coral and algal covered reef, or sand with moderate exposure, 0–44 m. Houtman Abrolhos, north to Ningaloo Reef, Montebello Islands and Dampier Archipelago, WA to Cape Grenville, Queensland. Common in suitable habitats.

TYPE LOCALITY

Nicobar Islands, Bay of Bengal.

FURTHER DISTRIBUTION

From the Red Sea, Maldive Islands, India, Sri Lanka, Cocos (Keeling) Islands, Indonesia, the Philippines and Papua New Guinea.

REFERENCES

Clark, H.L., 1921: 50, 52–53 (as *Nardoa mollis*).

Clark, A.M. and Rowe, 1971: 64 (as *Nardoa mollis*).

NEW CALEDONIAN SEASTAR
Nardoa novaecaledoniae
(Perrier, 1875)

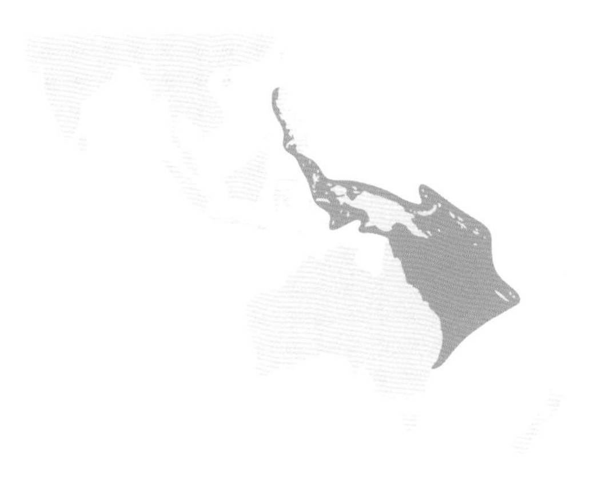

DESCRIPTION

Disc small, slightly convex; arms taper to fairly acute tips; largest abactinal plates at base of arm usually about 3 mm diameter, slightly to moderately convex, polygonal to rounded, not in regular rows; **one-third to halfway along the arm the abactinal plates decrease rapidly in size, become elongate and continue decreasing in size to the arm tip**; at the base of the arms there are usually nine abactinal plates across the arm between the superomarginals; the granules on the largest abactinal plates are polygonal, about three per mm, on the papular areas and between the plates granules are smaller, about seven per mm; papulae 8–22 per papular area, between abactinal plates, marginals and actinals; actinal plates in three rows in the arm angle; 3–4 furrow spines, 3–4 subambulacral spines, and 3–4 enlarged granules on the outer part of the adambulacral plate.

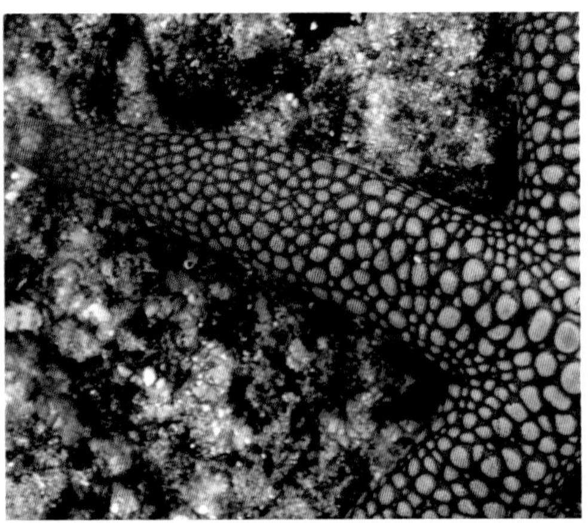

Walindi, Papua New Guinea [top], abactinal plate characteristics [bottom] (N. Coleman).

MAXIMUM SIZE

R/r = 110/17 mm (R = 6.5r, 7.4br).

COLOUR

Convex plates cream to yellow-brown to orange-brown, khaki between the larger plates.

HABITAT, DEPTH AND AUSTRALIAN DISTRIBUTION

On coral reef flats and shallow reefs often among rubble and seagrass, 0–10 m. Murray Islands, Torres Strait, to Bunker Group, Great Barrier Reef, Queensland, and Bungan Head, NSW.

TYPE LOCALITY

New Caledonia.

FURTHER DISTRIBUTION

Eastern Indonesia, New Britain, Papua New Guinea, Solomon Islands and Vanuatu to New Caledonia.

REFERENCES

Clark, H.L., 1946: 115.

Clark, A.M. and Rowe, 1971: 64.

Solomon Islands [top] (H. Beral); Great Barrier Reef, Queensland [bottom] (F. Michonneau).

TUBERCULATE SEASTAR
Nardoa tuberculata
Gray, 1840

Ashmore Reef, Timor Sea (L. Marsh).

Nardoa tuberculata forma *tuberculata*
Gray, 1840

DESCRIPTION
Disc small; arms nearly cylindrical or slightly tapering to a blunt tip; abactinal plates not in rows except at the base of the arm where there are usually about nine plates across the arm; **there are scattered tubercles on the arms,** usually about 34 per arm; tubercles **usually about 2.2 mm diameter, slightly convex to hemispherical; abactinal plates irregular in shape but uniform in size along the length of the arm not decreasing distally**; abactinal granules hexagonal to polygonal, their upper surface slightly convex, usually 4–5 per mm; granules smaller in papular areas and between plates (6–7 per mm); papulae about 13 per area, abactinally, intermarginally and a few on the actinal surface; usually two rows of actinal plates in the arm angle; 3–4 furrow spines, 3–4 subambulacral spines, and 2–3 enlarged granules.

MAXIMUM SIZE
R/r = 135/21 mm (R = ~6.4r, ~6br),
R usually 80–90 mm.

COLOUR

Cream to beige, grey or light brown, usually with 2–5 darker transverse bands on the arms. Papular areas and granules between the plates brown or grey.

HABITAT, DEPTH AND AUSTRALIAN DISTRIBUTION

On sand, rubble or coral, on coral reefs (reef flats, back reef and lagoons) and inner shelf, 0–84 m; common. Adele Island and nearshore Kimberley reefs, Rowley Shoals and Scott Reef, WA and Ashmore Reef, Timor Sea.

TYPE LOCALITY

Sual, Luzon, the Philippines.

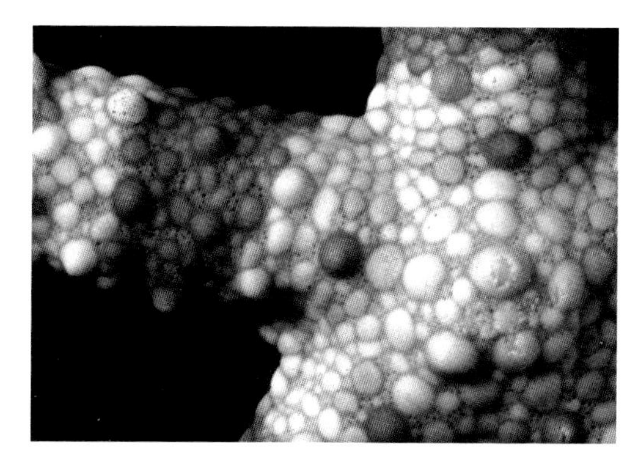

Abactinal plate characteristics [top] (L. Marsh); Ashmore Reef, Timor Sea [bottom] (C. Bryce).

FURTHER DISTRIBUTION

From Cocos (Keeling) Islands, Christmas Island, Moluccas, Indonesia and the Philippines, to Palau and Papua New Guinea, north to the Ryukyu Islands.

REFERENCES

Clark, A.M. and Rowe, 1971: 63; pl. 8 fig. 1.
Marsh, 1977: 242–263.
Hayashi, 1938: 430.
Rowe and Gates, 1995: 88.

REMARKS

In Palau a full range of characters from tuberculate *Nardoa tuberculata* f. *tuberculata* to small plated, almost non tuberculate *N. tuberculata* f. *pauciforis* is found in the same habitat. Through Indonesia to north-western Australia most are *N. tuberculata* f. *tuberculata* while on the Great Barrier Reef, Queensland, almost all are non-tuberculate *N. tuberculata* f. *pauciforis*. The two are therefore regarded as forms of *N. tuberculata*. Hayashi (1938) came to the same conclusion after studying the species in Palau. Rowe (in Rowe and Gates, 1995) recognises a single species, *N. tuberculata*.

Nardoa tuberculata forma *pauciforis*
(Von Martens, 1866)

DESCRIPTION
Disc small; arms long, nearly cylindrical; 11–13 (rarely 14) small irregular plates across the base of each arm with tubercles usually absent but some have small low tubercles; other characters as for *N. tuberculata* f. *tuberculata*.

MAXIMUM SIZE
R/r = 140/15 mm (R = 9.3r, 9.3br).

COLOUR
Usually uniform yellow-brown with darker brown papular areas.

Great Barrier Reef, Queensland [top] (J. Keesing); abactinal plate characteristics [bottom], Milne Bay, Papua New Guinea (N. Coleman).

HABITAT, DEPTH AND AUSTRALIAN DISTRIBUTION
In the open on coral reef flats, 0–70 m. From Torres Strait to the Capricorn Group, Great Barrier Reef, Queensland.

TYPE LOCALITY
Near Flores, Indonesia.

FURTHER DISTRIBUTION
Palau, Caroline Islands, Flores and Ambon, Indonesia, the Philippines, Papua New Guinea, Solomon Islands.

REFERENCES
Clark, H.L., 1946: 115 (as *Nardoa pauciforis*).

Clark, A.M. and Rowe, 1971: 64 (as *Nardoa pauciforis*).

Marsh, 1977: 262–263.

REMARKS
In Australia the two forms do not overlap geographically and the colour of each is distinctive but, in Palau, both subspecies and intermediates are found. In Indonesia the two forms appear to be distinct.

GENUS *OPHIDIASTER* L. Agassiz, 1836

Species of this genus have a small disc and more or less cylindrical arms. The abactinal plates are in regular longitudinal series, with **papular areas in eight series**, of which one series is below the inferomarginals on each side. All the plates are completely covered with granules, rarely with a few spines or tubercles. The adambulacral armature is in two, rarely three, unequal series, the outer one heavier than the inner, and distinctly spaced.

Type species: *Ophidiaster ophidianus* (Lamarck, 1816)
(north-eastern Atlantic, Mediterranean Sea)
Seven species are known from shallow Australian waters, one is currently an undescribed species confined to south-eastern Queensland.

Ophidiaster duncani, Great Barrier Reef, Queensland (J. Keesing).

KEY TO THE SPECIES OF *OPHIDIASTER* IN SHALLOW AUSTRALIAN WATERS

Couplet	Character	Go to / Species
1	Tubercles or enlarged granules on abactinal plates with or without a small spine on marginal plates.	2
	No tubercles or spines except for furrow and subambulacral spines.	3
2	Coarse granules in centre of abactinal plates sometimes becoming a small tubercle, no spines; $R \leq 45$ mm.	***O. granifer*** page 347
	Central granules become rounded tubercles and marginal plates bear a small spine on outer one third of arms; $R \leq 130$ mm.	***O. armatus*** page 341
3	Arms 4–6, regularly autotomous, reproducing asexually, comets common; often two madreporites; abactinal granules polygonal forming a smooth mosaic; pedicellariae numerous, broad sugar-tongs shape; $R \leq 55$ mm.	***O. cribrarius*** page 343
	Not autotomous.	4
4	Skin fairly loose, wrinkled.	5
	Skin not loose or wrinkled.	***O. hemprichi*** page 349
5	Polygonal granules 4–6 per mm on abactinal plates, 9–10 per mm in large papular areas, which have up to 60 pores; pedicellariae abundant, sugar-tongs shaped with 2–8 teeth fitting into a notched alveolus; paired furrow spines separated from each other by granules on the furrow face, outer side covered by granular skin; a large species; $R \leq 160$ mm.	***O. confertus*** page 342
	Arms readily autotomised but not reproducing asexually; skin covered by very small squamiform granules, 7–9 per mm on abactinal plates and in papular areas; pores up to 20 per papular area; pedicellariae few to many, distinctive, narrow, not toothed but wider at the tip, nearly straight to boomerang shaped, fitting into a narrow thin walled alveolus; paired furrow spines not separated by granules; $R \leq 70$ mm.	***O. duncani*** page 345

ARMOURED SEASTAR
Ophidiaster armatus
Koehler, 1910

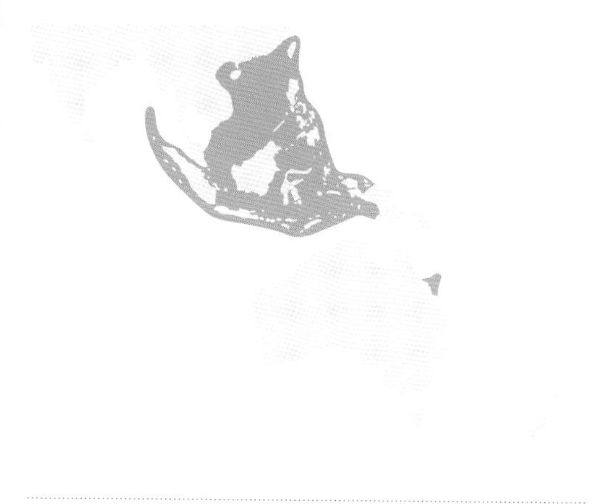

DESCRIPTION
Disc small; arms nearly cylindrical but tapering to a rounded tip; papular areas are large, rounded, in eight longitudinal rows; abactinal plates on the arms form regular longitudinal and transverse rows, **the plates are covered with fairly coarse, rounded granules in the centre with smaller granules towards the edge of each plate; on the last third of the arm one central granule becomes a rounded tubercle on the abactinal arm plates and a small spine on the marginal plates**; the actinal surface of the arms is covered with rounded, convex granules, similar to or larger than those on the abactinal plates, obscuring the plate outlines; furrow spines are paired, subequal, vertical, flattened with a blunt tip, while at some distance from the furrow spines is a row of large, blunt conical subambulacral spines; there are no pedicellariae.

Hong Kong [top], abactinal arm plate characteristics [bottom] (L. Marsh).

MAXIMUM SIZE
R/r = 130/~18 mm (R = ~7r).

COLOUR
Dark, yellowish violet with 2–3 lighter bands on the outer third of the arm. Actinal surface greyish with irregular yellow violet spots on some inferomarginal plates.

FOOD
Probably grazes the substrate biofilm.

HABITAT, DEPTH AND AUSTRALIAN DISTRIBUTION
On rock or rubble, 5–40 m. In Australia known only from Keeper Reef, off Townsville to Lindeman Island, Queensland.

TYPE LOCALITY
Aru Islands, Indonesia.

FURTHER DISTRIBUTION
Indonesia, Andaman Islands, Taiwan.

REFERENCES
Chao, 2001: 6, figs 21–24.

Clark, A.M. and Rowe, 1971: 60.

Koehler 1910b: 277–278; pl. 15 fig. 8; pl. 17 fig. 6.

ORANGE SEASTAR
Ophidiaster confertus
H.L. Clark, 1916

DESCRIPTION

Disc small; five arms, long, cylindrical and blunt-ended; abactinal and marginal plates convex, **body covered by a thick densely granular skin wrinkled transversely; granules polygonal, tightly packed, not squamous, 4–6 per mm on abactinal arm plates, 9–10 per mm in papular areas**; eight rows of papular areas, each area with up to 60 pores; not autotomous, anus and madreporite single; furrow spines paired, flattened, round ended, subequal, all separated by granules; outer face covered by granular skin, only granule tips protrude; subambulacral spines are almond-shaped, round-ended, lower part embedded in skin on each plate proximally, on every second plate distally or irregularly; **pedicellariae abundant, embedded, broad to narrow sugar-tongs shaped, valves with 3–5 (2–8) teeth fitting into a weakly notched alveolus with thick walls**; pedicellariae abundant on actinal surface.

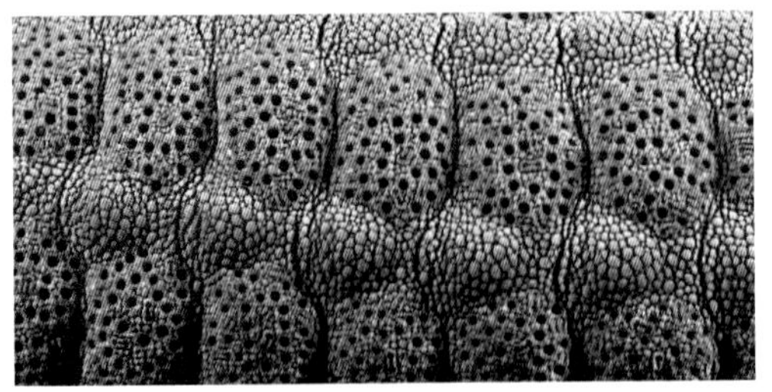

Lord Howe Island, Tasman Sea [top], abactinal arm plate and pedicellariae characteristics [bottom] (N. Coleman).

MAXIMUM SIZE

R/r = 160/20 mm (R = 7–9r).

COLOUR

Uniform tawny yellow, orange-red, orange-brown to red-brown.

FOOD

Sessile invertebrates, algae.

HABITAT, DEPTH AND AUSTRALIAN DISTRIBUTION

Rock platforms and coral reefs, among coral heads and rock fragments, 0–50 m. Found in south-eastern Australia, from the Capricorn group, Great Barrier Reef, Queensland, south to Shell Harbour, NSW, Lord Howe and Norfolk Islands, Elizabeth and Middleton Reefs, Tasman Sea.

TYPE LOCALITY

Lord Howe Island, Tasman Sea.

FURTHER DISTRIBUTION

Uncertain.

REFERENCES

Clark, H.L., 1938: 138–139; pl. 10 figs 2–3.

Clark, H.L., 1946: 121–122.

Coleman, 1994: 27 (food).

SIEVE SEASTAR
Ophidiaster cribrarius
Lütken, 1871

Scott Reef, Western Australia (C. Bryce).

DESCRIPTION
Disc small; 4–6 arms, cylindrical, distal half tapering slightly to a blunt tip; **arms readily autotomised resulting in comets, usually two madreporites and two anal apertures**; the abactinal plates are covered in a fairly smooth mosaic of unequal sized convex granules, larger in the centre of the plates (4–6 mm) than in the papular areas (9–10 per mm); papulae in eight rows, 1–12 pores per area; pedicellariae numerous, in the papular areas and near the anus, **broad sugar-tongs form with 4–6 well developed teeth that fit into a notched alveolus**; the paired furrow spines are separated by 1–2 granules on the furrow face (between pairs of spines in small animals); **almond-shaped subambulacral spines are separated from the furrow spines by a strip covered in squamous granules**.

MAXIMUM SIZE
R/r = 60/6 mm (R = 5–10.5r, 3–8.5br),
R usually < 40 mm.

COLOUR
Cream to yellow-brown variously mottled with brown, purple, green, or yellow, pink or red.

FOOD
Substrate biofilm of detritus and microorganisms.

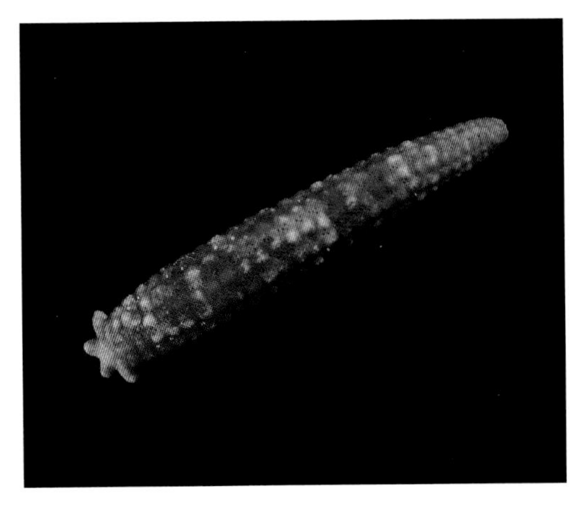

Comet development, Scott Reef, Western Australia (L. Marsh).

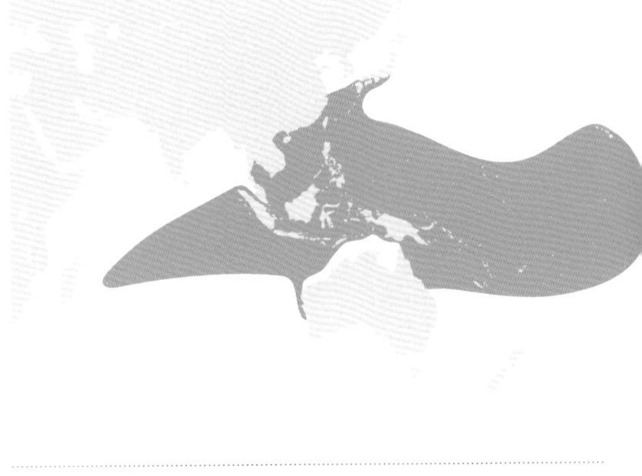

REPRODUCTION

Induced spawning took place in early summer, the eggs are small and non-yolky suggesting planktotrophic development. Autotomous asexual reproduction probably plays a major part in the maintenance of populations, as comets are commonly found among large populations in fairly isolated situations.

HABITAT, DEPTH AND AUSTRALIAN DISTRIBUTION

Under boulders and coral slabs or rubble on coral reefs, or among dead branching corals, 0–5 m. From the Houtman Abrolhos, Ningaloo Reef, Rowley Shoals and Scott Reef, WA and the Great Barrier Reef south to the Capricorn Group, Queensland.

TYPE LOCALITY

Tonga.

FURTHER DISTRIBUTION

Widespread in the tropical Indo-Pacific from Mauritius (as *O. robillardi*) to Guam, Taiwan and Japan, and east to Hawaii and Pitcairn Island (as *O. lorioli*).

REFERENCES

Clark, A.M. and Rowe, 1971: 61; fig. 15a.
Fisher, 1906: 1077; pl. 31 fig. 4a–d.
Marsh, 1974: 78–80 (as *Ophidiaster lorioli* (part)).
Yamaguchi, 1975: 20 (as *Ophidiaster robillardi*).

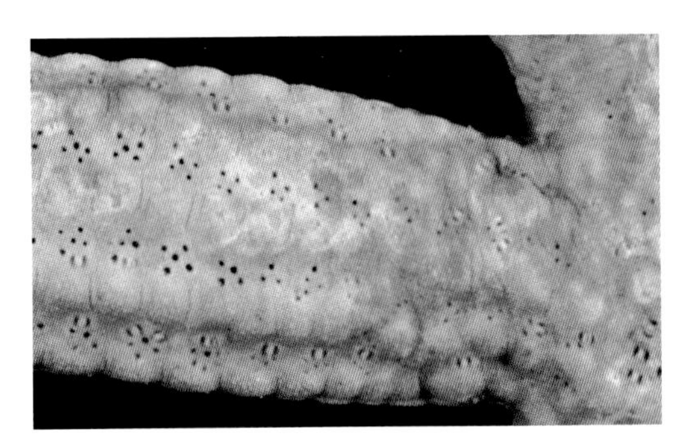

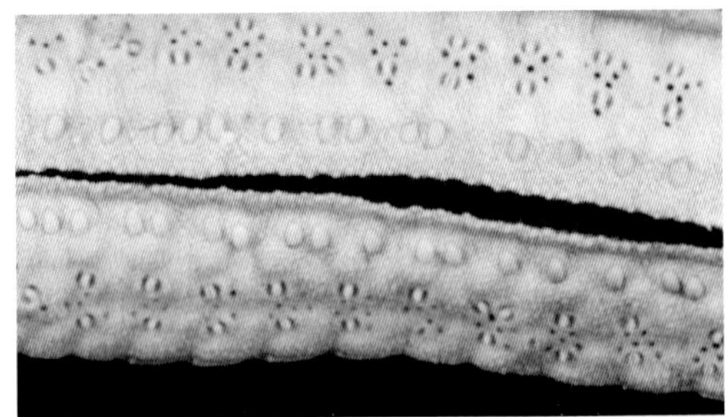

Abactinal plate and pedicellariae characteristics [top], furrow characteristics [bottom], Western Australian Museum (WAMZ19970).

DUNCAN'S SEASTAR
Ophidiaster duncani
de Loriol, 1885

Loloata Island, Papua New Guinea (N. Coleman).

DESCRIPTION

Disc small; arms cylindrical for half to three quarters of their length, then tapering to a fairly narrow tip; the arms readily autotomise; **the skeleton is covered by a loose skin, often wrinkled in preserved specimens, covered with very small, squamiform granules (7–9 linear mm)**, in small specimens they are slightly larger in the centre of the plates than elsewhere, otherwise the granules are the same size in the papular areas as on the plates; papulae are in eight rows with 12–20 pores per area (size dependent); **pedicellariae are usually present, few to many, embedded, narrow, nearly straight to very strongly curved, valves wider at the tip than the base**, not toothed; furrow spines are paired, flat or slightly grooved on the furrow face, round-ended, subequal, no granules between spines; subambulacral spines slightly flattened, tapering to a fairly sharp point, one per plate proximally, more widely spaced distally; the shape of the pedicellariae and the fairly thick, very finely granulated skin are the main diagnostic characters of this species.

MAXIMUM SIZE

R/r/br = 70/9/8.8 mm (R = 7.2–10r).

COLOUR

Dull light brownish red, sometimes blotched with darker brown.

FOOD

Probably grazes on substrate biofilm.

HABITAT, DEPTH AND AUSTRALIAN DISTRIBUTION

Under boulders on the reef crest of coral reefs and in the open at night on reef slopes, 0–15 m; rare. In Australia it has only been found on the Great Barrier Reef, from the Murray Islands to Heron Island (23°24'S), Queensland.

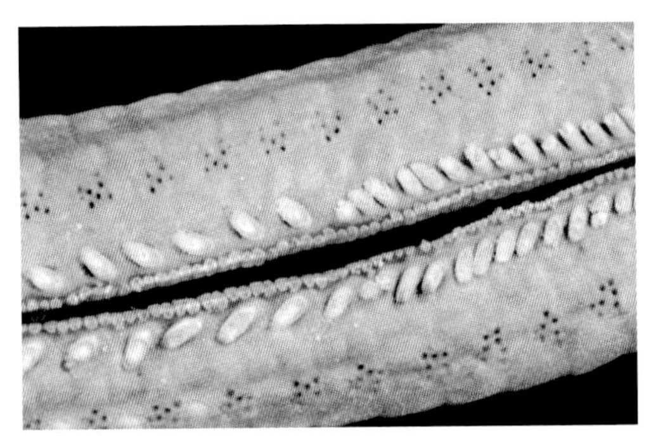

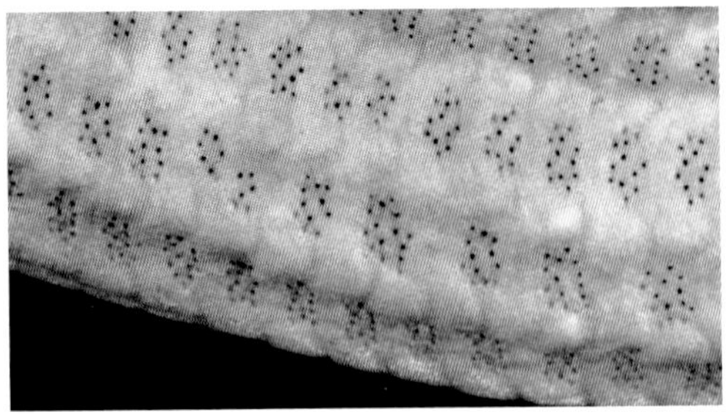

Furrow characteristics [top] and abactinal plate characteristics [bottom], Western Australian Museum (WAMZ99828).

TYPE LOCALITY

Mauritius.

FURTHER DISTRIBUTION

East Africa and Madagascar, the Philippines and the Cook Islands, South Pacific.

REFERENCES

Clark, H.L., 1921: 80–81; pl. 27 figs 3–4 (as *Ophidiaster lioderma*).

Clark, A.M. and Rowe, 1971: 60; fig. 15b.

de Loriol, 1885: 15–17; pl. 11 fig 2.

Jangoux, 1985: 21.

REMARKS

Clark and Rowe (1971) noted that a specimen from Mauritius had both nearly straight and strongly curved pedicellariae. The shape of the pedicellariae was the only character separating *Ophidiaster duncani* from *O. lioderma* H.L. Clark 1921 from Torres Strait. The two species were synonymised by Jangoux (1985).

GRAINY SEASTAR
Ophidiaster granifer
Lütken, 1871

Ashmore Reef, Timor Sea (C. Bryce).

DESCRIPTION

A small species with a small disc; five fairly short, stout, more or less cylindrical arms tapering to a blunt tip; **abactinal plates with several coarse central granules, sometimes one is enlarged to a low tubercle; sugar-tongs pedicellariae with toothed sockets are common**, papulae are in groups of up to nine, in eight rows; furrow spines are in unequal pairs, without any granules between them (on the inner face); subambulacral spines are almond-shaped.

MAXIMUM SIZE

R/r = 45/7 mm (R = 4–6.6r).

COLOUR

Base colour cream to beige or grey, more or less blotched and/or banded with dark brown or dark grey, occasionally all dark or all light coloured; sometimes pale grey with violet markings.

FOOD

The substrate biofilm of detritus and microorganisms.

REPRODUCTION

Spawning takes place in late summer to autumn (February to April in Queensland) at full moon. Development is parthenogenetic, no males have been found at any locality. The large (600 µm diameter) yolky, bright orange eggs are buoyant but adhesive, some attach to the substrate under rocks and metamorphose without a free swimming stage, others develop into a lecithotrophic swimming brachiolaria that settles and metamorphoses 9–15 days after spawning, but larval life can be prolonged to 7 weeks. Juveniles grow to R = 12 mm in one year and reach sexual maturity at two years.

HABITAT, DEPTH AND AUSTRALIAN DISTRIBUTION

Found on coral reefs, under boulders and coral slabs on reef flats or among dead basal branches of coral thickets, 0–100 m. From Shark Bay and Ningaloo Reef, inshore Pilbara reefs and the offshore atolls, Rowley Shoals and Scott Reef, WA, Ashmore Reef, Timor Sea and on the Great Barrier Reef and coastal reefs to the Capricorn Group (24°S) and Lady Elliott Island, Queensland.

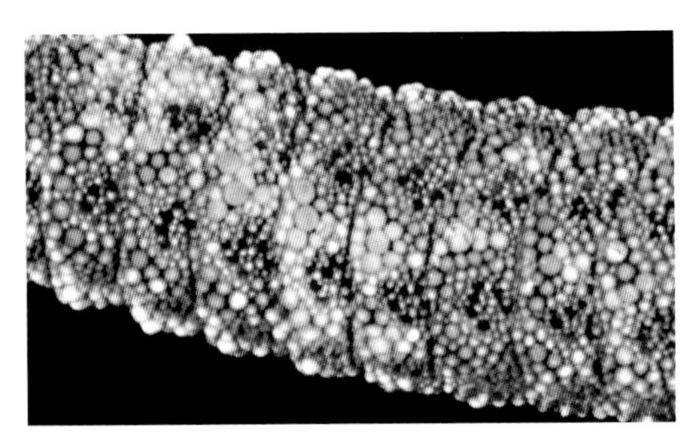

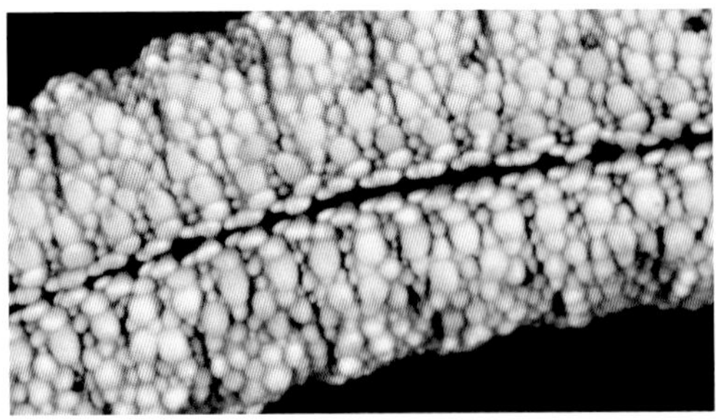

Abactinal plate characteristics [top] and furrow characteristics [bottom], Western Australian Museum (WAMZ19999).

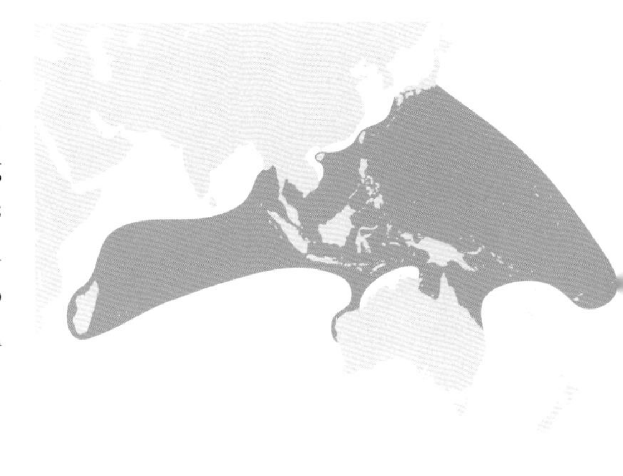

TYPE LOCALITY

Tonga.

FURTHER DISTRIBUTION

Widely distributed in the tropical Indo-West Pacific from Madagascar and the Seychelles to the Marshall Islands and Samoa.

REFERENCES

Fisher, 1919: 390–392; pl. 84 fig. 5; pl. 95 fig. 6a–d; pl. 103 figs 2–3; pl. 107 fig. 6.

Yamaguchi, 1975: 20.

Yamaguchi and Lucas, 1984: 33–42 (reproduction).

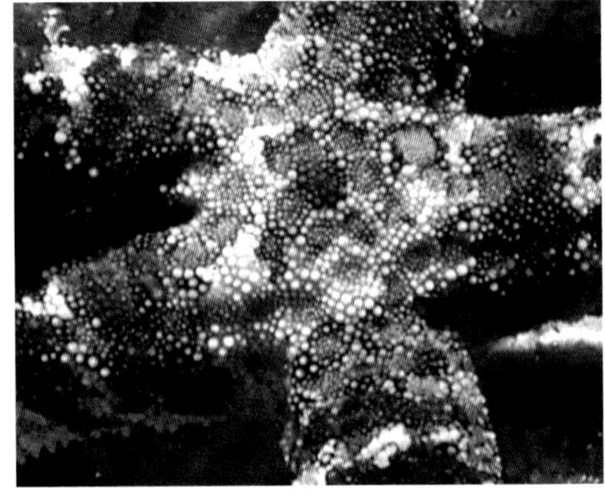

Abactinal plate characteristics, Ashmore Reef, Timor Sea (C. Bryce).

HEMPRICH'S SEASTAR
Ophidiaster hemprichi
Müller and Troschel, 1842

Ashmore Reef, Timor Sea (C. Bryce).

DESCRIPTION

A small species with a small disc; five arms, often unequal in length, cylindrical for most of their length, then tapering slightly to a blunt tip; **abactinal plates slightly convex, with thin skin completely covered in squamous granules, irregularly larger in the centre of plates (3–4 per mm) than in pore areas (7–8 per mm)**; papular areas in eight rows, up to 15 pores per area; pedicellariae present or absent, embedded, of sugar-tongs form, fairly narrow with 2–4 ill-defined teeth; furrow spines in subequal pairs, separated by one or

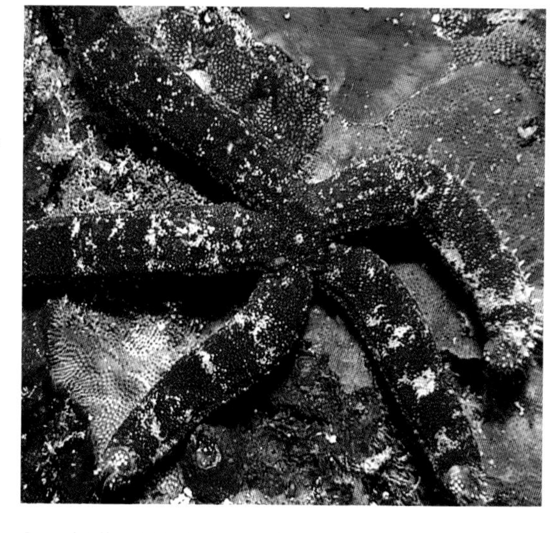

Seychelles (N. Coleman).

more granules, subambulacral spines almond-shaped or tapering to a blunt point.

MAXIMUM SIZE

R/r = 73/7 mm (R = 7.1–10.7r).

COLOUR

Reddish brown mottled with old rose, cream or grey, sometimes blotched with dark brown.

FOOD

The substrate biofilm of detritus and microorganisms.

REPRODUCTION

Induced spawning took place in autumn, the eggs are small and after fertilisation develop as a planktotrophic bipinnaria larva.

HABITAT, DEPTH AND AUSTRALIAN DISTRIBUTION

On coral reefs, under boulders on reef flats and reef crests, 0–10 m (rarely to 276 m). In Australia it is found on the offshore reefs off north-western Australia (Scott Reef, Rowley Shoals and Ashmore Reef) and on the Great Barrier Reef, Queensland, south to Lady Elliott Island and on Middleton Reef, Tasman Sea. Recorded from Christmas Island by the Western Australian Museum in the 1960s and 1970s, not found in 1987 (Marsh, 2000).

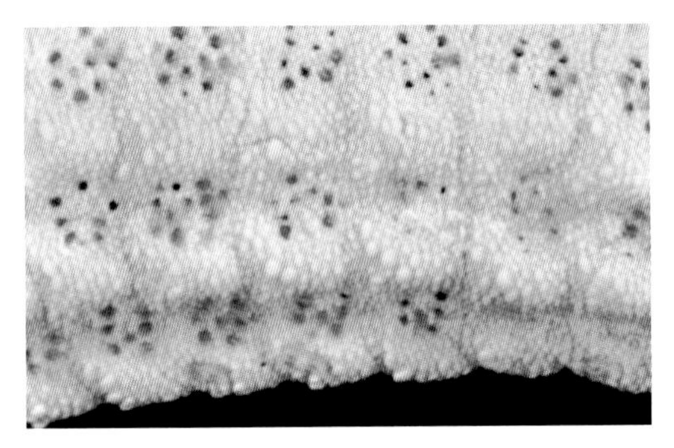

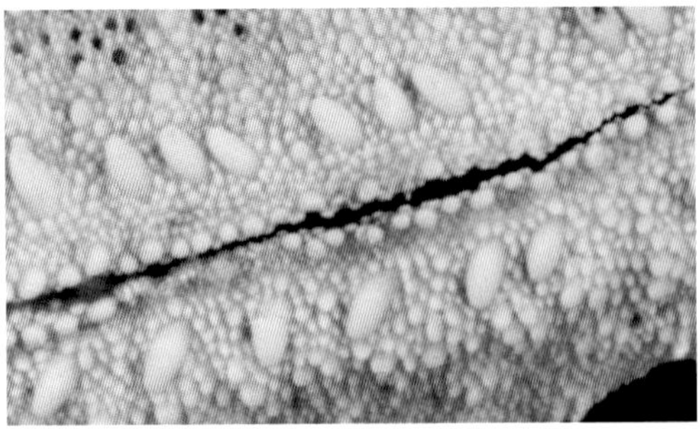

Abactinal plate characteristics [top], furrow characteristics [bottom], Western Australian Museum (WAMZ19994).

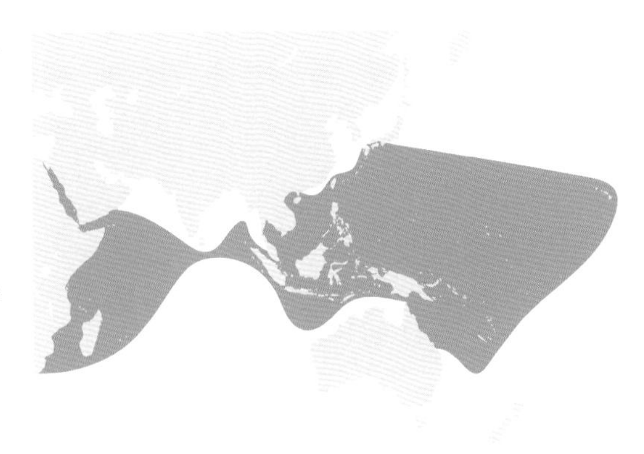

TYPE LOCALITY

Red Sea.

FURTHER DISTRIBUTION

This is one of the most widely distributed tropical Indo-West Pacific seastar species, found from east Africa and the Red Sea, through South-East Asia and the Pacific to Hawaii, north to Taiwan and southern Japan.

REFERENCES

Clark, A.M., 1993: 348, 350 (synonymy).

Clark, A.M. and Rowe, 1971: 59, 61; fig. 14f; pl. 8 fig. 3 (as *Ophidiaster squameus*).

Fisher, 1906: 1079; pl. 31 fig. 6a–b; pl. 37 fig. 4 (as *Ophidiaster squameus*).

Yamaguchi, 1975: 20 (food, as *Ophidiaster squameus*).

Lane et al., 2000: 121.

Marsh, 2000: 99.

GENUS *TAMARIA* Gray, 1840

Species of this genus have a small disc and more or less cylindrical arms. The abactinal plates are in regular longitudinal series and the **papular areas are in only six series** (eight in *Ophidiaster*). The adambulacral armature is similar to that of *Ophidiaster*. Pedicellariae, when present, are of the sugar-tongs type.

Type species: *Tamaria fusca* Gray, 1840 (Indo-West Pacific)

Four species are known from shallow Australian waters.

Tamaria tumescens, Port Hedland, Western Australia (N. Coleman).

KEY TO THE SPECIES OF *TAMARIA* IN SHALLOW AUSTRALIAN WATERS

1	Tubercles or small spines on at least the marginal plates, often on other plates as well.	2
	No tubercles; carinal plates twice as broad as long; all abactinal plates strongly convex; R < 117 mm.	***T. tumescens*** page 357
2	Abactinal plates rounded, strongly convex with 1–6 rounded, very prominent convex granules standing out from the general granulation; inferomarginals each with a large central granule or small tubercle; pedicellariae with 5–10 teeth, alveolus not notched for teeth.	***T. fusca*** page 353
	Abactinal plates not strongly convex with no enlarged convex granules on plates; tubercles and/or spines variously developed.	3

3 Abactinal plates flat, covered with slightly convex polygonal granules; a single pointed tubercle usually on some carinal and marginal plates; large papular areas; pedicellariae with 4–5 teeth fitting into a notched alveolus; northern Queensland. ***T. megaloplax*** page 356

Abactinal plates convex; each arm plate carries a conical spine as do superomarginal and inferomarginal plates; pedicellariae present or absent, they have five poorly developed teeth in a weakly notched alveolus; north-western Australia. ***T. hirsuta*** page 354

Tamaria fusca, New Caledonia (P. Laboute).

SPINDLED SEASTAR
Tamaria fusca
Gray, 1840

DESCRIPTION
Small disc; five arms tapering evenly to a narrow tip; abactinal plates strongly convex, skeleton covered with close fitting slightly convex polygonal granules, 6–8 per mm on the periphery of the plates and 8–10 per mm of mixed sizes in the papular areas; there are **1–6 greatly enlarged, strongly convex granules (two per mm) in the centre of each abactinal plate on the arms**, up to 15 per plate on the disc; papulae in six rows with 6–10 pores per area near disc, one per papular area near arm ends; **inferomarginal plates bear a single large conical central granule or very small tubercle distally**; furrow spines in unequal pairs, flattened, truncate, no granules between them, subambulacral spines almond-shaped proximally, conical, pointed distally; on the actinal plates granules are squamous, about seven per mm; pedicellariae of sugar-tongs shape bearing 5–10 poorly formed teeth in a moderately thick walled, non-notched alveolus.

MAXIMUM SIZE
R/r = 50/9 mm (R = 5.5r).

COLOUR
Purplish brown with 2–4 darker bands on arms. Actinal surface paler.

FOOD
Sponges, algal turf and epibenthic biofilm (Birtles, personal communication).

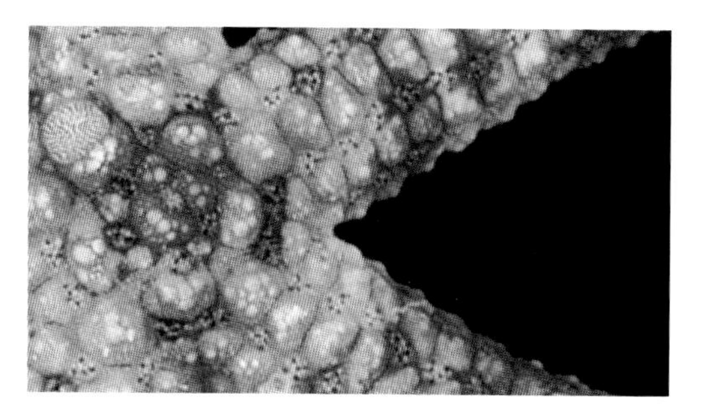

Abactinal plate characteristics, Western Australian Museum (WAMZ99850).

HABITAT, DEPTH AND AUSTRALIAN DISTRIBUTION
Inner continental shelf, on sand or dead coral with shell gravel or stones, 0–70 m. Off Direction Island to Lindeman Island, Queensland, Darwin, NT.

TYPE LOCALITY
Migupou, the Philippines.

FURTHER DISTRIBUTION
Philippines and Sulawesi, Indonesia, New Caledonia, Japan and the South China Sea.

REFERENCES
Lane et al., 2000: 473.
Livingstone, 1932a: 257–259; pl. 9 figs 4–7; pl. 11 figs 1–4, 7–8; pl. 12, figs 1, 4, 6–7, 10, 15, 17, 19.
VandenSpiegel et al., 1998: 435.

REMARKS
Tamaria fusca described by Studer (1884) from 91 m on the North West Shelf, WA is not conspecific with *T. fusca* Gray, and probably represents *T. hirsuta*. Livingstone (1932a) redescribed *T. fusca* from Gray's type specimen and clarified the synonymy, removing misidentified records.

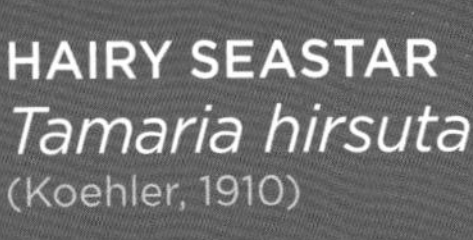

HAIRY SEASTAR

Tamaria hirsuta

(Koehler, 1910)

Western Australian Museum (WAMZ8802).

DESCRIPTION

Small disc; cylindrical arms, slightly flattened on the lower side and tapering to blunt points; **the disc plates are irregular, some bearing spines, the dorsalaterals carry two conical tubercles**, some of the interradial plates have tubercles, the carinals have enlarged granules at their centre; **the arm plates are in seven fairly regular rows (carinal, dorsolateral, superomarginal and inferomarginal) of convex plates, covered with large granules (3–4 per mm) in the centre, and small on the periphery and between the plates; each plate bears a conical spine, equal in length to the diameter of the plate; the terminal plate also carries 3–4 tubercles**; papular areas have up to 20 pores, granules 7–8 per mm, the papular areas are in six rows characteristic of the genus *Tamaria*, there are also 1–2 pores of an additional actinal row;

superomarginal and inferomarginal plates both carry a spine on each plate; actinal plates are granule-covered; furrow spines are paired, one large, one small, with no granules between them; 1–2 rows of granules separate the furrow spines from the large, conical, pointed subambulacral spines, one per plate basally, of irregular occurrence distally; actinal granules three per mm, tightly packed, slightly imbricating; pedicellariae have five poorly developed teeth, the alveolus is weakly notched, pedicellariae may be absent.

MAXIMUM SIZE

R/r/br = 35/5/4.5 mm (R = 7r).

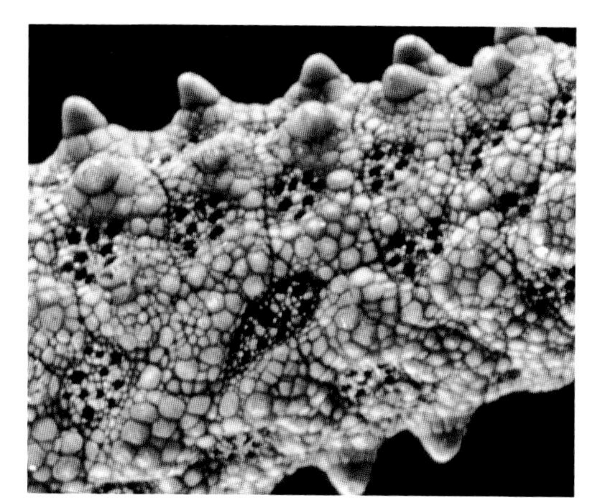

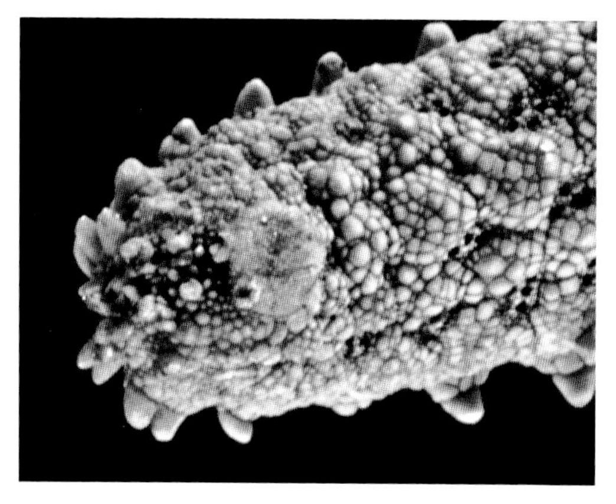

Abactinal plate charcateristics [top], terminal plate characteristics [bottom], Western Australian Museum (WAMZ8802).

COLOUR

Abactinal surface greenish yellow. Actinal surface white.

HABITAT, DEPTH AND AUSTRALIAN DISTRIBUTION

Inner continental shelf, 18–84 m. Off Red Bluff (24°S), south of Ningaloo Reef, North West Shelf from off Dampier Archipelago to Holothuria Banks off Kimberley, WA.

TYPE LOCALITY

Andaman Islands, Bay of Bengal.

FURTHER DISTRIBUTION

Andaman Islands.

REFERENCES

Clark, H.L., 1921: 89–90 (as *Tamaria fusca* (part)).

Koehler, 1910b: 149–151; pl. 18 figs 5–6 (as *Ophidiaster hirsutus*).

Livingstone, 1932a: 260; pl. 10 figs 5–7; pl. 11 figs 9–10; pl. 12 fig. 20.

LARGE-PLATED SEASTAR
Tamaria megaloplax
(Bell, 1884)

DESCRIPTION

Small disc; five nearly cylindrical arms tapering slightly to a blunt tip; **abactinal granules large on plates and uneven in size, polygonal, flat-topped (4–5 per mm), granules in papular areas smaller**; papular areas fairly large, with 10–20 pores in six rows; tubercles variously developed, may be on disc, carinal and dorsolateral plates as well as marginals, or they may only be on the marginals; furrow spines, with rounded tips, in unequal pairs; subambulacral spines tapering, pointed, usually another row of rounded spines between the furrow and subambulacral series; sugar-tongs pedicellariae near subambulacral spines and irregularly on abactinal plates; pedicellariae are broad with 4–5 teeth that fit into thick walled, notched alveoli.

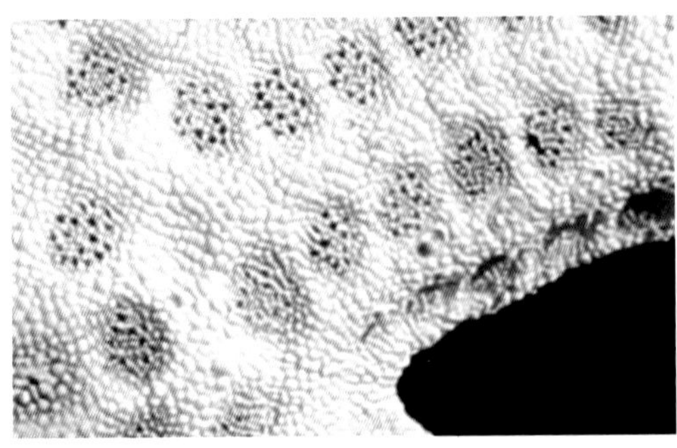

Australian Museum (AMJ12264), abactinal plate characteristics [bottom].

MAXIMUM SIZE

R/r = 67/9 mm (R = 7–7.5r).

COLOUR

Light brown, blotched with dark brown, papular areas dark brown. Actinal surface cream to beige.

FOOD

Zoantharians and epibenthic biofilm (Birtles, personal communication).

HABITAT, DEPTH AND AUSTRALIAN DISTRIBUTION

On the inner continental shelf, on coralline mud with coral isolates, 13–117 m. Found from the Gulf of Carpentaria south to off Townsville, Queensland.

TYPE LOCALITY

Albany Island, Cape York, Queensland.

FURTHER DISTRIBUTION

Andaman and Nicobar Islands.

REFERENCES

Livingstone, 1932a: 259; pl. 9 figs 1–3; pl. 12 figs 8, 12, 14 (synonymy of *Ophidiaster tuberifer* with *Tamaria megaloplax*).

Sladen, 1889: 404; pl. 65 figs 1–4 (as *Ophidiaster tuberifer*).

SWOLLEN SEASTAR
Tamaria tumescens
(Koehler, 1910)

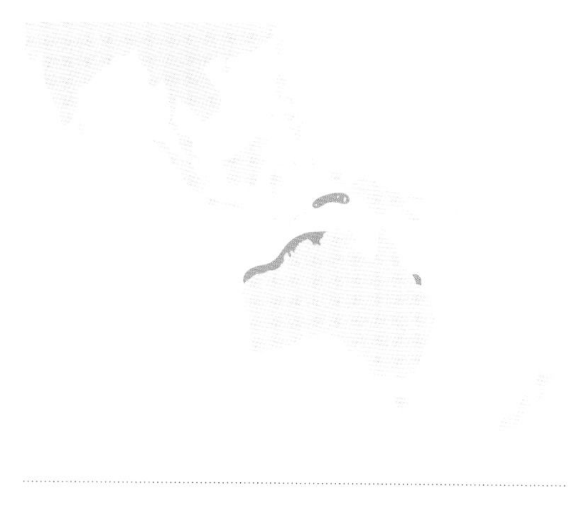

DESCRIPTION
Disc small and high; five arms taper evenly to narrow tips; **abactinal disc plates are uneven, usually strongly convex, primary radial and interradial plates are prominent, separated by deep grooves; the carinal plates are strongly convex, broader than long and much broader than the dorsolateral plates**; papulae are in six rows, in groups of up to 20; the abactinal plates are covered by close packed, polygonal convex granules, larger on the plates than between them or in the pore areas; sugar-tongs pedicellariae are common, usually on the superomarginal plates; furrow spines are paired, equal, flat with a rounded tip, no granules between them; subambulacral spines are single.

MAXIMUM SIZE
R/r = 117/15 mm (R = 6–8r).

COLOUR
Pink to old rose with light grey-brown papular areas, sometimes all grey-brown.

HABITAT, DEPTH AND AUSTRALIAN DISTRIBUTION
On the inner continental shelf, on sand or mud with shells, sponges, bryozoans and rubble, 15–130 m. North West Shelf, WA from the Montebello Islands to Parry Shoal, NT and off Keppel Island to Heron Island, Queensland.

TYPE LOCALITY
Aru Islands, Indonesia.

FURTHER DISTRIBUTION
Indonesia (Aru Islands and Dammer Island, Banda Sea.)

REFERENCES
Clark, A.M., 1967: 194–195.
Clark, H.L., 1946: 123–124.
Clark, A.M. and Rowe, 1971: 58.
Koehler, 1910a: 281–283; pl. 16 figs 3–4.

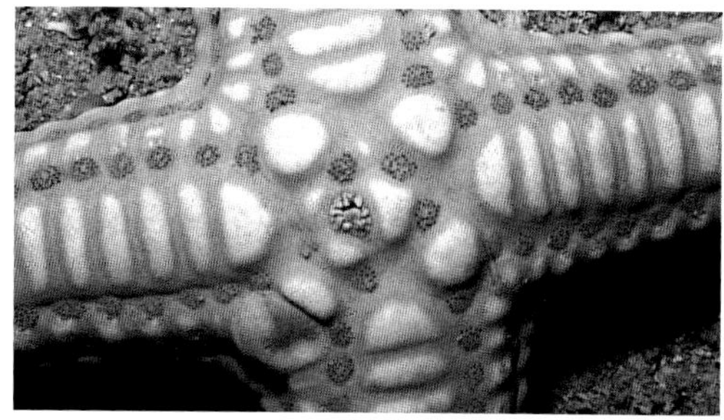

Kimberley, Western Australia (J. Keesing) [left]; abactinal plate characteristics [above], Port Hedland, Western Australia (N. Coleman).

OREASTERIDAE

Oreasterids usually have a **large body that is stellate to pentagonal** with five arms. The **abactinal skeleton is reticulate** with secondary plates linking the primaries and leaving conspicuous papular areas. The marginal plates are well developed but not always conspicuous. There are no actinal papulae and the **interbrachial septum is calcareous**. The tube feet have spicules in their walls and terminal discs.

The Oreasteridae is a tropical and temperate family of 19 genera, 15 of which (comprising 54 species) are represented in Australia; 12 genera and 38 species are found in depths of less than 30 m. The number of species in the genera *Anthenea* and *Pentaceraster* cannot be determined accurately as species of both genera are highly variable and some may be synonymised when more is known of their biology.

Culcita schmideliana, Montebello Islands, Western Australia (C. Bryce).

Pseudoreaster obtusangulus, Dampier Archipelago, Western Australia (S. Morrison).

KEY TO THE GENERA OF OREASTERIDAE IN SHALLOW AUSTRALIAN WATERS

1	Each of the larger actinal interradial plates has one conspicuously large valve-shaped pedicellaria, surrounded by a single or double circle of granules.	2
	Numerous actinal interradial plates without conspicuous pedicellaria.	3
2	The whole abactinal surface covered with a fused pavement of polygonal granules, abactinal tubercles numerous.	***Anthaster*** page 361
	Abactinal surface densely or sparsely covered with small tubercles, single or arranged in groups.	4
3	Marginal plates not visible, covered on the outside.	5
	Marginal plates visible, not covered.	6

4	Abactinal surface covered with smooth skin with widely spaced small tubercles; upper surface of superomarginal plates horizontal, bare except for one large granule on the upper edge and 3–4 small granules on the lateral face; a single row of subambulacral spines.	***Gymnanthenea*** page 405
	Abactinal surface fairly densely covered by small tubercles; superomarginals more or less covered by granules; two rows of subambulacral spines.	***Anthenea*** page 364
5	Round or pentagonal; granules large.	***Culcita*** page 399
	Stellate; entire surface covered by very finely granulated skin, appearing smooth.	***Choriaster*** page 364

Anthenea conjungens, Dampier Archipelago, Western Australia (C. Bryce).

Couplet	Character	Go to
6	All carinal plates similar, forming a regular row, the individual plates of which may diminish in size distally.	7
	Some carinal plates noticeably enlarged, bolster-shaped, nodular, or spine-like, the intermediate ones are smaller and inconspicuous.	11
7	Stellate; body flat or convex, granule-covered.	8
	Pentagonal; disc flat or convex.	10
8	Central pentagonal area limited by a sharp margin of large granules.	***Bothriaster*** page 364
	No sharp margin surrounds the central area.	9
9	Very prominent usually domed carinal plates.	***Pseudoreaster*** page 433
	Carinal plates not domed.	***Goniodiscaster*** page 405
10	Body flat; pore areas on the abactinal surface not very sharply defined; tubercles on abactinal surface; R < ~70 mm.	***Culcita*** (juv.) page 399
	Abactinal surface convex, with very irregular sharply defined usually triangular pore areas; no tubercles; R < 160 mm.	***Halityle*** page 405
11	Actinal interradial plates and the last marginal plates with polygonal plates of uneven size, always smooth and flat.	***Protoreaster*** page 433
	Actinal interradial plates usually with rounded granules, in larger specimens with uneven surface; marginal plates usuallywith even small granules.	12
12	Pore areas in the interradial area between upper and lower marginal plates.	***Poraster*** page 424
	No pore areas between marginal plates (except in the arm angle in *Pentaceraster gracilis*); abactinal surface with spines or nodules in regular rows, usually with a bare point.	***Pentaceraster*** page 424

(Modified from Döderlein, 1935.)

GENUS *ANTHASTER* Döderlein, 1915

Type species: *Anthaster valvulatus* (Müller and Troschel, 1843) (Australia)

Monotypic genus. Diagnosis as for the species (page 362).

VALVULATE SEASTAR
Anthaster valvulatus
(Müller and Troschel, 1843)

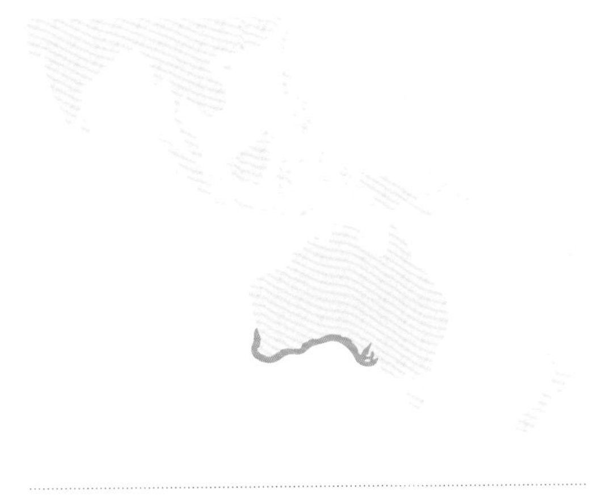

DESCRIPTION

Abactinal surface convex, arms fairly long, tapering to a broad tip; **the whole surface is covered with small polygonal granules, abactinal plates are convex and have a large, bare rounded tubercle**; interradial plates may have a medium sized pedicellaria instead of a tubercle; tubercles of the primary radial plates are 6 mm diameter, those of the disc centre and carinal row are up to 4 mm wide; superomarginal plates are convex with various sized tubercles and a pedicellaria, inferomarginals near the arm tip are similar, the rest lack tubercles but have 1–2 bivalved pedicellariae (up to 4 mm long); **most inner actinal plates have a very large pedicellaria (up to 4 mm long)**; adambulacral armature is in three rows; there are 9–10 slender furrow spines, subambulacral spines consist of three thicker spines and a thick pedicellaria, 3–5 irregular spinelets outside the subambulacral spines.

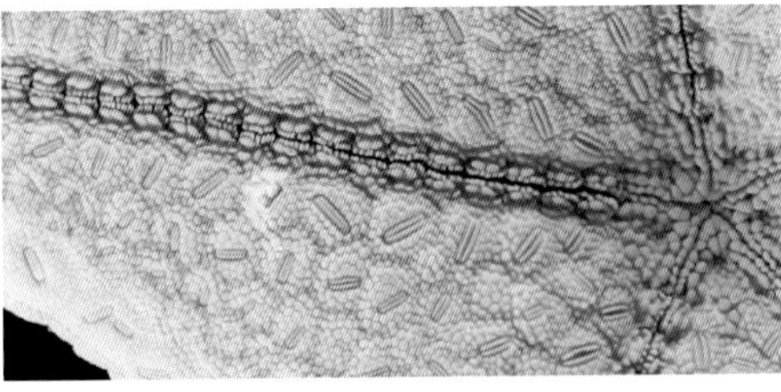

Abactinal plate characteristics [top], South Australian Museum (SAMAK3374); furrow and actinal plate characteristics [bottom], South Australian Museum (SAMAK3351).

MAXIMUM SIZE

R/r = 153/52 mm (R = 2.9r).

COLOUR

Orange or crimson, typically blotched with large or small areas of white. Actinal surface paler with crimson pedicellariae.

FOOD

Encrusting sponges and ascidians.

HABITAT, DEPTH AND AUSRALIAN DISTRIBUTION

On rock, sand, seagrass beds, or on mud, 3–40 m. From Encounter Bay, SA to near Lancelin (30°50'S), WA. South-western Australian endemic.

TYPE LOCALITY

'South-western Australia'.

FURTHER DISTRIBUTION

Australian endemic.

REFERENCES

Clark, H.L. 1928: 386.

Clark, H.L. 1946: 98–99.

Shepherd, 1968: 743 (food).

Zeidler and Shepherd, 1982: 411; figs 5e–f, 10.

Geographe Bay, Western Australia [top] (S. Morrison);
Penguin Island, Western Australia [bottom] (M. Rule).

GENUS *ANTHENEA* Gray, 1840

The body is flat to convex and the arms are short (R = 1.5–2.1r). The abactinal surface is skin-covered with close or sparse small tubercles and bivalved pedicellariae. The plates are joined by connecting ossicles leaving large papular spaces, occasionally the plates have crystal bodies. **The marginal plates are well developed and have a dense or sparse covering of granules often making a pattern that is a useful diagnostic feature for some species.** The inferomarginals have few or many pedicellariae. The actinal plates each have a large pedicellaria surrounded by a single or double row of granules, and the furrow and subambulacral spines form three rows. Most species are inflated but may become flat when preserved.

Type species: *Anthenea pentagonula* (Lamarck, 1816) (Australia, the Philippines)

Anthenea is a tropical genus confined to southern Asia, from India to China, and the northern, eastern and western coasts of Australia. Seventeen nominal species have been recorded from Australia of which 15 are found in depths of less than 30 m. Several of the species are very hard to separate; Rowe (in Rowe and Gates, 1995) notes that 'without doubt a rationalisation of species will occur when a much needed revision of the genus is carried out, integrating morphological and genetic characters, species and their distributions recorded herein should be treated with reservation'. It is also possible that hybridisation between species occurs. In this work we focus on details of superomarginal and inferomarginal plates, tubercle arrangement on the abactinal surface, and pedicellariae type and abundance on both surfaces.

A.M. Clark (in Liao and Clark, 1995) noted that specimens of *A. chinensis* are not conspecific with specimens of *A. pentagonula*, that a neotype for the latter species should be described, and *A. chinensis* reinstated as a valid species and the type species of the genus. This action has not been taken yet.

Several species show little variation in their characters and can be fairly reliably identified, for example *A. australiae*, *A. conjungens*, *A. sidneyensis*, *A. polygnatha* and *A. tuberculosa*. H.L. Clark (1938) recognised 14 species from the coasts of Australia including five that he described as new. He noted that small individuals of any species (with R < 35 mm) could not be reliably identified and that the variation in many species made it difficult to provide a reliable key.

GENUS *BOTHRIASTER* Döderlein, 1916

Type species: *Bothriaster primigenius* Döderlein, 1916 (Indo-West Pacific)

Monotypic genus. Diagnosis as for the species (page 395).

GENUS *CHORIASTER* Lütken, 1869

Type species: *Choriaster granulatus* Lütken, 1869 (Indo-West Pacific)

Monotypic genus. Diagnosis as for the species (page 397).

KEY TO THE SPECIES OF *ANTHENEA* IN SHALLOW AUSTRALIAN WATERS

1	Arms relatively long, narrow distally and more or less pointed; R = 2r (1.9–2.45r, rarely 1.5r).	2
	Arms short and rounded at tip, form sometimes pentagonal; R = 1.6–1.8r.	9
2	Small bivalved pedicellariae extraordinarily abundant on actinal surface and inferomarginals, fewer on superomarginals; north-western Australia.	***A. polygnatha*** page 384
	Pedicellariae not excessively abundant on actinal surface, rarely more than 3–4 on any inferomarginal plate.	3
3	Superomarginal plates in interradial arc more or less horizontal forming a conspicuous part of the abactinal surface.	4
	Superomarginals in interradial areas more or less vertical, forming an inconspicuous part of the abactinal surface.	6
4	Abactinal surface with numerous bluntly pointed tubercles forming 5–9 distinct longitudinal series on each arm; superomarginal plates completely covered by granules; inferomarginals closely covered with small granules and sometimes a small pedicellaria; north-western Australia.	***A. elegans*** page 378
	Abactinal surface diverse; inferomarginals with coarse granules and usually two or more larger pedicellariae.	5
5	Abactinal tubercles numerous and coarse; upper half of superomarginals bare on each side with a few coarse granules in a narrow or single vertical series; Queensland.	***A. crassa*** page 376
	Abactinal tubercles few, scattered; superomarginals rather uniformly covered with granules which may be fewer and coarser at the upper end; Queensland (see also *A. pentagonula*).	***A. aspera*** page 370
6	Disc more or less elevated, reticulate skeleton may be visible; superomarginal plates covered with granules to their upper margin; southern Queensland and NSW.	***A. sidneyensis*** page 389
	Disc not very elevated or reticulated; upper end of superomarginals more or less bare.	7

7	Superomarginals low, wide, with tubercles only on lower half; abactinal tubercles low and more or less flattened; Western Australia.	***A. australiae*** page 372
	Superomarginals high, narrow, with a large granule near the top, below it a swollen pair or trio, the lower half of the plate more or less covered with smaller but often coarse granules; abactinal tubercles high, not flattened; north-western Australia and NT.	***A. conjungens*** page 374
8	Abactinal plates near arm tips enlarged and more or less conspicuous; abactinal surface usually with more or less numerous tubercles.	**10**
	Abactinal plates near arm tips not much enlarged or conspicuous; abactinal surface with few (or more numerous) tubercles or large spinelets, commonly with numerous pedicellariae and small spinelets (animals with R < 65 mm may have smooth abactinal plates and few pedicellariae).	**9**
9	Abactinal tubercles few, irregularly scattered, seldom capitate; superomarginals with small granules, mostly on lower half of plates; one or more pedicellariae on inferomarginals, arms often somewhat pointed; north-western Australia.	***A. pentagonula*** page 382
	Abactinal tubercles more or less numerous, especially on arms, somewhat capitate; marginal plates uniformly covered by small granules; arms wide and rounded; usually ocurrs in depths of more than 30 m; north-western Australia.	***A. regalis*** page 385
10	Enlarged abactinal plates near arm tips each with several (2–8) large granules or small tubercles.	**11**
	Enlarged abactinal plates near arm tips each with one or two large granules or tubercles.	**14**
11	Whole abactinal surface covered with large pointed tubercles that form 9–13 series on each arm; superomarginals with coarse spaced tubercles, the uppermost largest; usually ocurrs in depths of more than 30 m.	***A. acanthodes*** page 368
	Whole abactinal surface covered in small, blunt or capitate tubercles.	**12**

12	Abactinal tubercles small, numerous in about nine parallel series on each arm; superomarginals have a row of small granules at their base.	***A. godeffroyi*** page 379
	Abactinal tubercles fewer, larger, in not more than five series on each arm.	**13**
13	Abactinal surface with few, small spinelets and many pedicellariae; superomarginals with bare margins.	***A. sibogae*** page 387
	Abactinal surface with numerous, more or less capitate spines and few pedicellariae; superomarginals have a vertical row of 5–6 tubercles with a row of small granules at their base.	***A. mertoni*** page 380
14	Tubercles on abactinal surface and superomarginals few and coarse, smaller on lower half.	***A. tuberculosa*** page 391
	Tubercles on abactinal surface and superomarginals small, like large granules.	***A. viguieri*** page 393

(Modified from H.L. Clark, 1938.)

Anthenea mertoni, Robroy Reef, Western Australia (C. Bryce).

RIBBED CUSHION STAR
Anthenea acanthodes
H.L. Clark, 1938

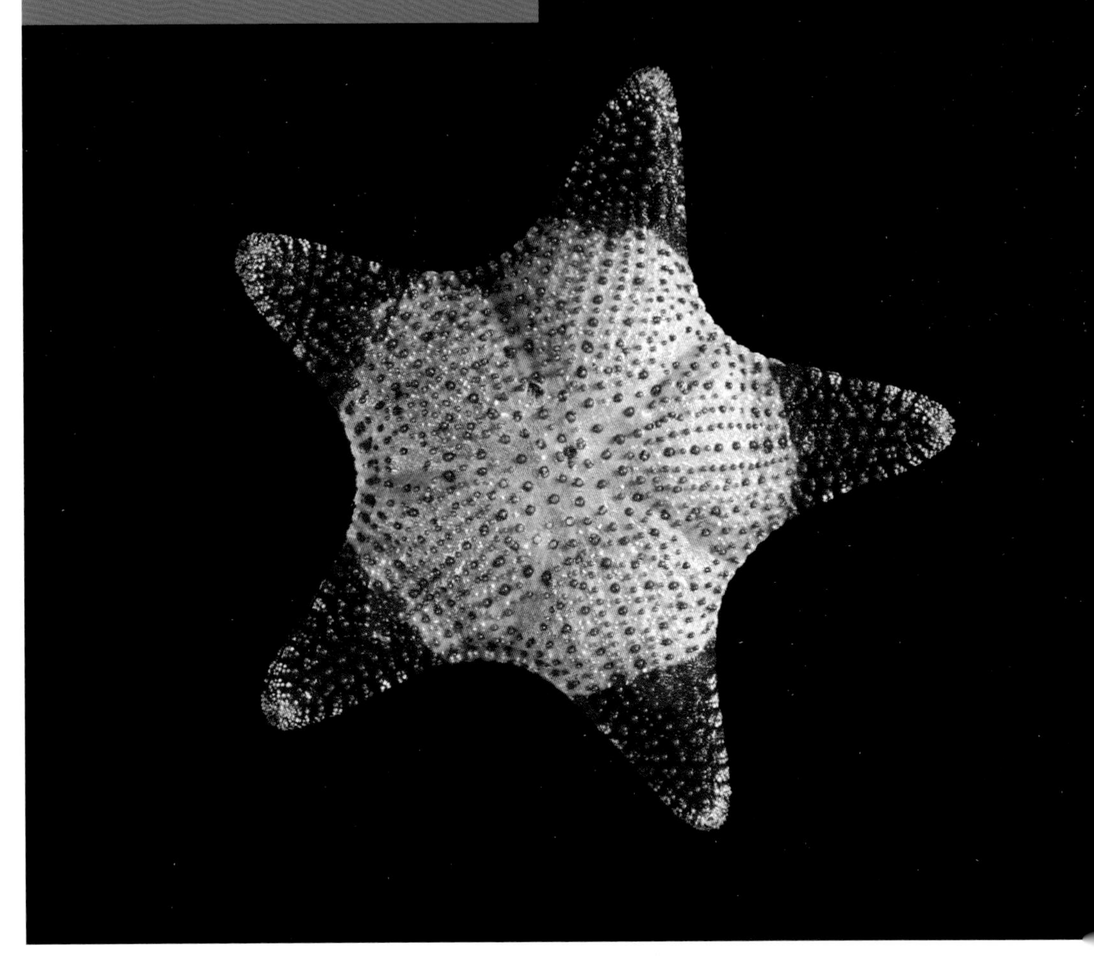

North West Shelf, Western Australia (L. Marsh).

DESCRIPTION

Disc and arms very convex with the arms flatter near the tips; the abactinal surface of the disc and basal part of arms is covered with large, **bluntly pointed tubercles or low stout spines that form 9–13 nearly parallel series on the basal part of the arms**; the distal part of the arms covered with irregular tuberculated plates, similar to those of *A. mertoni*, the tubercles on these plates are coarse and rounded, similar to those on the superomarginal plates;

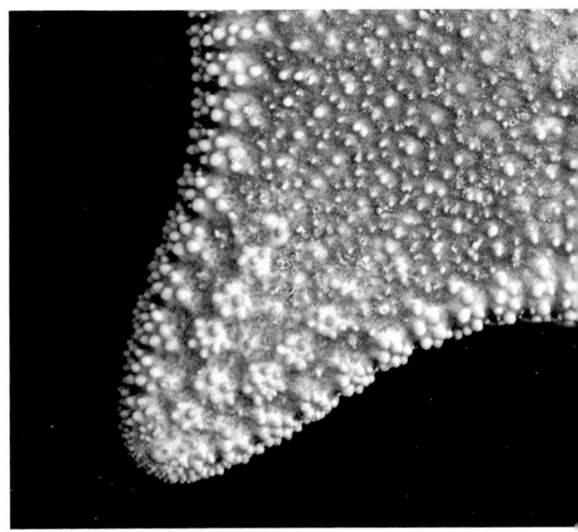

Distalmost abactinal plate characteristics, Western Australian Museum (WAMZ13764).

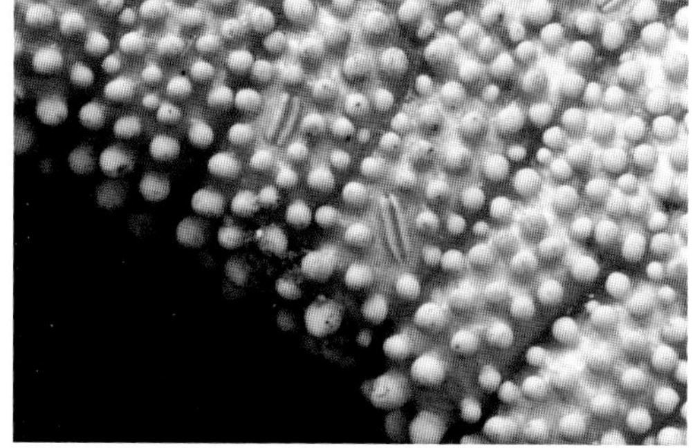

Superomarginal [above], inferomarginal [middle] and actinal plate characteristics [bottom], Western Australian Museum (WAMZ13764).

superomarginal plates are vertical in position, higher than long, covered except at the margins with coarse granules, the uppermost largest; inferomarginal plates are covered in coarse granules; there may be a few small pedicellariae on superomarginal and inferomarginal plates; adambulacral armature stout, similar to other species.

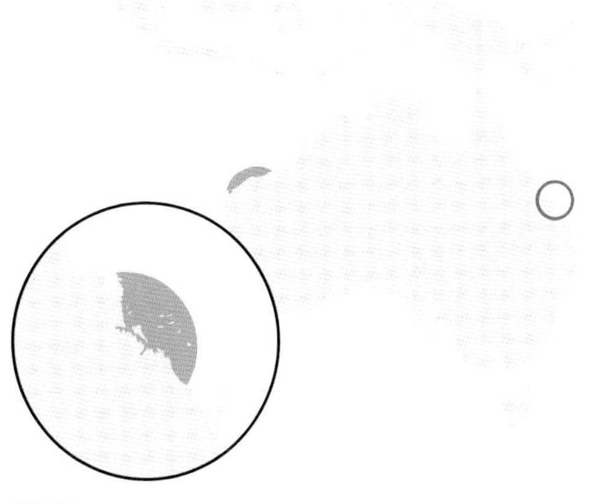

MAXIMUM SIZE

R/r = 97/55 mm (R = 1.76r).

COLOUR

Mottled light and dark brown or red-brown.

HABITAT, DEPTH AND AUSTRALIAN DISTRIBUTION

Habitat unknown, probably soft substrate, ~20–106 m. Known from the type locality and from the North West Cape to Port Hedland, WA. Northern Australian endemic.

TYPE LOCALITY

Port Curtis, Queensland.

FURTHER DISTRIBUTION

Australian endemic.

REFERENCES

Clark, H.L., 1938: 124; pl. 18 fig. 2.

ROUGH CUSHION STAR

Anthenea aspera

Döderlein, 1915

DESCRIPTION

Disc large; arms short, broad and tapering to a fairly acute tip; abactinal surface densely covered with papular pores except on the outer part of the arms and along the interradial line; surface covered with scattered, blunt, conical tubercles, the largest about 1 mm in diameter and height, smaller on the arms; between the tubercles are numerous bivalved pedicellariae up to 1.5 mm wide; the surface between the pores, pedicellariae and tubercles is covered with delicate, very small spinelets, sometimes in groups; **superomarginals prominent, convex, three times as wide as long** in the arm angle, about twice as wide as long distally, **completely covered with fairly small granules** and a few pedicellariae (up to 2 mm wide); inferomarginals with coarser granules on their lower half than on their upper part, and several bivalved pedicellariae up to 3 mm wide; actinal plates each have a large pedicellaria (up to 3 mm wide) surrounded by a single or double ring of granules and a bare edge to the plates; 6–7 slender furrow spines, two rows of 2–3 much thicker subambulacral spines, near the second row there is usually a short thick, upright pedicellaria.

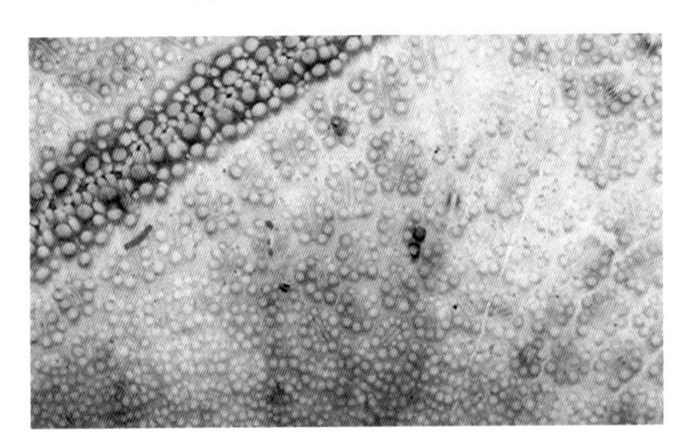

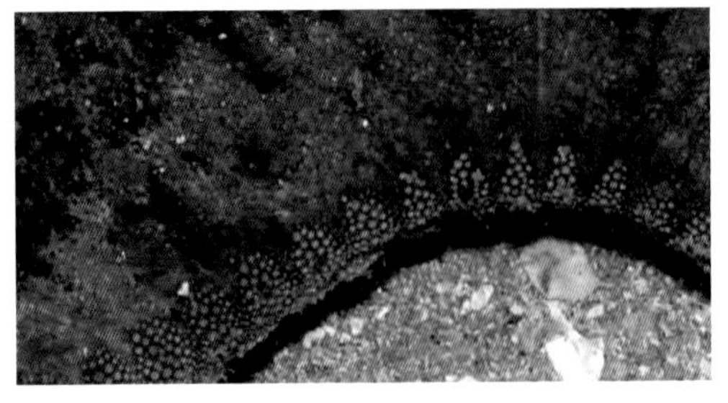

Actinal [top] and superomarginal plate characteristics [bottom], Singapore (R. Tan).

MAXIMUM SIZE

R/r = 115/65 mm (R = 1.5–2.1r).

COLOUR

Irregularly mottled with brown, red and sage green. Actinal surface yellow.

HABITAT, DEPTH AND AUSTRALIAN DISTRIBUTION

On muddy sand, 8–40 m. Rockhampton to Capricorn Group, Queensland, Dampier Archipelago, WA and Arafura Sea.

TYPE LOCALITY

'Australia'.

FURTHER DISTRIBUTION

Indonesia, Singapore and Hong Kong.

REFERENCES

Clark, H.L., 1946: 102.

Döderlein, 1915: 35–37; pl. 4 figs 1–2.

Purwati and Lane, 2004: 89–102.

Singapore [top and bottom] (R. Tan).

Cockburn Sound, Western Australia (A. Brearley)

DESCRIPTION

Stellate seastar with a large disc and triangular arms tapering to acute tips; skin obscures the abactinal plates which have fairly small truncate or hemispherical tubercles 0.6–1.7 mm wide and about 3–4 mm apart; **tubercles form irregular longitudinal rows, about seven extending along the arms** and small bivalved pedicellariae are present between the tubercles; tiny spicules project from the skin (visible when dry); **the granules on the superomarginal plates are fairly coarse and cover the lower half of the plates leaving a flat, nearly horizontal bare upper half of the plates**; the inferomarginal plates are densely granulated and have 1–2 very large pedicellariae up to 3.5 mm long; no pedicellariae on the superomarginal plates, actinal plates each with a large pedicellaria surrounded by a ring of granules; there are 6–7 furrow spines in a webbed fan, 2–3 thick subambulacral spines, and 1–3 shorter spines

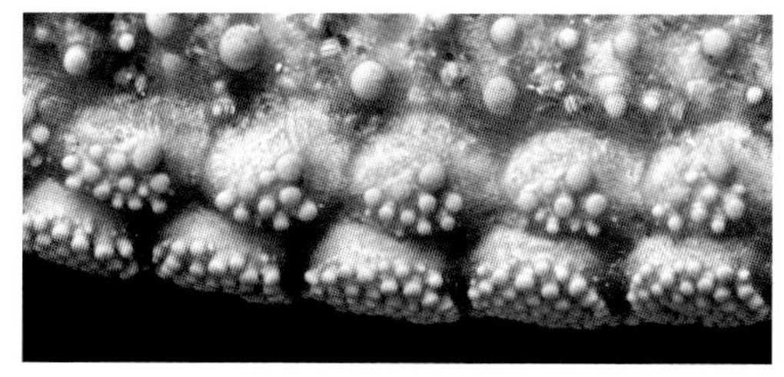

Abactinal [top], superomarginal [middle] and inferomarginal plate characteristics [bottom], Western Australian Museum (WAMZ13710).

outside the subambulacral spines; a short upright pedicellaria is sometimes present on the edge of the adambulacral plates.

MAXIMUM SIZE

R/r = 120/55 mm (R = 2.2r).

COLOUR

Variable, often yellow with red blotches, all orange, cream with yellow and brown blotches, crimson disc, pinkish arms with darker mottling.

FOOD

Likely feeds on substrate biofilm or scavenges.

HABITAT, DEPTH AND AUSTRALIAN DISTRIBUTION

On coral, algal covered rocks, among seagrass or on silty sand, 1–15 m. From Cockburn Sound near Fremantle and the Swan River estuary to the North West Shelf and Broome. Western Australian endemic.

TYPE LOCALITY

Fremantle and Shark Bay, WA.

FURTHER DISTRIBUTION

Australian endemic.

REFERENCES

Clark, H.L., 1946: 102.

Clark, A.M. and Rowe, 1971: 51.

Swan River estuary, Western Australia (G. Edgar).

CONJOINED CUSHION STAR
Anthenea conjungens
Döderlein, 1935

Shark Bay, Western Australia (S. Morrison).

Dampier Archipelago, Western Australia (C. Bryce).

DESCRIPTION

One of the least variable of the *Anthenea* species; **the superomarginal plates in the interradial arc are high and narrow, forming an inconspicuous part of the abactinal surface, they have a large granule near the bare top, crowning a wedge-shaped area of progressively less coarse granules**, and sometimes a pedicellaria; most abactinal plates bear a smallish high, rounded tubercle, not flattened; small bivalved pedicellariae numerous in the papular areas; with 3–4 pedicellariae on inferomarginals and a large bivalved pedicellaria on each actinal plate;

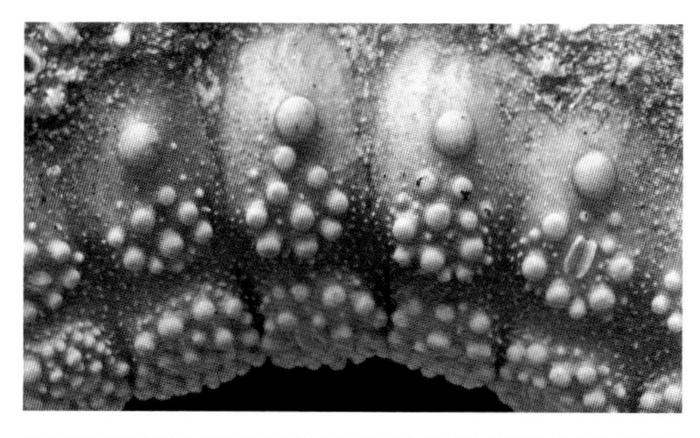

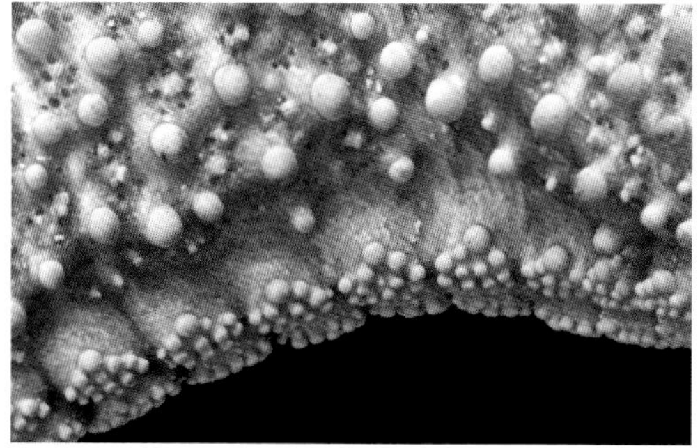

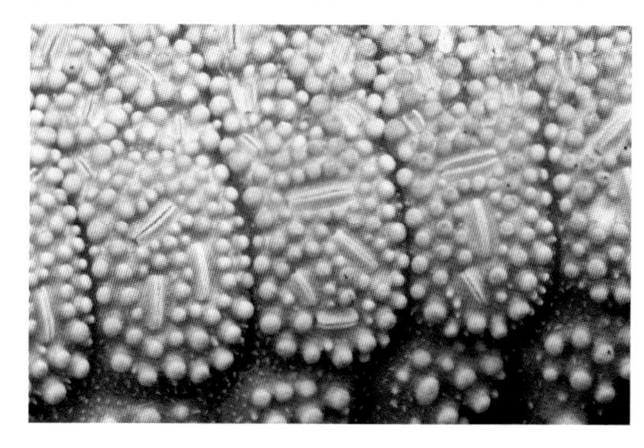

Superomarginal [above], abactinal [middle] and inferomarginal plate characteristics [bottom], Western Australian Museum (WAMZ13932).

there are 6–8 furrow spines, three (sometimes 2–4) coarse, rounded subambulacral spines and two (sometimes 1–3) spines outside the subambulacral spines; sometimes there is a small upright pedicellaria between the furrow and subambulacral spines.

MAXIMUM SIZE

R/r = 150/72 mm (R = 2.1r),
R usually < 100 mm.

COLOUR

Variable, commonly orange, red, purple or brown often with darker blotches.

HABITAT, DEPTH AND AUSTRALIAN DISTRIBUTION

On soft substrates, 0–76 m. Fremantle, WA to Darwin, NT. Also found in the Gulf of Carpentaria. Common in Shark Bay and off Broome and the Kimberley coast, WA.

TYPE LOCALITY

'Australia'.

FURTHER DISTRIBUTION

Australian endemic.

REFERENCES

Clark, H.L., 1946: 103.

Clark, A.M. and Rowe, 1971: 52 (as *Anthenea conjugens*).

Döderlein, 1935: 107; pl. 27 figs 3, 3a.

COARSE CUSHION STAR

Anthenea crassa

H.L. Clark, 1938

Holotype, Australian Museum (AMJ5368).

DESCRIPTION

Stellate seastars with abactinal surface fairly flat to convex and covered with fairly thick skin bearing **coarse to fine tubercles or stout, blunt spines in nine rows at the base of arms** (rows sometimes indistinct); tubercles may be more than 2 mm high and wide at the base; superomarginal plates wider than long interradially, length and width more equal distally; at the base of the arms **granules on the upper part of the superomarginals are in a single vertical series of 2–4 large granules**, distally the plates are more or less covered with coarse well spaced granules;

Stradbroke Island, Queensland (L. Newman and A. Flowers).

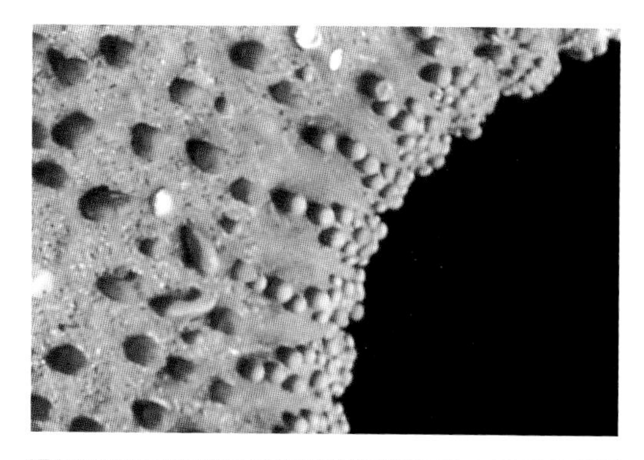

Superomarginal [top] and abactinal plate characteristics [bottom], holotype, Australian Museum (AMJ5368).

inferomarginal plates closely covered with rounded granules, coarsest at the outer end, with 2–6 pedicellariae; few to numerous small bivalved pedicellariae on the abactinal surface; actinal plates with moderate sized bivalved pedicellariae surrounded on most plates by two rows of granules; 6–8 fairly coarse furrow spines, two stout, rounded subequal to unequal subambulacral spines, and 2–3 smaller, similar spines outside the subambulacral spines; a stout upright pedicellaria on proximal side of each adambulacral plate between the two rows of subambulacral spines.

MAXIMUM SIZE

R/r = 115/55 mm (R = 2.1–2.5r).

COLOUR

Abactinal surface brown, orange or olive green. Actinal surface cream.

HABITAT, DEPTH AND AUSTRALIAN DISTRIBUTION

On soft substrates, 15–73 m. Weipa, Gulf of Carpentaria to Moreton Bay, Queensland. North-eastern Australian endemic.

TYPE LOCALITY

Rat Island, Port Curtis, Queensland.

FURTHER DISTRIBUTION

Australian endemic.

REFERENCES

Clark, H.L., 1938: 124–126; pl. 18 fig. 1.

REMARKS

The name *crassa* is from the Latin *crassus,* meaning 'coarse', in reference to the large tubercles and granules of the abactinal surface.

ELEGANT CUSHION STAR

Anthenea elegans

H.L. Clark, 1938

DESCRIPTION

Disc large; arms taper to a fairly acute tip, the disc is covered with small, rounded or bluntly pointed tubercles, with small bivalved pedicellariae and minute spinelets embedded in the skin; **tubercles may be up to 2 mm high, in 9–11 distinct rows on the arms; superomarginals in the arm angle about twice as wide as long**, distally length and width are about equal, **superomarginals completely covered with granules**, largest on the convexity of the plate, smallest close to the edges, distally some or many granules are enlarged on the upper half of the plate; abactinal plates near the arm tip of large specimens are enlarged and carry several tubercles and granules; inferomarginal plates, large, closely granulated with or without a bivalved pedicellaria; most actinal plates with a large bivalved pedicellaria surrounded by granules, 5–7 furrow spines, 2–4 thick, flat subambulacral spines and 2–3 slightly shorter spines outside the subambulacral spines.

Exmouth Gulf, Western Australia [top], marginal plate characteristics [bottom] (S. Morrison).

MAXIMUM SIZE

R/r = 120/60 mm (R = 2r).

COLOUR

Orange-brown.

HABITAT, DEPTH AND AUSTRALIAN DISTRIBUTION

On silty sand, 9–24 m. Shark Bay, WA to Weipa, Queensland. Northern Australian endemic.

TYPE LOCALITY

Broome, WA.

FURTHER DISTRIBUTION

Australian endemic.

REFERENCES

Clark, H.L., 1938: 126; pl. 18 fig. 4.

GODEFFROY'S CUSHION STAR
Anthenea godeffroyi
Döderlein, 1915

DESCRIPTION

Disc large; arms short, broad and rounded to pointed; abactinal surface of disc and arms covered with low blunt tubercles, less than 1 mm high and wide, in rows on the arms, nine of which reach the outer half of the arms; in larger specimens abactinal plates near the margin have 2–6 small tubercles of mixed sizes; the skin of the disc and arms is bare except for minute spinelets and small bivalved pedicellariae; **superomarginal plates are more vertical than horizontal, upper half is bare except for minute spinelets and one or two coarse granules at the apex of a group of spaced granules of mixed sizes,** and sometimes with a pedicellaria on the lower half of the plate; inferomarginals have a dense cover of granules of similar size to those on the actinal plates and several bivalved pedicellariae (up to 1.5 mm wide); actinal plates each have a pedicellaria up to 2.5 mm wide ringed by 1–2 rows of granules; 6–7 furrow spines, 2–3 subambulacral spines, and 2–3 spines outside the subambulacral spines; there may be a short upright pedicellaria next to the subambulacral spines.

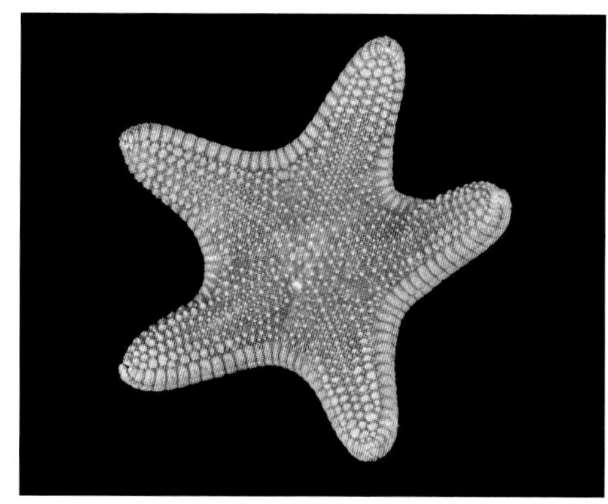

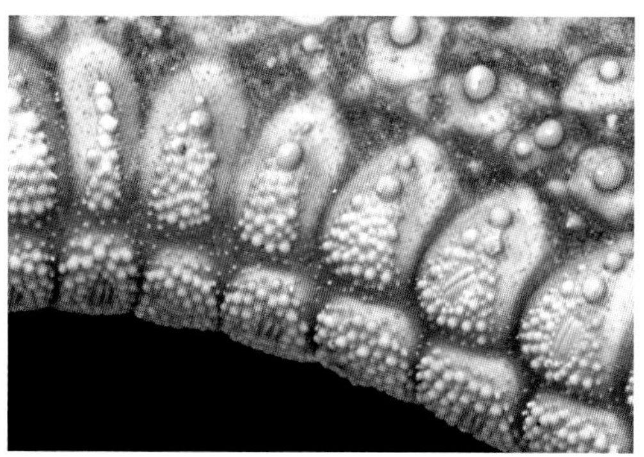

Western Australian Museum (WAMZ17943), superomarginal characteristics [bottom].

MAXIMUM SIZE

R/r = 79/44 mm (R = 1.8r).

COLOUR

Not recorded. Dried specimens are reddish grey to fawn.

HABITAT, DEPTH AND AUSTRALIAN DISTRIBUTION

On soft substrates, 9–63 m. Exmouth Gulf, WA to Queensland. Northern Australian endemic.

TYPE LOCALITY

'Samoa' (erroneous).

FURTHER DISTRIBUTION

Australian endemic.

REFERENCES

Clark, A.M. and Rowe, 1971: 52.
Döderlein, 1915: 45–47; pl. 11 figs 1–2.

MERTON'S CUSHION STAR
Anthenea mertoni
Koehler, 1910a

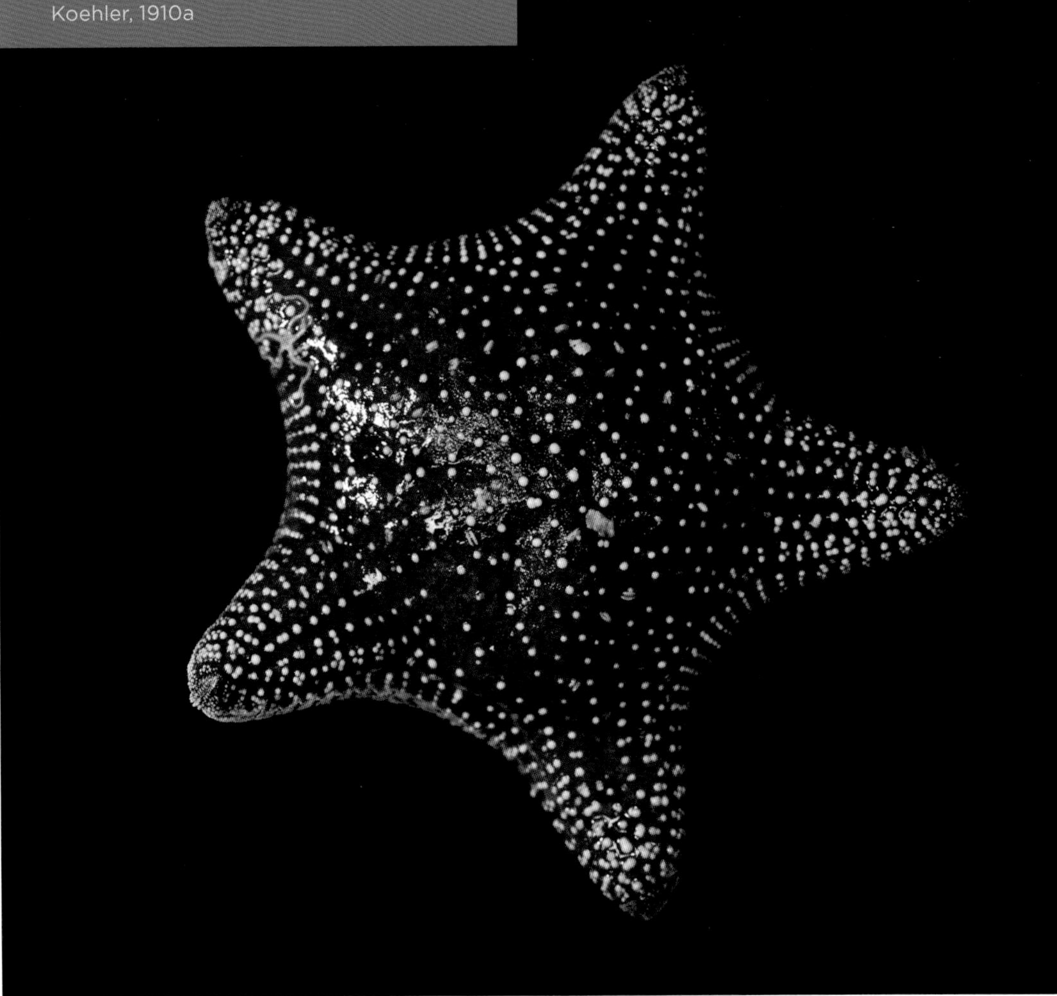

King George River, Western Australia (J. Keesing).

DESCRIPTION

Body stellate, robust; arms broad, triangular, tapering to a blunt tip; the body is very thick, covered with thick skin obscuring the plate outlines; the abactinal surface has well spaced rounded tubercles (up to 1.2 mm across the base) arranged in poorly defined longitudinal rows except for a distinct carinal row; **on the outer third of the arm each plate bears a group (up to 12) of small tubercles on five rows of raised plates across the arm near the tip**; among the abactinal tubercles are scattered smaller tubercles, and numerous split-granule and small bivalved pedicellariae (up to 1.5 mm); the superomarginal plates in the arm angle appear about twice as wide as long, length and width about equal distally; **superomarginals in the arm angle are bare except for a vertical row of 5–6 large granules**, the uppermost largest; granules are smaller and more numerous on the lower part of the plate, distally the granules are more numerous and the bare margins of the plates narrower; some plates may have a

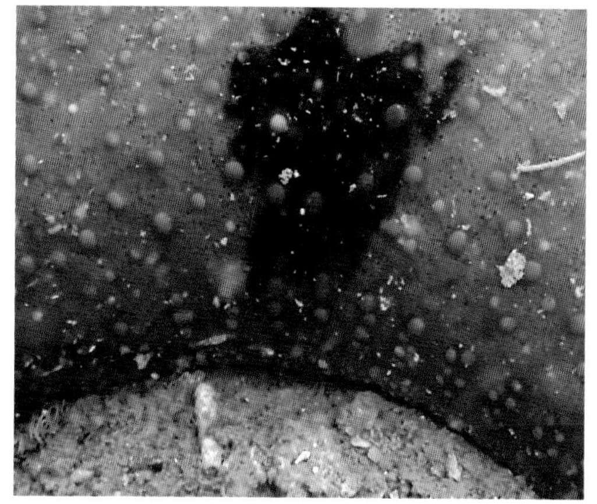

Dampier Archipelago, Western Australia, superomarginal plate characteristics [bottom] (C. Bryce).

small bivalved pedicellaria; the inferomarginal plates are closely granulated and each has 1–2 bivalved pedicellariae (up to 2 mm); most actinal plates have a large bivalved pedicellaria (up to 3 mm) surrounded by a double row of granules; there are seven furrow spines, three subambulacral spines (two distally) and usually three irregular spines outside the subambulacral spines; there is usually a pedicellaria distal to the subambulacral row.

MAXIMUM SIZE

R/r = 95/47 mm (R = 2r).

COLOUR

Variable, for example cream with maroon blotches in the arm angles and dark brown disc; orange with dark brown patches in the arm angle; or brown.

HABITAT, DEPTH AND AUSTRALIAN DISTRIBUTION

On soft substrates, 9–23 m. Dampier Archipelago, WA to Sir Edward Pellew Group, Gulf of Carpentaria.

TYPE LOCALITY

Aru Islands, Indonesia.

FURTHER DISTRIBUTION

Indonesia

REFERENCES

Clark, H.L., 1938: 121; pl. 7.

Koehler, 1910b: 268; pl. 16 figs 1–2.

PENTAGONAL CUSHION STAR
Anthenea pentagonula
(Lamarck, 1816)

DESCRIPTION

Body stellate; arms taper to a rounded tip; abactinal plates not obscured by the skin, tubercles low, widened at the apex or truncate (0.5–1 mm wide), in irregular radial rows on the disc, absent from the outer part of the arms and bands in the interradial area; bivalved pedicellariae of mixed sizes lie between the tubercles; the skin is covered in minute spicules and crystal bodies are visible through the skin on the main disc plates and the upper end of the superomarginals; **superomarginals are more than twice as wide as long in the arm angle**, slightly wider than long distally, the plates are convex, **the upper edge bare of granules, the remainder covered by small sparse granules and 1–2 small bivalved pedicellariae**; the inferomarginals are closely granulated, granules coarser on the upper half and with one or more small pedicellariae; actinal plates have large pedicellariae (up to 2 mm) surrounded by granules; there are six furrow spines, 2–3 thicker subambulacral spines and 2–3 shorter spines outside the subambulacral spines; there is often a thick upright pedicellariae next to the subambulacral spines (Döderlein's figures show superomarginals with a uniform covering of close packed granules).

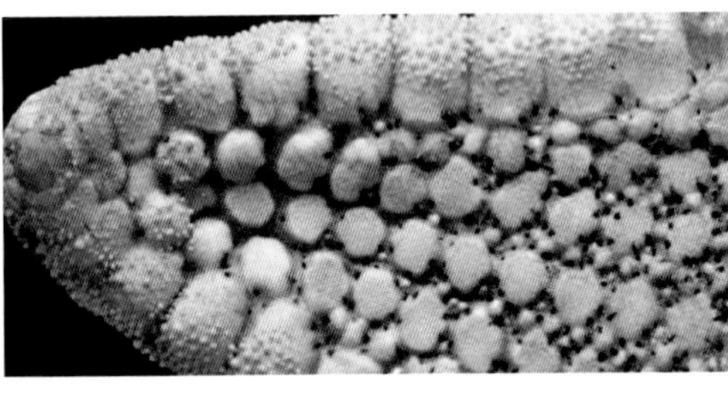

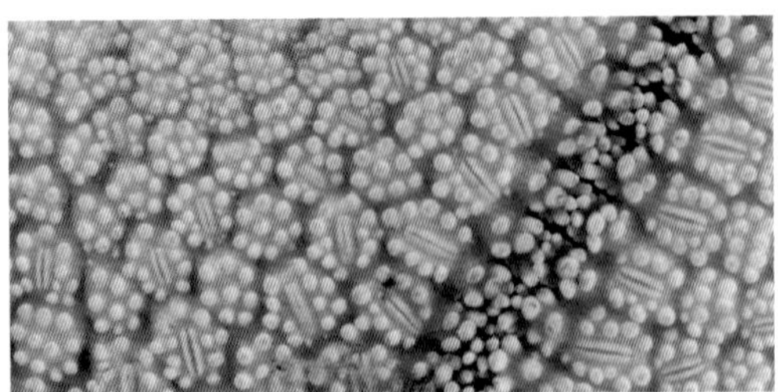

Abactinal [top] and actinal plate characteristics [bottom], Western Australian Museum (WAMZ48495).

MAXIMUM SIZE

R/r = 73/41 mm (R = 1.8r).

COLOUR

Not recorded.

HABITAT, DEPTH AND AUSTRALIAN DISTRIBUTION

On the inner continental shelf on soft substrates, 9–90 m. Dampier Archipelago, WA to Darwin, NT.

TYPE LOCALITY

'Australia Seas'.

FURTHER DISTRIBUTION

South-East Asia.

REFERENCES

Clark, A.M., 1993: 296 (validity doubtful, requires neotype (syntypes in Muséum National d'Histoire Naturelle)).

Liao and Clark, A.M., 1995: 98–99.

Döderlein, 1915: 32–33; pl. 4 figs 3–4; pl. 5 fig. 1.

Western Australian Museum (WAMZ48495).

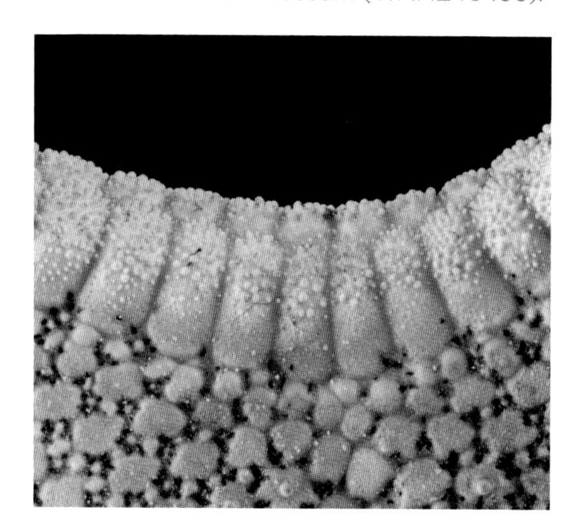

Superomarginal characteristics, Western Australian Museum (WAMZ48495).

REMARKS

Döderlein (1915) described and provided figures of two of Müller and Troschel's specimens of *Anthenea pentagonula* but it is not known if they are the same as Lamarck's type specimens. Liao and Clark (1995) suggested that *A. chinensis* Gray, 1840, long believed to be synonymous with *A. pentagonula*, is a valid species, thus all records of *A. pentagonula* from southern China should be referred to *A. chinensis*. Other records from South-East Asia are questionable.

MULTI-JAWED CUSHION STAR
Anthenea polygnatha
H.L. Clark, 1938

DESCRIPTION
Disc large; arms taper to a fairly acute tip; **easily recognised by the abundance of bivalved pedicellariae especially on the inferomarginal and actinal plates**; abactinal surface with spaced, small rounded or pointed tubercles, between these the skin is bare except for fairly numerous, quite small split-granule to protruding bivalved pedicellariae; tubercles sometimes in rows on the arms; **superomarginal plates in the arm angle have granules in a triangular pattern with one or two larger granules at the apex and 1–5 mixed sized bivalved pedicellariae**; on the bare upper and lateral parts of the plate there are minute calcareous grains; the inferomarginals have numerous (up to 20) smallish bivalved pedicellariae with coarse granules around them; the actinal plates have similar granules with up to five pedicellariae of various sizes per plate; there are 6–7 furrow spines, 2–3 stout subambulacral spines and 2–3 smaller spines outside the subambulacral spines.

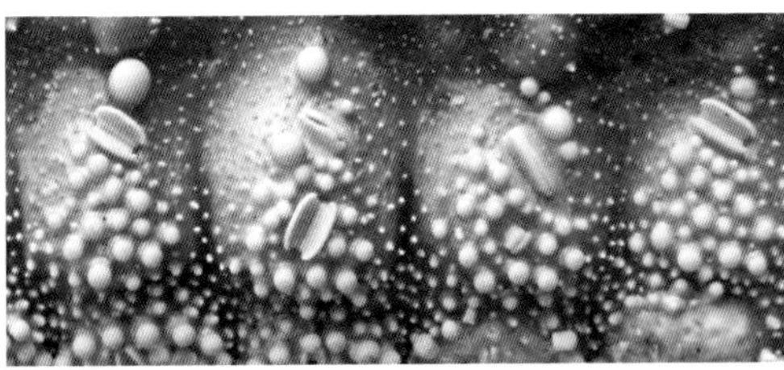

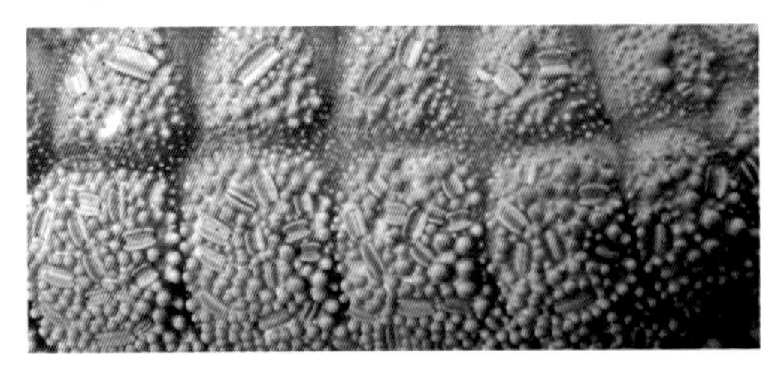

Western Australian Museum (WAMZ17962), superomarginal [middle] and inferomarginal characteristics [bottom].

MAXIMUM SIZE
R/r = 135/70 mm (R = 1.9–2.1r).

COLOUR
Variable, commonly purple, violet or orange.

HABITAT, DEPTH AND AUSTRALIAN DISTRIBUTION
On sand or muddy sand, 9–105 m. From off Lancelin (31°S) to Broome, WA. Western Australian endemic.

TYPE LOCALITY
Broome, WA.

FURTHER DISTRIBUTION
Australian endemic.

REFERENCES
Clark, H.L., 1938: 128; pl. 18 fig. 3; pl. 19 figs 2–3.

REGAL CUSHION STAR
Anthenea regalis
Koehler, 1910b

Australian Museum (AMJ18406).

DESCRIPTION

Stellate seastar with rounded arm ends; the abactinal plates are obscured by fairly thick skin, with small cylindrical tubercles 1.5–2 mm high and wide; the tubercles form 1–3 rows on each side of the carinal series on the arms and there is a slightly larger tubercle at the base of each carinal row; each interradial area is marked by a strip 12–15 mm long, bare of tubercles; between the tubercles are a few small granules and bivalved pedicellariae 1.2–1.5 mm long; **superomarginal plates are covered with coarse granules, unequal in size but never forming tubercles**, they are smaller and more widely spaced near the disc plates, sometimes bare near their inner margin, some have very small bivalved pedicellariae; inferomarginal plates are nearly twice as wide as the superomarginals interradially and are more than twice as wide as long, length and breadth are nearly equal near the arm end, they are densely covered by small granules and have one or two small bivalved pedicellariae; actinal plates have a very large bivalved pedicellaria up to 4 mm long, surrounded by large granules; pedicellariae and granules decrease in size distally and towards the interradial margin; there are five furrow spines, three subambulacral spines and 2–3 shorter spines outside the subambulacral spines.

MAXIMUM SIZE

R/r = 115/60 mm (R = 1.9r).

COLOUR

Not recorded.

HABITAT, DEPTH AND AUSTRALIAN DISTRIBUTION

Inner continental shelf, 30–55 m. The only Australian record is from off Port Hedland, WA.

TYPE LOCALITY

Coast of Ganjam, northern India.

FURTHER DISTRIBUTION

Southern India and Bay of Bengal, Sri Lanka, Myanmar, West Java and the Philippines.

REFERENCES

Clark, A.M. and Rowe, 1971: 52; pl. 7 figs 1–2.
Koehler, 1910b: 82; pl. 9 figs 1–2.

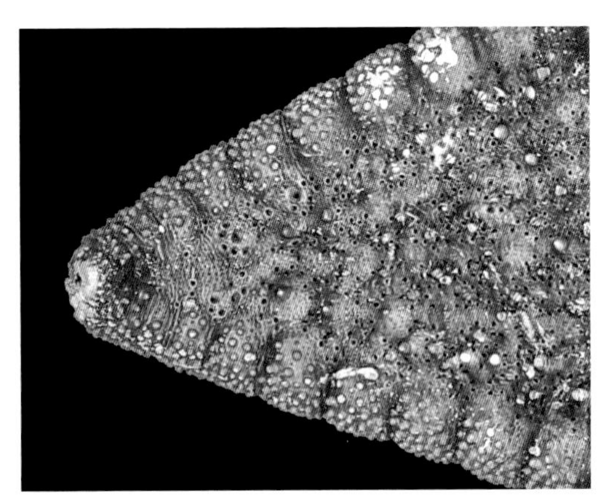

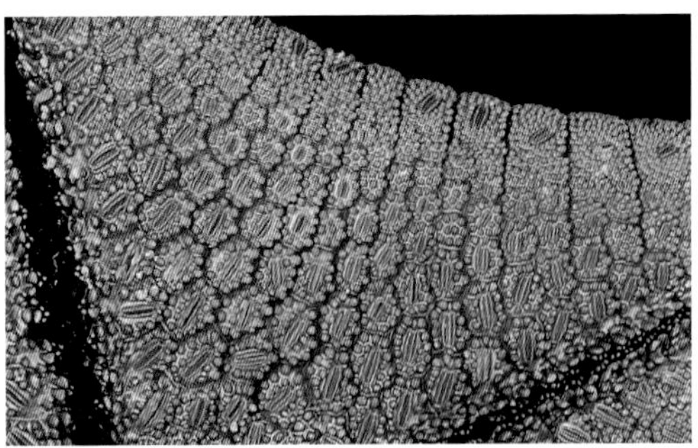

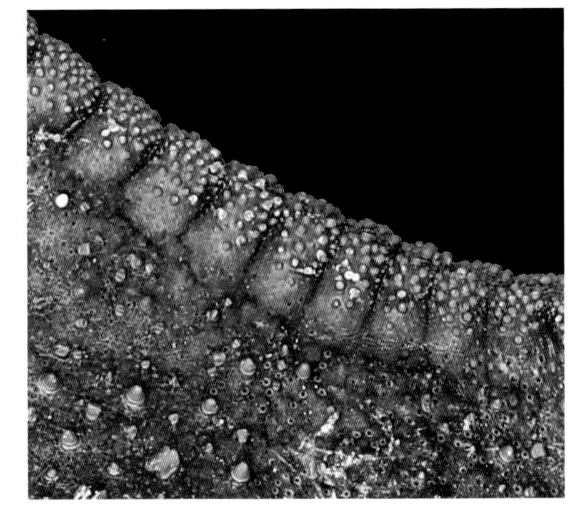

Abactinal arm tip [top], actinal and inferomarginal [bottom] and superomarginal plate characteristics [right], Australian Museum (AMJ18406).

SIBOGA CUSHION STAR

Anthenea sibogae

Döderlein, 1915

Western Australian Museum (WAMZ48445).

DESCRIPTION

Stellate seastar, arms not very long but moderately narrow, tapering to an acute tip; skin not completely obscuring the abactinal plates; abactinal tubercles conical, 1–1.6 mm wide and high, in fairly regular rows of which three reach the distal part of the arms; **plates near the arm tip are conspicuous, slightly raised and have groups of 6–20 granules**; between the disc tubercles are bivalved pedicellariae (up to 1.6 mm); **superomarginal plates slightly convex with a vertical band of moderately coarse granules, leaving bare margins to the plates**, distally granules cover more of the plate, smaller granules cover the lower part of the superomarginals, sometimes there is a small bivalved pedicellaria; inferomarginals covered with similar granules to those on the lower part of the superomarginals and 2–3

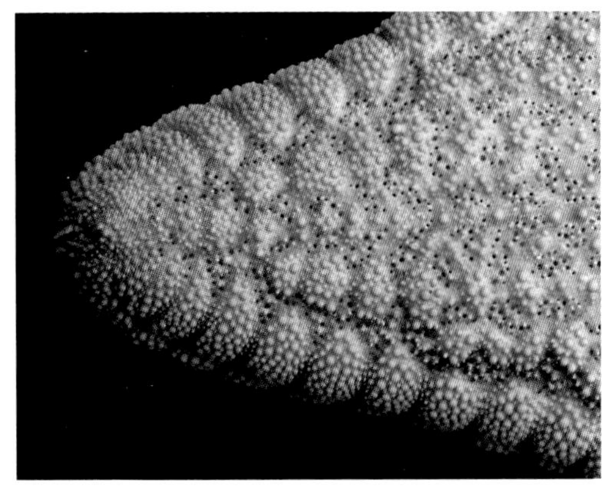

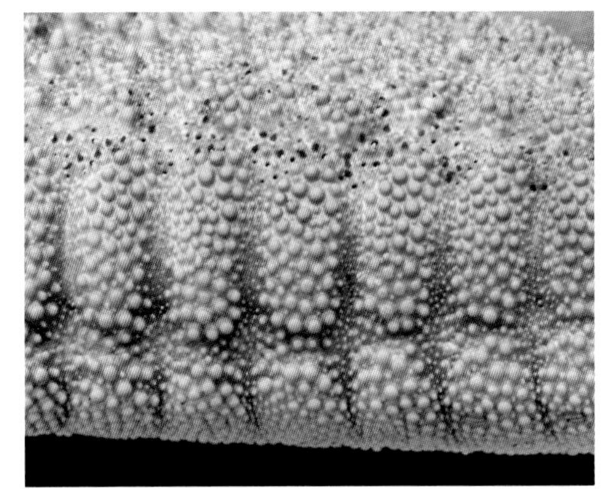

Distalmost abactinal [top], superomarginal and inferomarginal plate characteristics [bottom], Western Australian Museum (WAMZ13786).

pedicellariae (up to 1.5 mm long); actinal plates have very large pedicellariae (up to 3.5 mm, with largest near the furrow), surrounded by granules; there are 5–6 slender furrow spines, 2–3 thicker subambulacral spines and 2–3 shorter spines outside the subambulacral spines; there is usually a short upright pedicellaria next to each subambulacral spine.

MAXIMUM SIZE

R/r = 102/56 mm (R = 1.8r).

COLOUR

Not recorded.

HABITAT, DEPTH AND AUSTRALIAN DISTRIBUTION

Inner continental shelf on soft substrates, 9–50 m. Exmouth Gulf, WA to Albany Passage, Cape York, Queensland.

TYPE LOCALITY

North of Sumbawa, Indonesia.

FURTHER DISTRIBUTION

Indonesia.

REFERENCES

Döderlein, 1915: 47–48; pl. 10 fig. 5.

REMARKS

H.L. Clark (1946) suggested that *Anthenea mertoni, A. tuberculosa* and *A. viguieri* may be synonymous with *A. sibogae.*

SYDNEY CUSHION STAR

Anthenea sidneyensis

Döderlein, 1915

Port Hacking, New South Wales (J. Turnbull).

DESCRIPTION

Disc large; arms taper to an acute tip; similar to *A. crassa* but the abactinal tubercles are generally smaller; **granulation on superomarginals tapers to a point on the upper part of the plates but never forms a single vertical row, large bivalved pedicellariae are found on some superomarginal plates**; inferomarginals have very large bivalved pedicellariae, and fairly coarse rounded granules that obscure the boundaries of the plates and merge into the granulation of the actinal plates; the actinal plates have extremely large pedicellariae surrounded by granules that obscure the plate outlines; there are 5–6 moderately slender furrow spines followed by 1–2 coarse, rounded subambulacral spines and 2–3 slightly smaller spines outside the subambulacral spines.

MAXIMUM SIZE

R/r = 120/57 mm (R = 2.1r).

COLOUR

Mottled green or grey with rust coloured blotches and arm ends, the madreporite is orange.

FOOD

Probably feeds on substrate biofilm of microorganisms.

HABITAT, DEPTH AND AUSTRALIAN DISTRIBUTION

On sandy mud among rocks in bays and estuaries, 8–80 m. Lindeman Island, Queensland to Bass Strait. Eastern Australian endemic.

TYPE LOCALITY

Port Jackson, NSW.

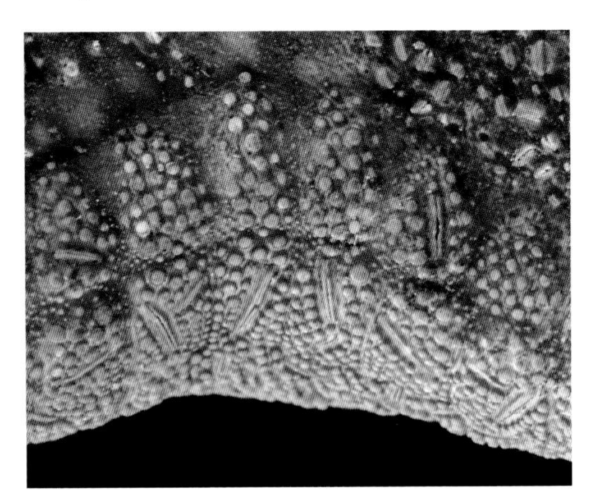

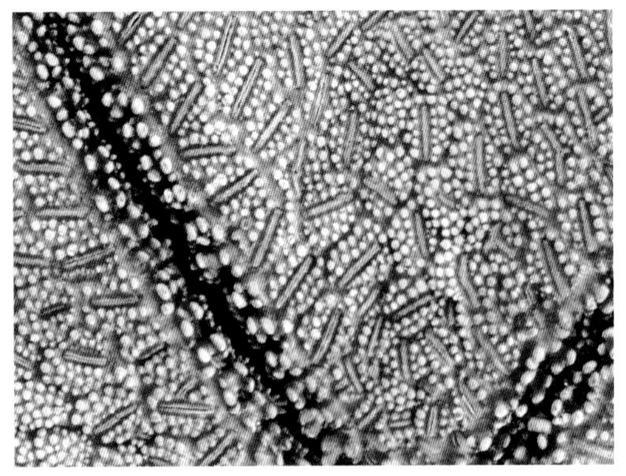

Superomarginal, inferomarginal [top] and actinal plate characteristics [bottom], Western Australian Museum (WAMZ88876).

FURTHER DISTRIBUTION

Australian endemic.

REFERENCES

Clark, A.M., 1970: 157; pl. 6 figs 1–2 (as *Anthenea edmondi*).

Clark, H.L., 1946: 102 (as *Anthenea acuta*).

Clark A. M. and Rowe, 1971: 51 (as *Anthenea edmondi*).

Rowe and Gates, 1995: 98.

Mah, 2019: as *Anthenea edmondi.*

REMARKS

A.M. Clark (1970) renamed *Anthenea acuta* (Perrier, 1869) as *A. edmondi* after finding the name was preoccupied. However Rowe (in Rowe and Gates, 1995) found there was a valid older name for the species (*A. sidneyensis* Döderlein, 1915). Mah (2019) uses *A. edmondi* for this species.

TUBERCULATE CUSHION STAR

Anthenea tuberculosa

Gray, 1847

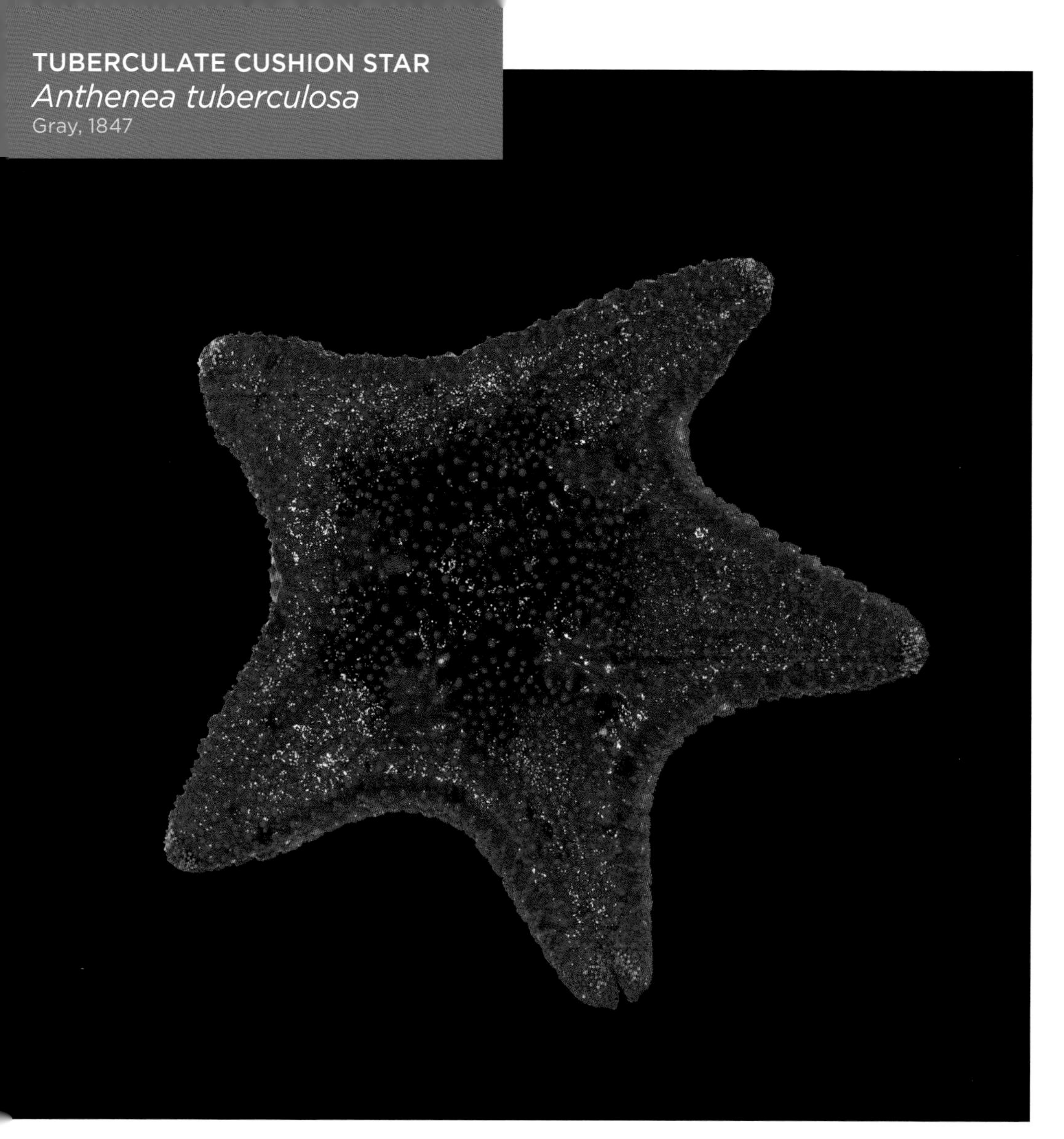

Ningaloo Reef, Western Australia (C. Bryce).

DESCRIPTION

Disc large; arms short and tapering to a rounded or slightly pointed tip; abactinal plates small, connected by elongate ossicles sometimes concealed by the skin; tubercles truncate to hemispherical (1–2 mm diameter), one per plate except near the arm ends where the lateral plates have 2–3, carinal tubercles slightly more prominent than those of the indistinct lateral rows; five rows of plates extend to near the arm tip; there are small scattered bivalved pedicellaria between the abactinal tubercles; **superomarginal plates are mostly aligned on the side of the body, not extending far onto the abactinal surface**, they are about 1.4 times wider than long in the arm angle and about one and a half times longer than wide distally; **upper parts of the superomarginals bare except for a large central granule, sometimes with a second one below it**, the lower half of the plate is covered with much smaller granules

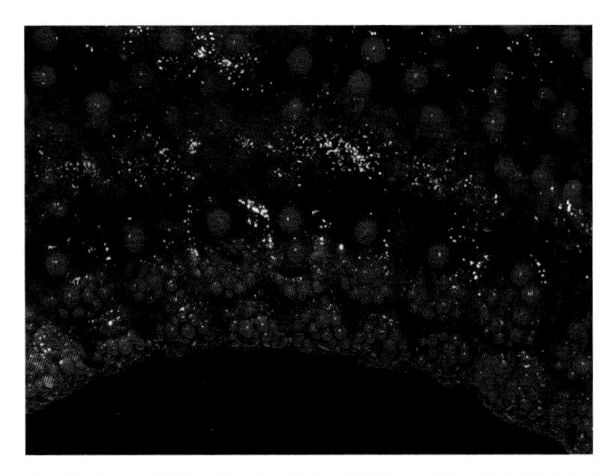

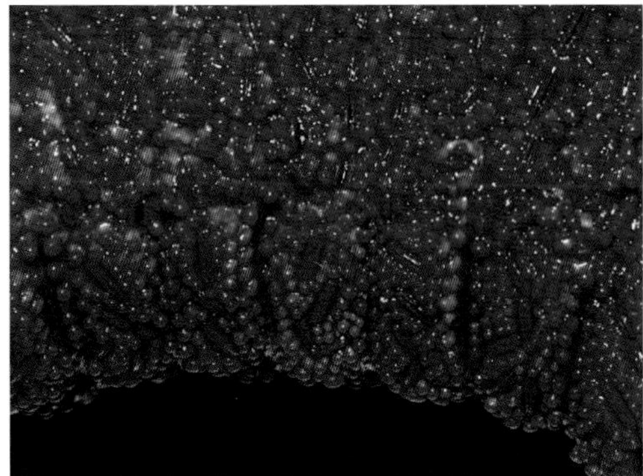

Superomarginal [top] and inferomarginal plate characteristics [bottom], Ningaloo Reef, Western Australia (C. Bryce).

forming a wedge-shaped pattern and there is often a small pedicellaria; the inferomarginal plates are closely granulated and have 2–6 pedicellariae (up to 2–3 mm); actinal plates have large pedicellariae (up to 3.5–4 mm) surrounded by granules; there are 4–5 slender furrow spines, two subambulacral spines and two (seldom three) thicker spines outside the subambulcral spines; there is a short upright pedicellaria next to each subambulcral spine.

MAXIMUM SIZE

R/r = 104/61 mm (R = 1.7r).

COLOUR

Orange-red with darker skin between tubercles in a specimen from Ningaloo.

HABITAT, DEPTH AND AUSTRALIAN DISTRIBUTION

Inner continental shelf on soft substrates, 11–82 m. North-western Australia to Lindeman Island, Queensland.

TYPE LOCALITY

Port Essington, NT.

FURTHER DISTRIBUTION

Indonesia (var. *vanstralaeni*).

REFERENCES

Clark, H.L., 1938: 122–123.

Clark, H.L., 1946: 104.

Döderlein, 1935: 106; pl. 27 figs 2, 2a.

Gray, 1866: 9; pl. 4 figs 1, 1a.

REMARKS

This is apparently a rare species but it has a valid name, being the first described of a group of related species, one or more of which may prove to be synonymous. It is closest to *Anthenea mertoni*.

LIVELY CUSHION STAR

Anthenea viguieri

Döderlein, 1915

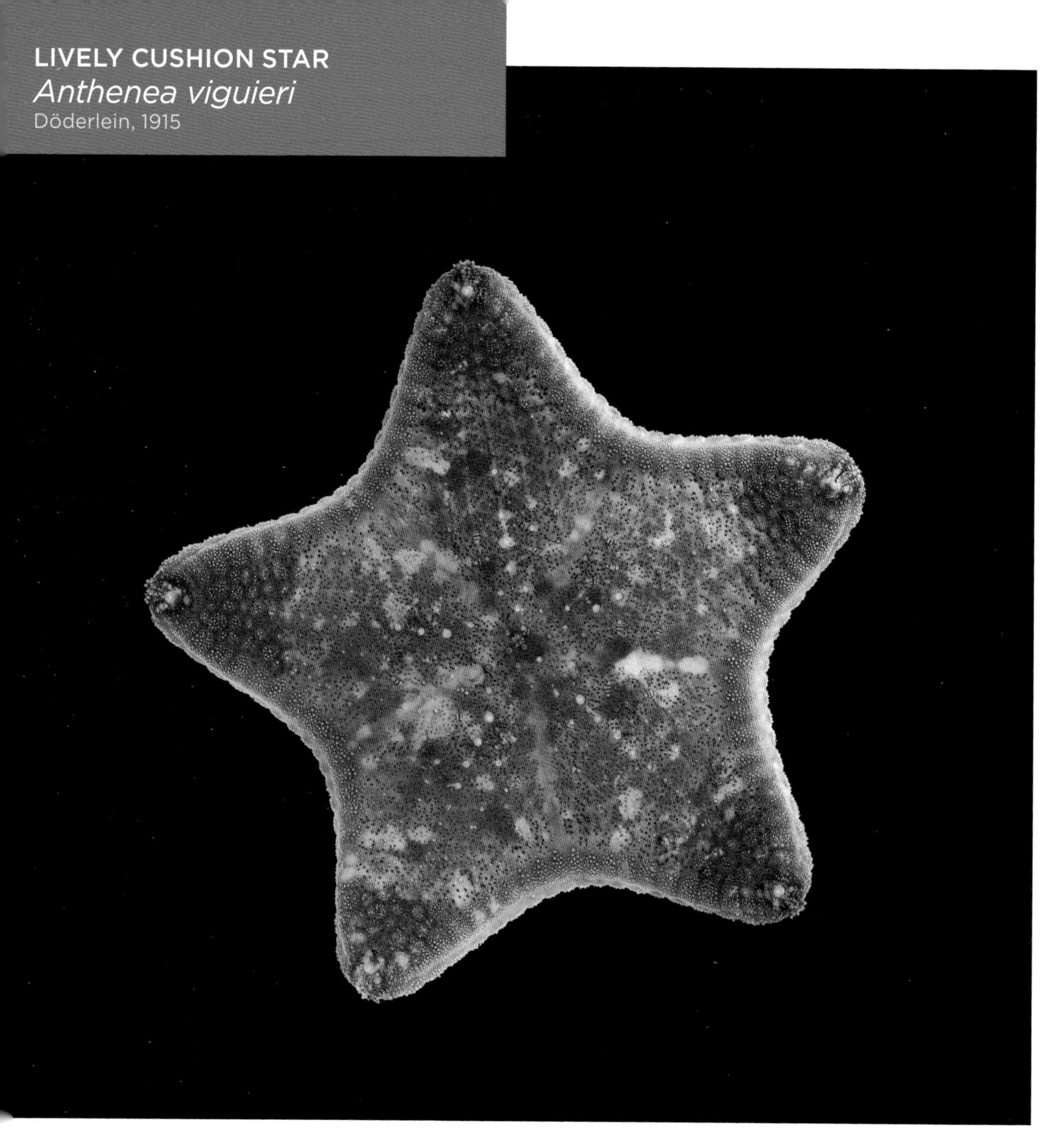

Exmouth Gulf, Western Australia (S. Morrison).

DESCRIPTION

Disc large; the arms are so short that the body is almost pentagonal; **abactinal surface bare**, skin partly conceals small plates with radiating connecting ossicles; tubercles (1 mm high and wide) broadly truncated, widely scattered, largest near the anus and on the first radial plate, usually absent on the outer part of the arms; small bivalved pedicellariae are fairly numerous, particularly in the papular areas; **the superomarginal plates are prominent, slightly more than twice as wide as long in the arm angle**, slightly less than that distally; **the upper margin of the plates is almost bare, the lateral margins are covered with small granules, the remainder by moderate sized granules** with occasionally a small pedicellaria; the lateral face of the inferomarginals has similar granules but the ventral surface of the inferomarginals has dense coarser granules and several pedicellariae (up to 1.5 mm); actinal plates have very large bivalved pedicellariae (up to 4 mm) with very few granules around them; there are five slender furrow spines, 3–4 thick subambulacral spines and 2–3 shorter spines outside the subambulcral spines; a short upright pedicellaria is often present next to each subambulacral spine.

MAXIMUM SIZE

R/r = 100/64 mm (R = 1.4–1.7r).

COLOUR

Mottled orange-red with pale blue patches.

HABITAT, DEPTH AND AUSTRALIAN DISTRIBUTION

Inner continental shelf on soft substrates, 8–60 m. Exmouth Gulf to Montague Sound, Kimberley, WA and Restoration Island, off Cape Weymonth, Queensland (12°38'S, 143°27'E).

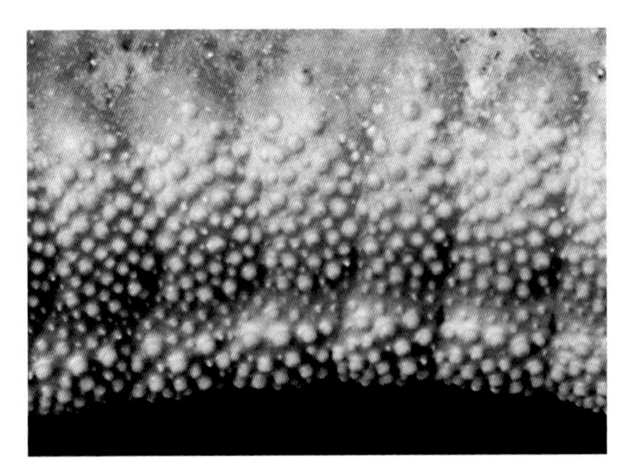

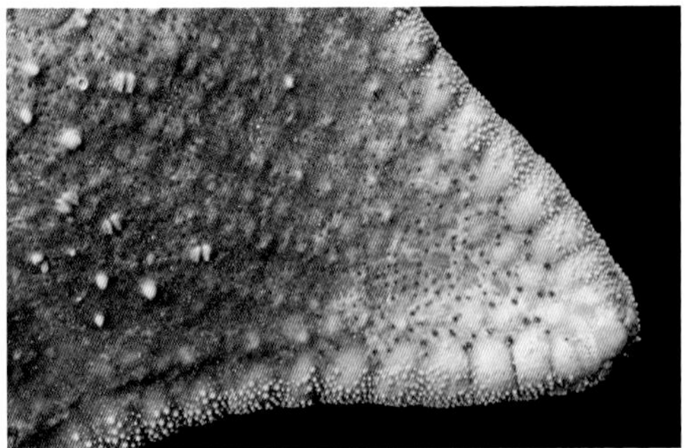

Superomarginal [top] and abactinal plate characteristics [bottom], Western Australian Museum (WAMZ13792).

TYPE LOCALITY

Ambon, Indonesia.

FURTHER DISTRIBUTION

Indonesia and southern China.

REFERENCES

Döderlein, 1915: 34–35; pl. 5 figs 2–3.

Liao and Clark, A.M., 1995: 102–103; pl. 11 fig. 1.

REMARKS

The Western Australian specimens identified as this species lack abactinal tubercles in large specimens but small tubercles are present in those with R < 80 mm. The abactinal pedicellariae are sparse, seldom with any on the superomarginal plates. Abactinal plates near the arm tip are enlarged and convex in large specimens, granule-covered in some, or with a small central tubercle in others.

PENTAGON CUSHION STAR
Bothriaster primigenius
Döderlein, 1916

Great Barrier Reef, Queensland (A. Hoggett).

DESCRIPTION
Similar in form to a small *Pentaceraster*, with fairly short arms; **disc raised, with a prominent central pentagon marked by a small tubercle or a group of enlarged granules on each primary radial plate**; carinal plates slightly more prominent than lateral plates; superomarginals wider than long interradially, more or less square (from above) until near the arm end where they are compressed; the carinal series reaches the terminal arm plate, one series each side of lateral plates extend to about the third last superomarginal; two further rows of lateral plates are confined to the arm angle; groups of up to 12 papulae lie between the abactinal plates; **there is a small deeply sunken area in each interradius**, half way between the primary pentagon and superomarginal plates which is unique to this species; there are five furrow spines and 2–3 subambulacral spines (*Pentacerastar regulus* of the same size have seven furrow spines).

MAXIMUM SIZE
R/r = 46/20 mm (R = 2.3r).

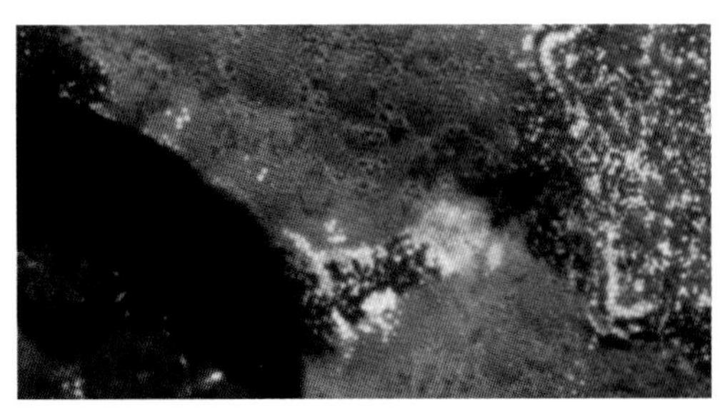

Abactinal interradial characteristics [top], Milne Bay, Papua New Guinea (R. Steene).

FURTHER DISTRIBUTION

Maldives, Indonesia, the Philippines, Papua New Guinea and the Solomon Islands.

REFERENCES

Coleman, 2007: 46.

Döderlein, 1916: 415.

Steene, 2014: 485, 491.

COLOUR

Highly variable, arms and radial areas of the disc may be orange, brown or dark green, interradial areas beige with a dark or light interradial spot, and superomarginals grey to brown, with some plates grey. In the Maldives a seastar with scarlet disc and arms, except for grey arm tips and a pale interradial spot has been recorded. All have a contrasting central pentagon on the disc.

FOOD

Scavenges.

HABITAT, DEPTH AND AUSTRALIAN DISTRIBUTION

On hard substrate on coral reefs, 2–45 m. In Australia it is only known from Broadhurst Reef off Townswille, Queensland.

TYPE LOCALITY

Timor, Indonesia.

REMARKS

This species has long been regarded to be of doubtful validity but recent specimens from Indonesia, Papua New Guinea, the Philippines and Australia pictured in Steene (2014) support Döderlein's opinion that this species differs from juvenile *Pentaceraster*.
The colour range of the species is highly variable, for example the disc centre may be brown and the outer disc and inner half of the arms sky blue, the arm bands brown, and beyond and outside this the arms are bright green, while the depressions on the disc margin are dark. The apical pentagon and the deeply sunken, dark interradial spot are unique to this species.

GRANULATED CUSHION STAR
Choriaster granulatus
Lütken, 1869

Scott Reef, Western Australia (C. Bryce).

DESCRIPTION

Arms stout, rubbery, arising from a strongly convex disc; **the abactinal surface is smooth, covered by a very finely granule-covered skin**, the actinal surface has a pavement of larger flat granules; there are 8–9 slender furrow spines and usually 3–4 (occasionally five) large flat, truncate subambulacral spines per plate; there may be small pincer-like pedicellariae near the furrow; papular pores are found only on the abactinal surface and, with the fine granulation, stop at the concealed superomarginal plates.

Cartier Island, Western Australia (C. Bryce).

MAXIMUM SIZE

R/r = 142/54 mm (R = 2.6r).

COLOUR

Dusty pink to brownish pink, rarely reddish with pale arm tips.

FOOD

Scavenges dead fish and other animals, may also feed on substrate biofilm of microorganisms.

REPRODUCTION

In Papua New Guinea, *Choriaster* spawns in April releasing eggs from the sides of the body wall (Coleman, personal communication). The larvae is planktotrophic.

HABITAT, DEPTH AND AUSTRALIAN DISTRIBUTION

Occurs on coral reefs, intertidal reef flats, lagoon slopes and outer slopes, in the open on sand, rubble or dead coral, 0–40 m. In Australia it is only found on the offshore atolls, Rowley Shoals, Scott and Ashmore Reefs off WA and on the Great Barrier Reef.

TYPE LOCALITY

Palau and Fiji.

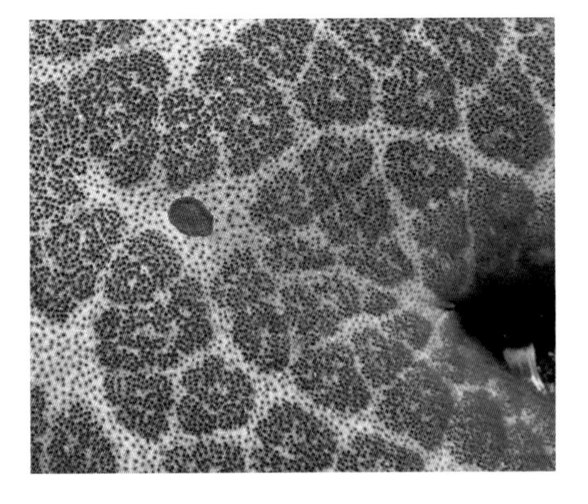

Granule-covered skin, Moalboal, the Philippines (H. Corneli).

FURTHER DISTRIBUTION

Widespread in the Indo-West Pacific, from east Africa and the Red Sea through Indonesia to Fiji, New Caledonia and the eastern Caroline Islands, and north to the Ryukyu Islands and southern China.

REFERENCES

Yamaguchi, 1975: 20–21 (food).

REMARKS

In Seychelles, some *Choriaster* have been observed harbouring pearlfish (Carapidae) in the coelom.

GENUS *CULCITA* L. Agassiz, 1836

Seastars of this genus have a massive pentagonal or almost circular body. The marginal plates are concealed by thickened skin, at least when $R > 60$ mm. Some small tubercles are often present on the upper side. Pore areas are usually rather irregular and sometimes indistinct or more or less continuous. The actinal granules are mostly coarse, often obscuring the limits of the plates. Juveniles have a flat pentagonal shape with prominent marginal plates and a tessellate skeleton similar to a goniasterid.

Type species: *Culcita schmideliana* (Bruzelius, 1805) (Indian Ocean)

Three species occur in the Indian and Pacific Oceans, two of which are found in shallow Australian waters; *C. novaeguineae* (page 400) and *C. schmideliana* (page 402).

Culcita novaeguineae, North West Cape, Western Australia (C. Bryce).

PINCUSHION STAR

Culcita novaeguineae

Müller and Troschel, 1842

Motupore Island, Papua New Guinea (S. Slack-Smith).

DESCRIPTION

A massive oreasterid, pentagonal to almost circular in outline, pincushion-shaped, abactinal and marginal plates concealed by a thick granule-covered skin; pore areas large and often confluent but there is a distinct pore free area on the lower part of the sides, near the margin; spines, **spinelets or tubercles are present in the pore areas** as well as on the reticulate skeleton; juveniles are flat, pentagonal, with prominent marginal plates and are covered with fine granules; they have bivalved pedicellariae on the abactinal and actinal surfaces; when R = ~35 mm the body tends to be convex and the tubercles become more prominent.

Bintan Island, Indonesia (C. Bryce).

MAXIMUM SIZE

R/r = 150/115 mm (R = 1.3r).

COLOUR

Variable, often cream, pale green or brown with patches of red-brown, dark green or dark brown corresponding with papular areas.

FOOD

Corals (such as *Acropora, Pocillopora* and *Porites*), echinoids and the substrate biofilm, everting its stomach and digesting the prey in situ.

REPRODUCTION

Sexes are separate, eggs small (190 µm), up to 7 million eggs may be spawned in one event. After fertilisation, a planktotrophic larva develops and swims near the surface for about 18 days before settling to the bottom on coral reef and metamorphosing into a juvenile seastar. It takes about two years from metamorphosis to attain the adult 'pincushion' form.

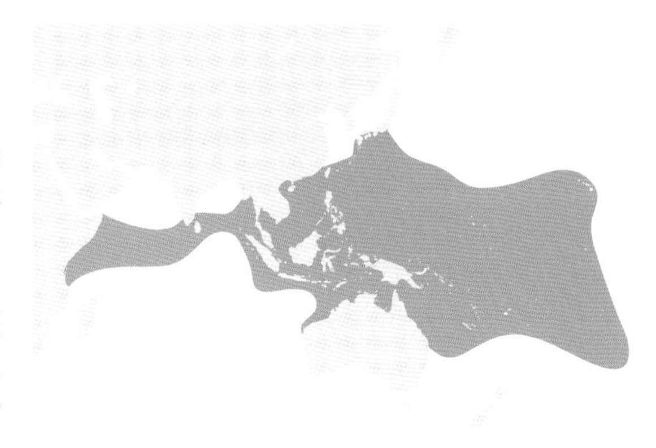

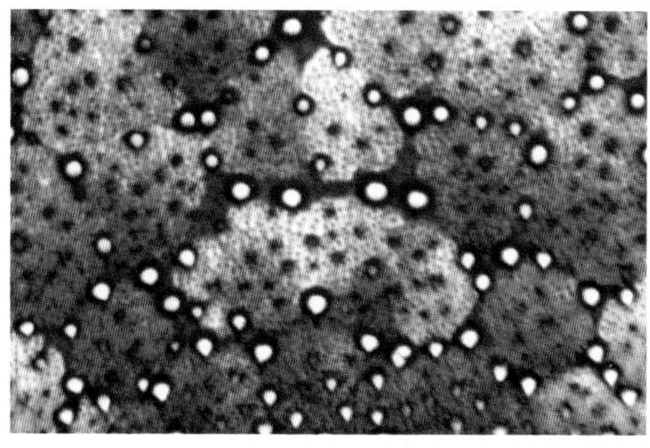

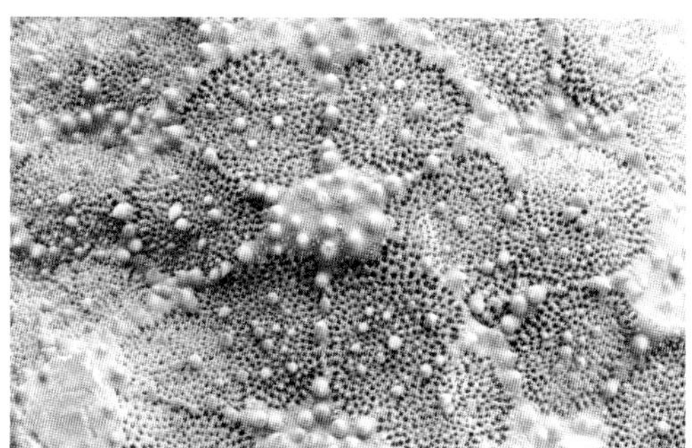

Abactinal tubercle characteristics, Scott Reef, Western Australia [top] (C. Bryce), Western Australian Museum [bottom] (WAMZ1888).

HABITAT, DEPTH AND AUSTRALIAN DISTRIBUTION

Coral reef flats and rubble bottoms, 0–30 m. Ningaloo, WA, Ashmore Reef, Timor Sea, to Lady Elliott Island, Queensland.

TYPE LOCALITY

Papua New Guinea.

FURTHER DISTRIBUTION

Widely distributed in the Indo-Pacific Ocean from east Africa, the Andaman Islands and South-East Asia to Taiwan, southern China, Japan, Hawaii, and south-eastern Polynesia to Pitcairn Island.

REFERENCES

Clark, H.L., 1946: 108.

Jangoux, 1974: 789–796.

Livingstone, 1932a: 265–270.

Yamaguchi, 1975: 20–21 (food).

Yamaguchi, 1976: 283–296 (reproduction).

REMARKS

Culcita novaeguineae have been observed harbouring pearlfish (Carapidae) in the coelom (Jangoux 1974).

SCHMID'S CUSHION STAR
Culcita schmideliana
(Bruzelius, 1805)

Dampier Archipelago, Western Australia (S. Morrison).

DESCRIPTION

Like *C. novaeguineae* this species has a distinct pore free area towards the lower side of the margins and extensive reticular areas on the upper side although pore areas may merge; **spines and/or tubercles are confined to the skeletal ridges between the pore areas, no spines or spinelets in the pore areas**; arms tend to be more evident than in *C. novaeguineae* and the tubercles more prominent; juveniles indistinguishable from those of *C. novaeguineae* are flat and *Tosia*-like.

MAXIMUM SIZE

R/r = 140/80 mm (R = 1.75r).

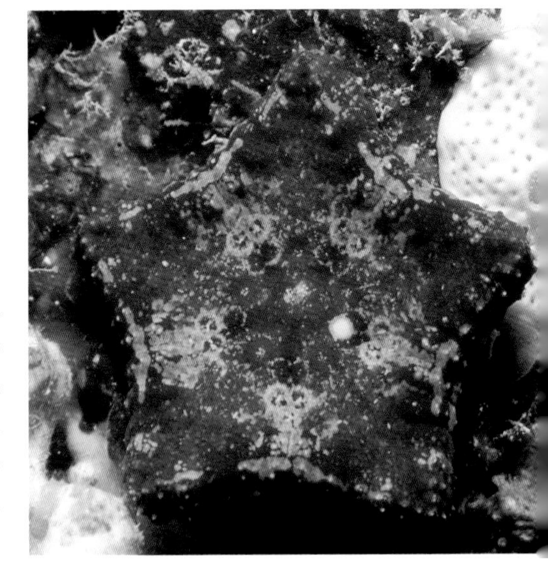

Juvenile, Dampier Archipelago, Western Australia (C. Bryce).

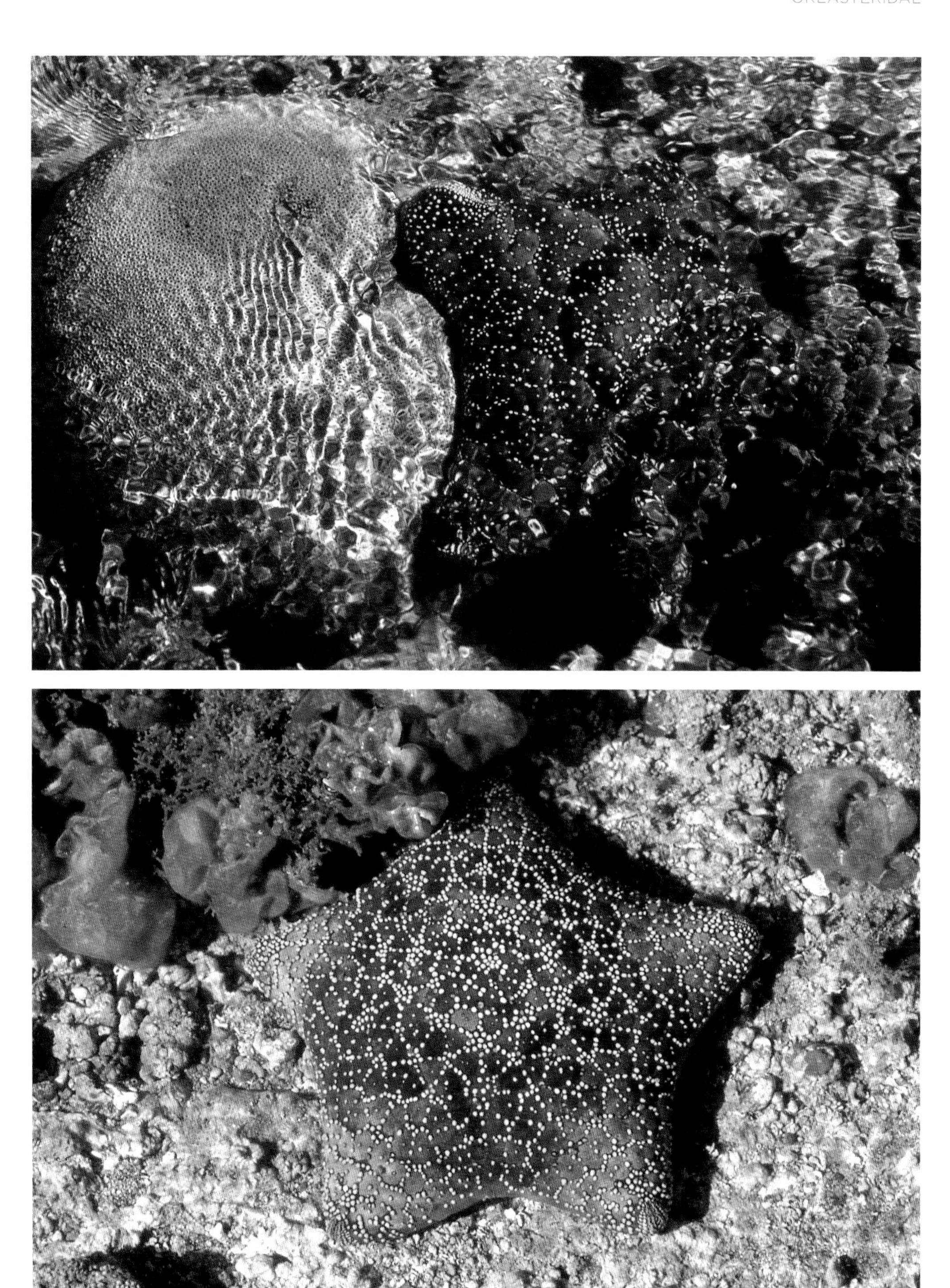

Encroaching on a colony of *Astreopora* [top],
Dampier Archipelago, Western Australia [bottom] (C. Bryce).

COLOUR

Variable, often light green or light brown with patches of darker colour or predominantly dark green or dark brown. Tubercles are usually contrasting, yellow, orange, red or purple.

FOOD

Hard and soft corals, echinoids, encrusting sponges and ascidians or the algal biofilm on reef, sand or seagrass substrate.

REPRODUCTION

In Mauritius, the gonads mature during spring and the planktotrophic larvae go through bipinnaria and brachiolaria stages, living for 38 days without metamorphosing.

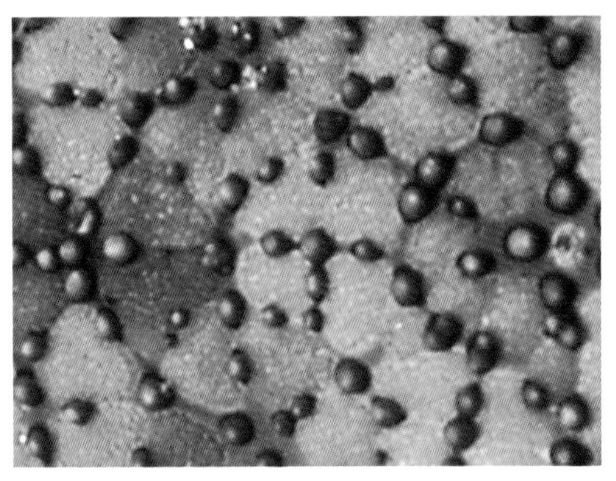

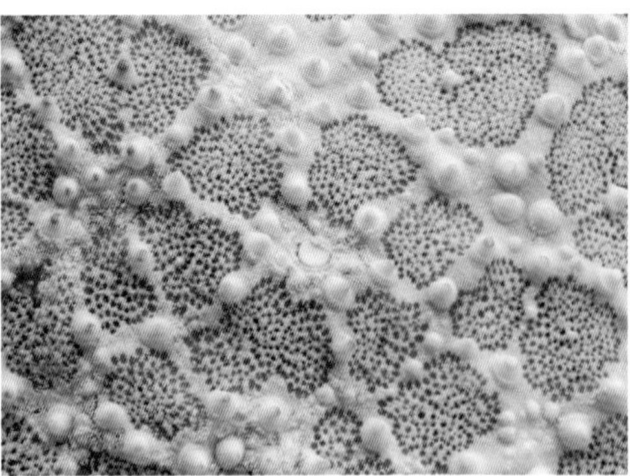

Abactinal tubercle characteristics, Dampier Archipelago, Western Australia [top] (C. Bryce), Western Australian Museum [bottom] (WAMZ99829).

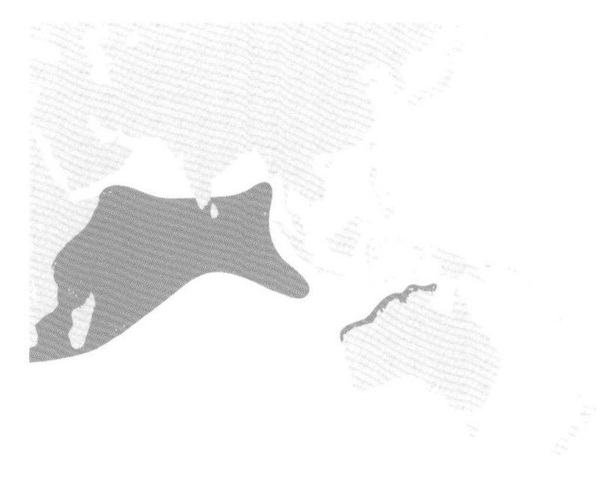

HABITAT, DEPTH AND AUSTRALIAN DISTRIBUTION

On coastal reefs and inshore islands, 0–92 m. Ningaloo Reef at Warroora, WA to Gove, NT and Cocos (Keeling) Islands.

TYPE LOCALITY

Not recorded.

FURTHER DISTRIBUTION

Indian Ocean, from east Africa to the Lakshadweep Islands.

REFERENCES

Mortenson, 1931: 30–31; pl. 3 fig. 6 (reproduction).

Sloan, 1980: 67 (food).

REMARKS

Culcita schmideliana have been observed harbouring pearlfish (Carapidae) in the coelom (Jangoux 1974).

GENUS *GONIODISCASTER* H.L. Clark, 1909b

Seastars of this genus are characterised by a flat or slightly convex disc with well developed arms (R = 1.9–3.4r), the tips of which are acute or rounded. The marginal plates are conspicuous and the abactinal plates are stellate, but not reticulate, and are in distinct longitudinal lines parallel to the carinal plates. The carinal plates are flat or slightly convex and lack spines except for a pentagon of five tubercles on the central disc in some species. There are no superomarginal spines and the abactinal plates are granule-covered with some coarse granules approaching small tubercles in size. **The pedicellariae are always small, bivalved, forceps-like or split-granule type**. Papular areas have 3–5 pores except near the arm ends where pores are absent; papulae are absent from the actinal surface. Furrow spines are followed by two or three rows of subambulacral spines and/or enlarged granules. **Large bivalved pedicellariae, typical of *Anthenea* are never found in *Goniodiscaster*.**

Type species: *Goniodiscaster pleyadella* (Lamarck, 1816) (Indonesia, Australia)
Like *Anthenea* some *Goniodiscaster* species are quite variable and several were described from single specimens. It is likely that some species will be synonymised when a range of specimens from northern Australia can be compared with those from the north-east and north-west. *Goniodiscaster* was formerly placed in the family Goniasteridae **but was transferred to the Oreasteridae because of the presence of a calcareous interbrachial septum** in contrast to the membranous septum found in goniasterids.

Ten species are recorded from Australia by Rowe and Gates (1995), of which seven are found in depths of 30 m or less. Excluded are *G. australiae* (considered to be a synonym of *G. rugosus* in this publication), *G. forficulatus* (deeper than 30 m) and *G. porosus* (from unknown depth). Döderlein (1936) noted that 'the species of *Goniodiscaster* are probably all very variable. The differentiation and division into species is very doubtful'. Therefore the following key should be used in conjunction with the species descriptions and illustrations.

GENUS *GYMNANTHENEA* H.L. Clark, 1838

Type species: *Gymnanthenea globigera* (Döderlein, 1915) (Australia)
One species is found in shallow Australian waters. Diagnosis as for the species (page 419).

GENUS *HALITYLE* Fisher, 1913

Type species: *Halityle regularis* Fisher, 1913 (Indo-West Pacific)
Monotypic genus. Diagnosis as for the species (page 422).

KEY TO THE SPECIES OF *GONIODISCASTER* IN SHALLOW AUSTRALIAN WATERS

1	Arms taper little to tip.	**2**
	Arms taper evenly to a blunt or acute tip.	**3**
2	Arms tips rounded; abactinal granulation variable.	***G. acanthodes*** page 408
	Arms tips rounded; abactinal granulation even.	***G. pleyadella*** page 415
3	Arms taper evenly to a blunt tip.	**4**
	Arms taper evenly to an acute tip.	**6**
4	Abactinal plates very convex, granulation evenly coarse or with a central group of enlarged granules to small tubercles; south-western Australia.	***G. seriatus*** page 418
	Abactinal plates flat with fine, even granulation.	**5**

Goniodiscaster seriatus, Marmion Lagoon, Western Australia (G. Edgar).

Goniodiscaster pleyadella, Port Hedland, Western Australia (G. Edgar).

5	Small pedicellariae abundant on actinal surface.	***G. foraminatus*** page 410
	Small pedicellariae absent or rare, usually on small tubercles on the inner carinal and adjacent disc plates.	***G. granuliferus*** page 412
6	Abactinal granules fairly even, slightly larger in centre of plates; a group of spine-like granules with bare conical tips on each primary radial plate; R = 2.3–2.7r.	***G. integer*** page 413
	Abactinal granules of mixed sizes with some emergent taller granules or small tubercles, or granules rounded, slightly spaced; R = 2.4–3.4r.	***G. rugosus*** page 416

RIB-EDGED SEASTAR

Goniodiscaster acanthodes

H.L. Clark, 1938

North West Shelf, Western Australia (L. Marsh).

DESCRIPTION

Disc large; arms fairly short; beyond the disc the arms barely taper until the end which is rounded and may be widened; this shape varies little among all specimens examined; the abactinal surface is covered in a dense coat of granules which may be smooth or evenly rounded (but smaller in the papular areas), or 1–8 central granules on some plates may be enlarged or become 1–5 small tubercles; **the five primary radial plates often have prominent conical tubercles, a central enlarged granule or small tubercle**, sometimes on only 2–3 arms; superomarginal plates in interradial area are more than twice as wide as long, distally they are slightly wider than long, they usually have close-set, convex polygonal granules, occasionally with one or more small tubercles at the upper end of the plate; 1–2 superomarginals near end of arm meet in the midline, inferomarginals with similar, slightly larger granules, no tubercles; actinal plates have larger, rounded, less closely packed granules; furrow spines 5–8, in a graduated fan, subambulacral spines paired broad, flat, round-ended, proximal one usually much larger than the distal one, then an outer series of 3–4 enlarged often prismatic granules; papulae are in small groups (~4–7) on abactinal surface only; few or many pedicellariae, small pincer-like only on abactinal plates, elongate split-granule type on actinal and marginal plates, and an upright bivalved pedicellaria next to each subambulacral spine.

MAXIMUM SIZE

R/r = 90/38 mm (R = 1.9–2.3r, 1.7–2.3br).

COLOUR

Variable, usually red and cream, blue-grey or grey and the actinal surface yellowish. The outer part of the arms, or just the marginal plates, interradial lines, carinal plates and all the tubercles or enlarged granules, may be red.

HABITAT, DEPTH AND AUSTRALIAN DISTRIBUTION

On muddy sand often with coral or rubble and sponges, 0–46 m. Houtman Abrolhos, WA to the Gulf of Carpentaria, Queensland. Northern Australian endemic.

TYPE LOCALITY

South-west of Broome, WA.

FURTHER DISTRIBUTION

Australian endemic.

Abactinal plate characteristics, North West Shelf, Western Australia (L. Marsh).

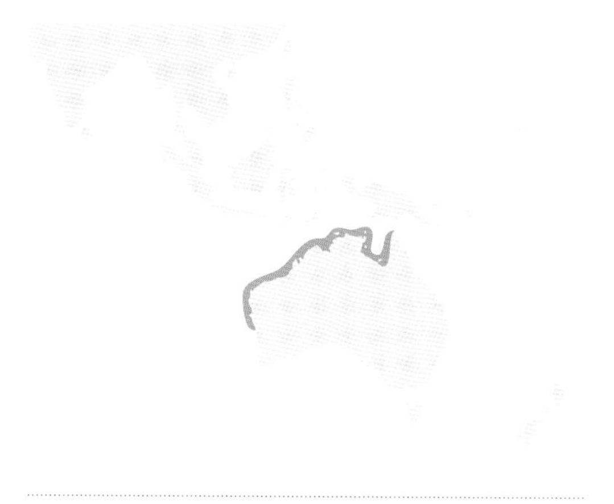

REFERENCES

Clark, H.L., 1938: 84; pl. 5 fig. 2 (as *Goniodiscaster acanthodes* and *G. bicolor*).

Rowe and Gates, 1995: 99–100 (synonymy of *Goniodiscaster bicolor* with *G. acanthodes*).

REMARKS

Clark and Rowe (1971) discuss the variability of specimens referred to *Goniodiscaster pleyadella, G. acanthodes, G. bicolor* and three other species, which in the authors' opinion are not closely related and will not be discussed here. Characters, such as primary radial tubercles, and the nature of the abactinal tubercles are highly variable within each of the nominal species. Examination of a large series of specimens from north-western Australia shows a complete range of these characters while the shape of the arms, marginal and actinal granulation and furrow armature shows little variation. *G. acanthodes* may prove to be a junior synonym of *G. pleyadella* (Lamarck, 1816); more specimens from the NT may clarify the relationship between *G. acanthodes* and *G. pleyadella.*

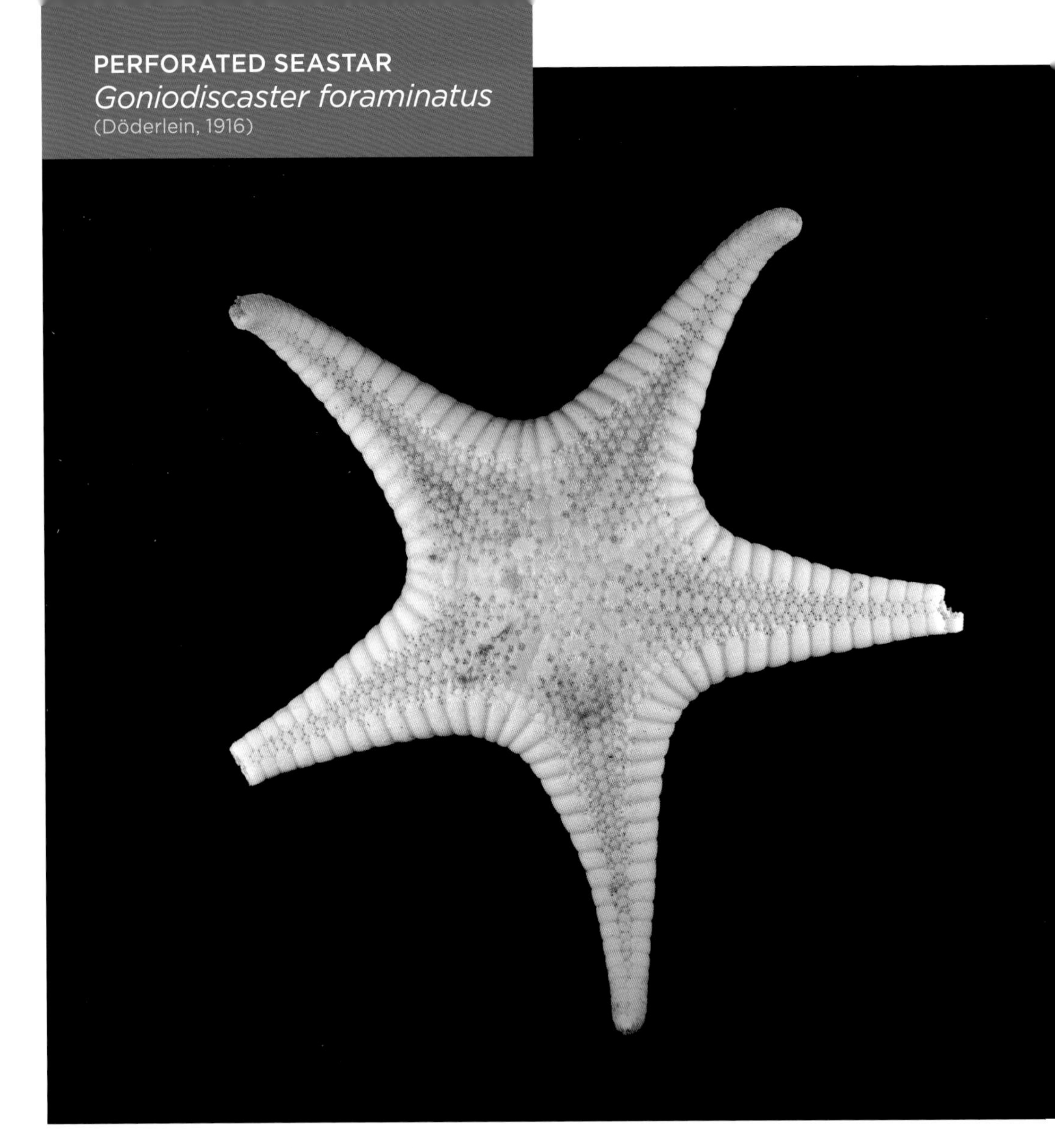

PERFORATED SEASTAR

Goniodiscaster foraminatus

(Döderlein, 1916)

Museums Victoria (NMVF240219).

DESCRIPTION

The disc is large; arms taper evenly to near the arm ends where they taper fairly abruptly immediately before the tip (this character is variable); the madreporite is very large, 2–3 times as large as the largest abactinal plates; the proximal radial plates are much broader than long, the plates adjoining the interradial line are also enlarged; **the abactinal and marginal plates are evenly covered by fairly fine granules**, between the abactinal plates the granules are very fine; papular areas between the angles of the plates are fairly large and covered by fine granules; pedicellariae are very numerous over the actinal plates; granulation on the actinal plates is slightly coarser than that on the abactinal plates but there are no enlarged granules or tubercles on any plate; the adambulacral armature is not described but from Döderlein's figure 4–6 furrow spines can be distinguished, followed by three subambulacral spines and 3–4 enlarged granules on the outer part of the plate.

MAXIMUM SIZE

R/r = 69/31 mm (R = 2.2r).

COLOUR

Not recorded, but the figure in Döderlein (1916) shows the plates outlined in a very dark colour.

HABITAT, DEPTH AND AUSTRALIAN DISTRIBUTION

Probably on a soft substrate, 12–20 m. Specimens are only recorded from Shark Bay and Joseph Bonaparte Gulf, NT.

TYPE LOCALITY

Shark Bay, WA (may be erroneous).

FURTHER DISTRIBUTION

Australian endemic.

REFERENCES

Döderlein, 1916: 414–415; fig. a. (as *Goniodiscus foraminatus*).

Döderlein, 1935: 77 (key), 79–80; pl. 20 fig. 6; pl. 21 fig. 5.

Clark, A.M. and Rowe, 1971: 50 (key).

Actinal plate characteristics, Museums Victoria (NMVF240219).

REMARKS

No specimens collected from Shark Bay since the type material match Döderlein's description. However, specimens from the northern part of the west coast continental shelf and off Ningaloo Reef (from 80–104 m) are close to the description apart from the presence of crystal bodies between the larger granules on the plates (not mentioned by Döderlein). This species seems to be distinguished by its smooth granulation and the presence of numerous pedicellariae. Döderlein (1916) describes the arm shape as resembling that of *Goniodiscaster pleyadella* but his figure shows more evenly tapered arms than that species or *G. acanthodes*, which has few pedicellariae. *G. foraminatus* was described with the 'Siboga' material from Indonesia so it is possible that labels were mixed and Döderlein's specimen may not have been from Australia or from shallow water.

GRANULAR SEASTAR
Goniodiscaster granuliferus
(Gray, 1847)

DESCRIPTION

The form is stellate with a large disc and arms tapering to an acute tip; abactinal plates form a carinal row of regular hexagonal plates on each arm, the remainder are irregularly arranged polygonal plates covered by a finely granulated thin skin**; denuded plates have embedded crystal bodies**; there is a small tubercle on the inner carinal and adjacent disc plates (absent in Western Australian specimens); the superomarginal plates are much wider than long interradially, becoming narrower distally but not meeting near the arm ends; inferomarginals are similar, both series are finely granulated; actinal plates have coarser granules; pore spaces, with 5–7 pores in groups or in a line of single pores around plates, are confined to the abactinal surface; a few minute bivalved pedicellariae occur on the abactinal surface and the marginals and slightly larger ones are present on the actinal plates; there are 5–8 furrow spines and three uneven subambulacral spines and 2–3 shorter spines and enlarged granules outside the subambulacral spines; there is often an upright two-bladed pedicellaria between the furrow and subambulacral spines.

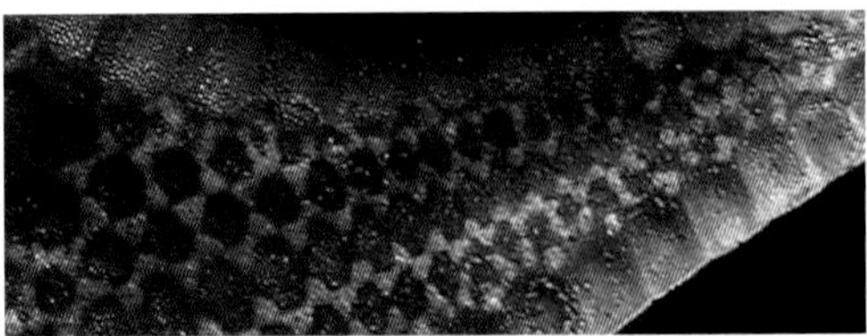

North West Shelf, Western Australia, abactinal plate characteristics [bottom] (L. Marsh).

MAXIMUM SIZE

R/r = 83/39 mm (R = 2.1r).

COLOUR

Beige to red.

HABITAT, DEPTH AND AUSTRALIAN DISTRIBUTION

Continental shelf, ~10–180 m. From off Jurien Bay to south of Rowley Shoals, WA and off Weipa, Queensland.

TYPE LOCALITY

Not recorded.

FURTHER DISTRIBUTION

Indonesia and Taiwan Strait.

REFERENCES

Clark, A.M. and Rowe, 1971: 49.

REMARKS

Another species, *Goniodiscaster forficulatus* (Perrier, 1875), has been recorded from 30–144 m off north-western Australia.

INTEGER SEASTAR
Goniodiscaster integer
Livingstone, 1931

Holotype, Australian Museum (AMJ5499).

DESCRIPTION

Stellate seastar, disc moderately elevated; arms taper gradually and evenly to a fairly sharp tip; interradial arcs rounded; papular areas confined to the abactinal surface, 3–9 pores in each papular area; **1–6 spine-like granules with bare conical tips mark each primary radial plate more or less prominently; carinal series distinct with plates broader than the lateral areas**; only the carinal series extends nearly to the arm tip, to the third and fourth last superomarginal plate; the first lateral row ends at the fifth last superomarginal while the third row extends only to about the third superomarginal in the arm angle, there may be one or two plates of a fourth series in the arm angle; the abactinal granulation is coarse and uneven, the larger granules pointed; granules in the papular areas are even-sized and smaller; the 15 superomarginals have fine smooth granulation except near the arm tip where some granules are larger; the inferomarginals, which project beyond the superomarginals in the arm angle, have similar granulation with a few split-granule pedicellariae and larger tubercle-like granules near the arm tips; actinal plates have larger granules and split-granule pedicellariae; there are six slender furrow spines, 2–3 large nearly equal subambulacral spines followed by 2–3 smaller spines or enlarged granules, and a fourth row of spine-like granules.

MAXIMUM SIZE

R/r = 99/37 mm (R = 2.3–2.7r).

COLOUR

Variable, the holotype had a sage green abactinal surface with granules of pentagon dark green, margins slate and the actinal surface brown with cream arms.

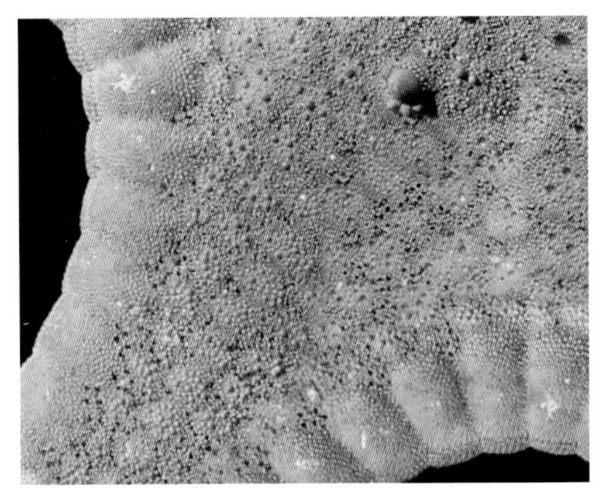

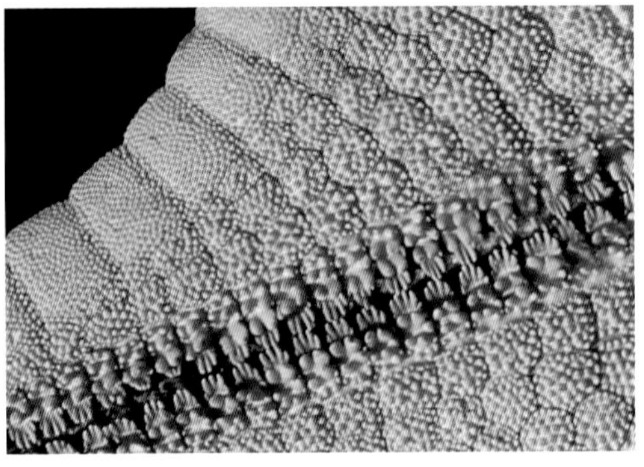

Abactinal [top] and actinal plate characteristics [bottom], holotype, Australian Museum (AMJ5499).

FOOD

The stomach is everted on compound ascidians and epibenthic biofilm, and it also scavenges damaged echinoids, for example *Breynia* (Birtles, personal communication).

HABITAT, DEPTH AND AUSTRALIAN DISTRIBUTION

On sand and rubble substrate, 0–22 m. From off Lizard Island, to Moreton Bay, Queensland. North-eastern Australian endemic.

TYPE LOCALITY

Port Curtis, Queensland.

FURTHER DISTRIBUTION

Australian endemic.

REFERENCES

Livingstone, 1931: 135; pl. 17 figs 1–2; pl. 18 fig. 34; pl. 19 figs 1–2.

REMARKS

Livingstone (1931) distinguished *Goniodiscaster integer* from *G. pleyadella* by its longer, tapering arms and coarse and uneven abactinal granulation, with granules larger on the plates than those in the papular spaces.

PLEYADELLA SEASTAR
Goniodiscaster pleyadella
(Lamarck, 1816)

DESCRIPTION
Disc large; arms fairly short, beyond the disc they barely taper until the ends, which are rounded; the primary radial plates are marked by a small tubercle at the head of a carinal series of plates, which are broader than the lateral plates; the carinals extend to the second last superomarginal plate; the first lateral series extends to the fourth last superomarginal, the third series extends from the arm angle to the fourth superomarginal plate, and two more radial series are confined to a few plates in the arm angle; the abactinal granulation is fine and even, slightly finer on the marginal plates; split-granule pedicellariae are found on both series of marginal plates and on the actinal plates; granules on the actinal plates are coarser than those on the abactinal surface; there are 5–6 slender furrow spines, and 2–3 unequal, stout subambulacral spines with enlarged granules on the outer part of the plate.

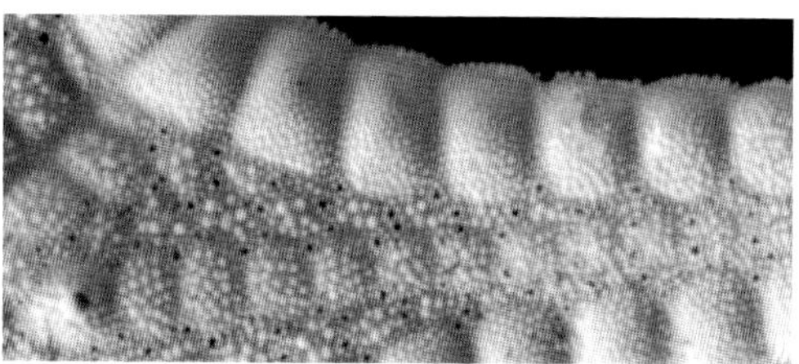

Western Australian Museum (WAMZ299830), abactinal plate characteristics [bottom].

MAXIMUM SIZE
R/r = 60/28 mm (R = 2.1r).

COLOUR
Not recorded.

HABITAT, DEPTH AND AUSTRALIAN DISTRIBUTION
Inner continental shelf, 0–37 m. North West Cape, WA to Tin Can Bay, Queensland.

TYPE LOCALITY
'Australian Seas'. Possibly Timor.

FURTHER DISTRIBUTION
Indonesia.

REFERENCES
Bell, 1884: 129 (as *Pentagonaster validus*).
Clark, H.L., 1921: 29.
Clark, H.L., 1946: 90–91.
Clark, A.M. and Rowe, 1971: 50.
Döderlein, 1896: 308; pl. 18.
Livingstone, 1931: 136–137; pl. 19 figs 3–4.

REMARKS
This species differs from *Goniodiscaster integer* in having shorter, more round-ended arms and finer, more even abactinal granulation. It is possible that further study will show that *G. pleyadella* intergrades with *G. acanthodes* in northern Australia.

WRINKLED SEASTAR
Goniodiscaster rugosus
(Perrier, 1875)

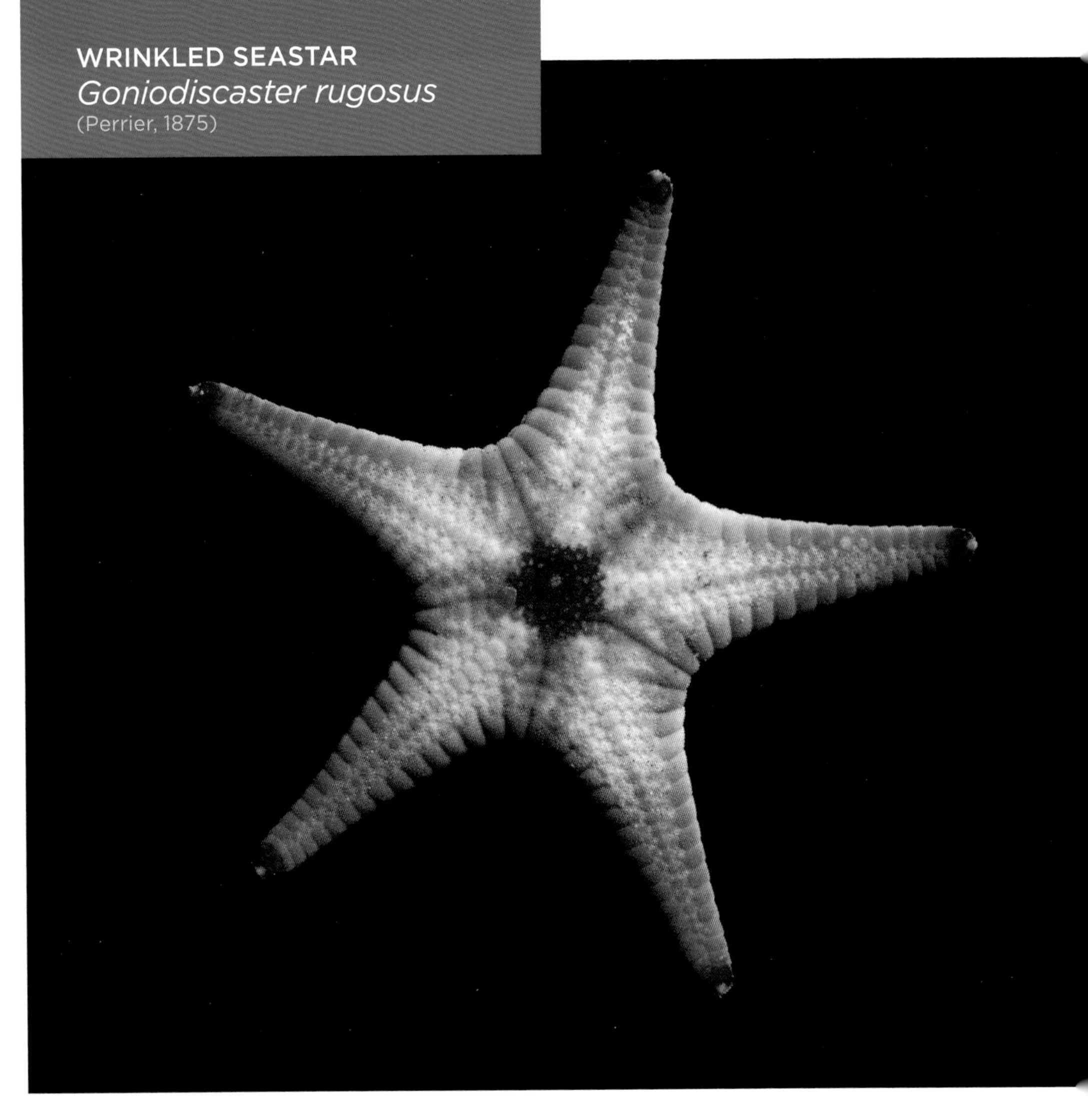

North West Shelf, Western Australia (L. Marsh).

DESCRIPTION

Disc of moderate size; arms taper evenly to a fairly narrow tip; abactinal surface covered with fairly even-sized, slightly spaced rounded granules or with granules of mixed sizes (3–5 per linear mm) with some emergent, taller granules, or in some cases small tubercles (1–4 per plate); there are a few minute, erect, pincer-like pedicellariae; superomarginal plates with close set, convex, rounded to polygonal granules, usually no tubercles but occasionally a vertical row of small tubercles on distal plates; 3–4 superomarginals are in contact at the arm tip; inferomarginals match superomarginals in number and granulation; actinal plates have coarser, spaced, rounded to polygonal granules with a few split-granule pedicellariae amongst them; there are 5–7 (up to nine) furrow spines, radiating, subequal except for a very small one at each end; 3–4 short, broad, truncate subambulacral spines and 2–3 similar spines and enlarged granules outside the subambulacral spines; elongate forceps-like pedicellariae may be near the subambulacral spines.

MAXIMUM SIZE

R/r = 95/40 mm (R = 2.4–3.4r, 3.4–5.2br).

COLOUR

Variable, for example grey or green with black or grey arm tips; grey with green interradial areas; or deep rose pink disc and marginal plates, arm tips and disc centre maroon with the rest of abactinal surface and interradial area cream.

HABITAT, DEPTH AND AUSTRALIAN DISTRIBUTION

Continental shelf and slope on silty sand, rubble or gravel substrate, 9–391 m. Exmouth Gulf, WA to Port Curtis, Queensland. Northern Australian endemic.

TYPE LOCALITY

Not recorded (*G. rugosus*). 'Western Australia' (*G. australiae*), Port Curtis, and Prince of Wales Channel, Queensland (*G. coppingeri*).

REMARKS

Comparison of a number of specimens with descriptions of the three species *Goniodiscaster rugosus* (Perrier, 1875), *G. coppingeri* (Bell, 1884) and *G. australiae* Tortonese,1937 supports the suggestion by Clark and Rowe (1971) that these names refer to a single species that varies within a population in the degree of roughness of the granulation and in the R/r ratio, but there is also some tendency for Queensland specimens to have slightly narrower arms than those from north-western Australia. Rowe (in Rowe and Gates, 1995) includes *G. coppingeri* in the synonymy of *G. rugosus* but retains *G. australiae* as a separate species. Mah (2019) recognises both *G. rugosus* and *G. australiae*.

FURTHER DISTRIBUTION

Australian endemic.

REFERENCES

Bell, 1884: 128 (as *Pentagonaster coppingeri*).
Clark, A.M. and Rowe, 1971: 50; pl. 5 fig. 7.
Clark, H.L., 1938: 80; pl. 8 fig. 2 (as *Goniodiscaster australiae*).
Rowe and Gates, 1995: 101
Tortonese, 1937: 1–5; fig. 1; pl. 1 (as *Goniodiscaster australiae).*
Mah, 2019: as *Goniodiscaster australiae* and *G. rugosus.*

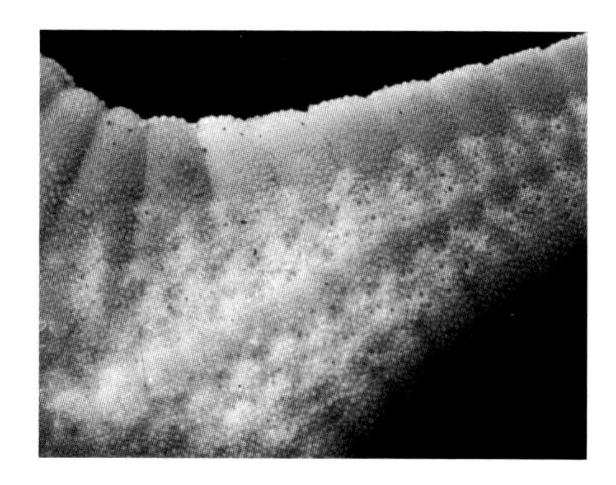

Abactinal plate characteristics, North West Shelf, Western Australia (L. Marsh).

SERIATUS SEASTAR
Goniodiscaster seriatus
(Müller and Troschel, 1843)

DESCRIPTION
Stellate seastar with arms tapering to a rounded tip; **abactinal plates very convex**, superomarginals moderately convex, abactinal plates appear quite smooth when covered with an even coat of fairly coarse granules, or rough when usually one, sometimes a group of central granules are enlarged, with the largest a small tubercle 1 mm in diameter; pore areas between the plates have ~12 pores; pedicellariae of the split-granule type are often numerous on abactinal and actinal surfaces, elongate pedicellariae with bent valves may also be present, near the ambulacral furrow there is often an upright pedicellaria; there are 5–8 furrow spines, 2–3 subambulacral spines, and 3–5 enlarged granules on each adambulacral plate.

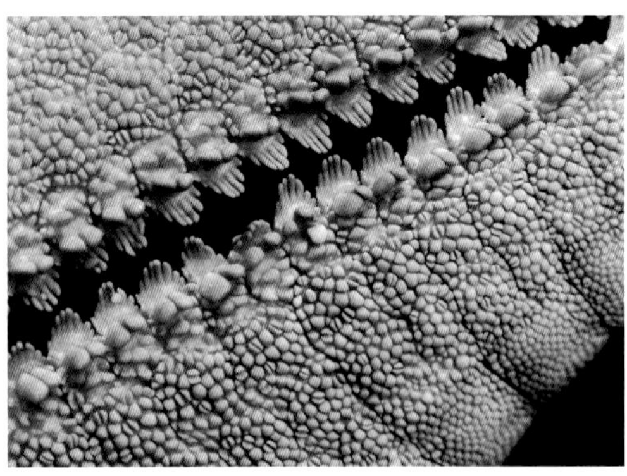

Shark Bay, Western Australia [top] (L. Marsh); actinal plate characteristics [bottom], Western Australian Museum (WAMZ31867).

MAXIMUM SIZE
R/r = 75/31 mm (R = 2.4r).

COLOUR
Often grey with a reddish cast and a pattern of red or tan plates, some are all red to tan, patterned in light and dark shades. Actinal surface pale cream to brown.

HABITAT, DEPTH AND AUSTRALIAN DISTRIBUTION
Sheltered waters among seagrass, rocks and algae and near jetty piles, 0–36 m; common. Cape Leeuwin to Shark Bay, WA. Western Australian endemic.

TYPE LOCALITY
'South-western Australia'.

FURTHER DISTRIBUTION
Australian endemic.

REFERENCES
Döderlein, 1935: 80; pl. 20 fig. 5; pl. 21 figs 4, 4a.
Gray, 1866: 6; pl. 6 fig. 3 (as *Pentaceros granulosus*).

GLOBE SEASTAR
Gymnanthenea globigera
(Döderlein, 1915)

Shark Bay, Western Australia (S. Morrison).

DESCRIPTION

Similar to *Anthenea* in appearance but the furrow armature is in two series (not three), there is a fan of about six webbed furrow spines and 2–3 subambulacral spines in a single row; the body is covered by a thick smooth skin with widely spaced small tubercles in a carinal and 3–4 dorsolateral rows, with few or many small bivalved pedicellariae between them; **upper surface of the superomarginal plates is horizontal and bare except for usually one large granule on the edge** and 3–4 (occasionally more) smaller granules on the lateral face of the plate; actinal plates each

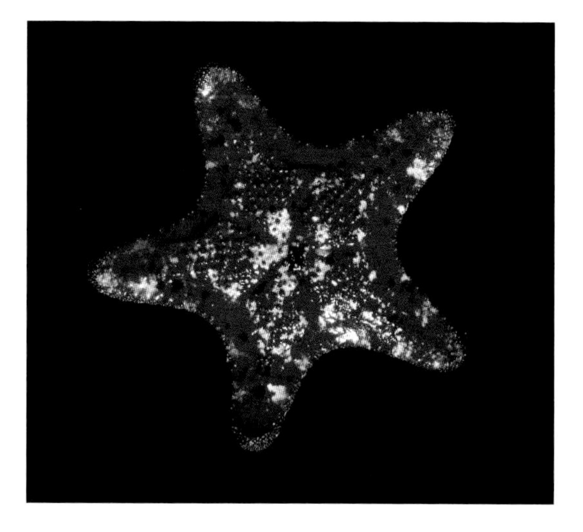

Houtman Abrolhos, Western Australia (B. Wilson).

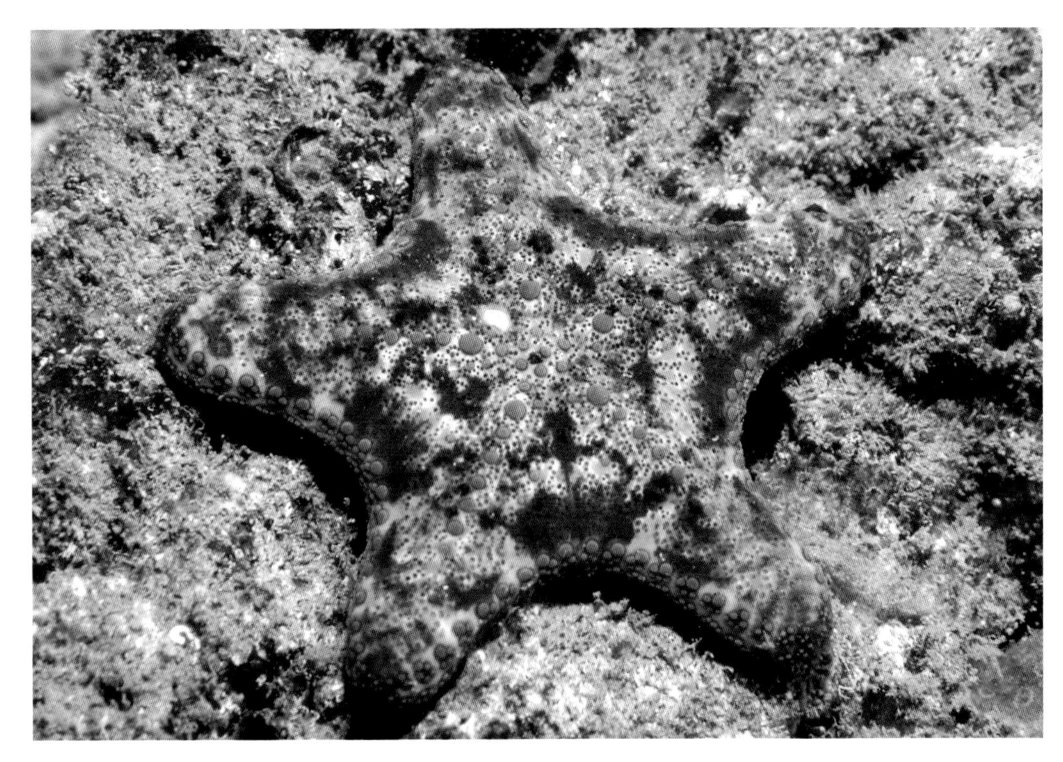

Dampier Archipelago, Western Australia (C. Bryce).

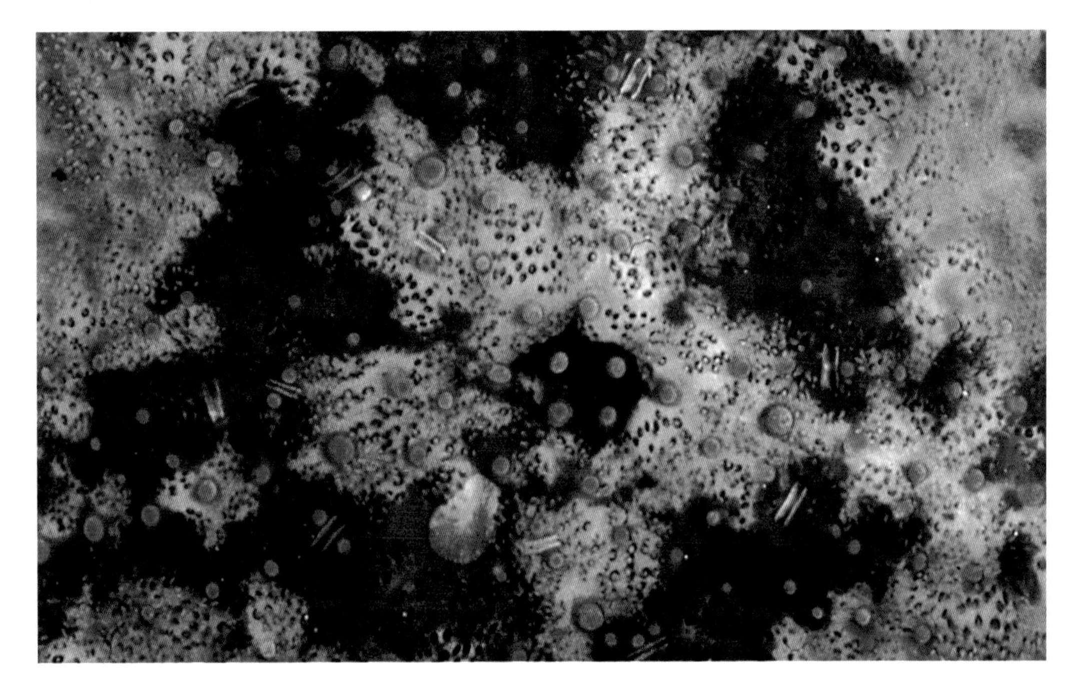

Abactinal pedicellariae and tubercle characteristics, Port Gregory, Western Australia (E. Hodgkin).

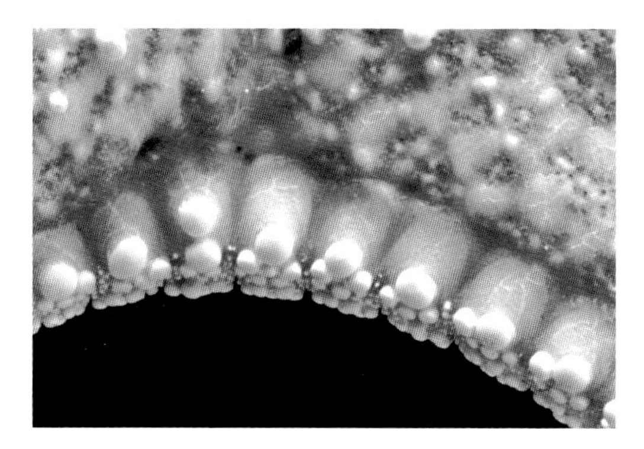

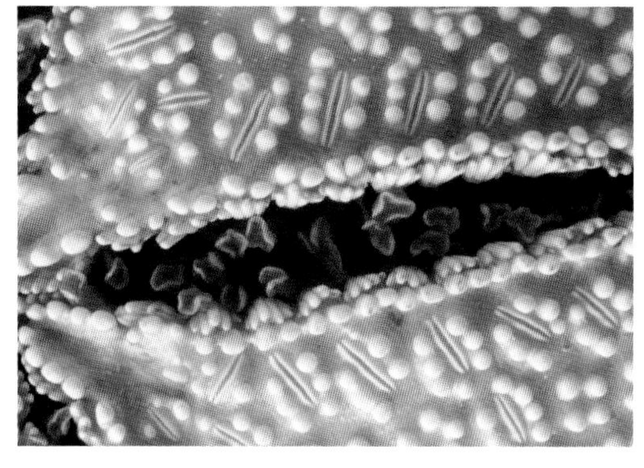

Superomarginal [top], furrow and actinal plate characteristics [bottom], Western Australian Museum (WAMZ5380).

have a large bivalved pedicellaria bordered by 3–4 granules each side; no pedicellariae on superomarginal or adambulacral plates.

MAXIMUM SIZE

R/r = 103/46 mm (R = 2.2r).

COLOUR

Orange, cream, grey or red patterned with varying shades of red or brown.

HABITAT, DEPTH AND AUSTRALIAN DISTRIBUTION

On sand and algal covered shore platforms, usually in the open, 0–104 m. From the Houtman Abrolhos (~28°S) and Port Gregory, WA to Darwin, NT on coastal and nearshore reefs and the inner continental shelf.

TYPE LOCALITY

Turtle Island, north-western Australia.

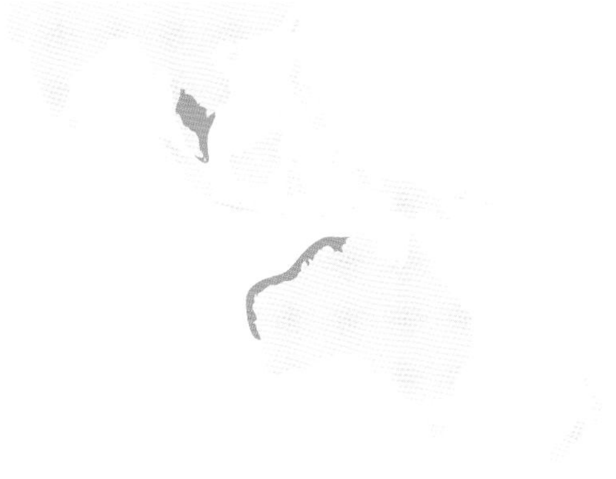

FURTHER DISTRIBUTION

Gulf of Thailand, southern South China Sea.

REFERENCES

Antokhina and Britayev, 2012: 897 (distribution).

Clark, H.L., 1938: 108; pl. 19 figs 4–5 (as *Gymnanthenea laevis*).

Clark, H.L., 1946: 99–100.

Clark, A.M. and Rowe, 1971: 52.

Döderlein, 1915: 50–51; pl. 8 figs 1–2.

Lane et al., 2000: 472 (distribution).

Marsh, 1976: 213–225 (synonymy).

Rowe and Gates, 1995: 102.

REMARKS

Marsh (1976) synonymised *Gymnanthenea laevis* (H.L. Clark, 1938) with *G. globigera*, this was followed by Rowe and Gates (1995) and is followed here. Lane et al. (2000) noted both species in the southern South China Sea and Antokhina and Britayev (2012) reported *G. laevis* from Mot Island, Vietnam. Mah (2019) recognises both species. Molecular analyses would determine if there are one or two species.

UNIFORM SEASTAR
Halityle regularis
Fisher, 1913

DESCRIPTION
This massive species is similar to *Culcita* but the marginal plates are clearly visible and it lacks tubercles; **the stellate plates are joined by slightly raised structures which divide the surface into regular triangular papular areas, arranged in hexagons**; the whole surface is covered by granular skin; the actinal surface is covered by large close fitting, nearly flat plates covered by flattish granules; there are 4–8 larger, rhombic plates near the mouth, these are usually violet, outlined by orange granules.

MAXIMUM SIZE
R/r = 162/102 mm (R = 1.18–1.66r).

COLOUR
Variable, often orange to maroon or purple with darker or yellow to orange papular areas. Actinal surface apricot to pink with the prominent plates near the mouth orange or violet outlined with orange granules.

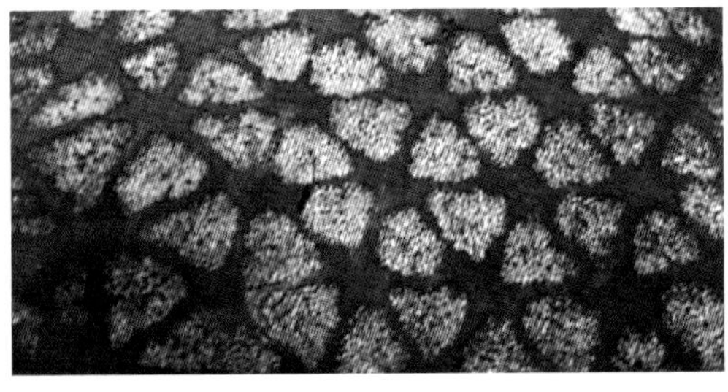

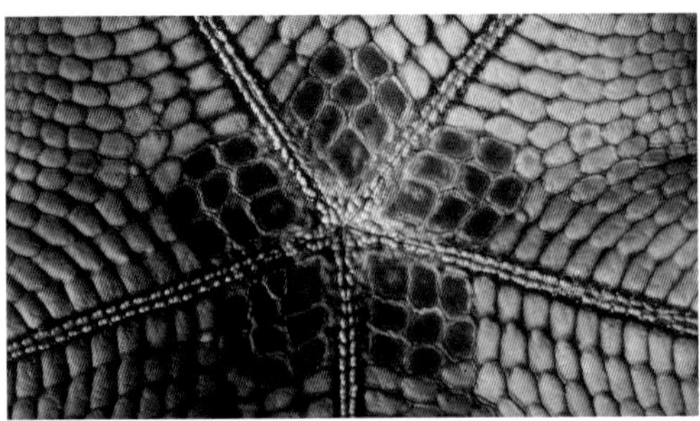

Abactinal [top] and actinal plate characteristics [bottom], Dampier Archipelago, Western Australia (N. Sinclair).

HABITAT, DEPTH AND AUSTRALIAN DISTRIBUTION
Found on sandy rubble, coral or coralline algae, 3–275 m; rare. In Australia known only from off Ningaloo Reef, Exmouth Gulf, the North West Shelf WA, Townsville, Keppel Bay and Heron Island (23°30'S), Queensland.

TYPE LOCALITY
Tawi Tawi Group, Sulu Archipelago, the Philippines.

FURTHER DISTRIBUTION
Madagascar and east Africa, southern India and the Lakshadweep Archipelago, the Philippines, north to Japan and east to New Caledonia.

REFERENCES
Baker and Marsh, 1976: 107–116; pl. figs 1a–d.
Clark, H.L., 1946: 109 (as *Halityle anamesus*).

REMARKS
Despite extensive trawling on the North West Shelf only one specimen has been taken there, two from Exmouth Gulf, and a juvenile from off Ningaloo Reef.

Sulawesi, Indonesia (B. Dupont).

Malapascua Island, the Philippines (S. Taykor).

GENUS *PENTACERASTER* Döderlein 1916

Species of this genus are characterised by having distal marginals and other plates covered with individually distinct, even-sized, usually projecting granules. The dorsolateral areas are distinctly reticulate with well defined pore areas. The primary plates at the nodes often have rounded or conical tubercles arranged in longitudinal series. Distal inferomarginal plates usually have an enlarged spine or conical projection. A few intermarginal plates may be present but there is no extensive development of intermarginal pore areas. R is rarely > 3r.

Type species: *Pentaceraster mammillatus* (Audouin, 1826) (Indo-West Pacific)

Twelve species are recorded from shallow water by Clark and Rowe (1971), only four have been found in Australian waters, all from shallow water. There is great variability amongst species of *Pentaceraster* and intermediate forms between species throw doubt on the reliability of any key. One species, *P. gracilis*, is distinctive with little variation.

GENUS *PORASTER* Döderlein, 1916

Type species: *Poraster superbus* (Möbius, 1859) (Indo-West Pacific)

Monotypic genus. Diagnosis as for the species (page 432).

Pentaceraster gracilis, Dampier Archipelago, Western Australia (C. Bryce).

Pentaceraster regulus, Houtman Abrolhos, Western Australia (C. Bryce).

KEY TO THE SPECIES OF *PENTACERASTER* IN SHALLOW AUSTRALIAN WATERS

1	A number of carinal spines well developed and at least some distal superomarginal spines also, no intermarginal plate; R = 2–3r.	2
	Carinal and distal superomarginal spines not strongly developed, intermarginal plates present.	3
2	Abactinal and marginal spines slender and conical; arm tips relatively narrow; R = 2.4–3r.	***P. alveolatus*** page 426
	Abactinal and marginal spines stout; arm tips rounded; R = 2.2–2.5r.	***P. multispinus*** page 429
3	Superomarginal and inferomarginal plates arching apart from each other interradially, with well developed intermarginal plates, which may bear spines.	***P. gracilis*** page 427
	Intermarginal plates, if present, not separating the two series of marginal plates but are simply interstitial.	***P. regulus*** page 430

(Modified from Clark and Rowe, 1971.)

HONEYCOMB SEASTAR
Pentaceraster alveolatus
(Perrier, 1875)

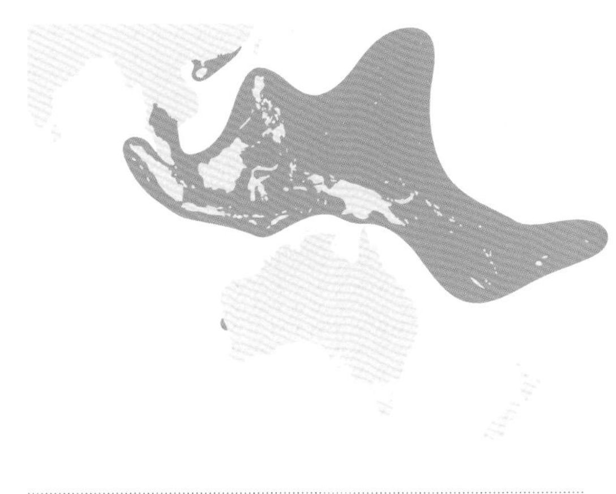

DESCRIPTION
Massive disc with disc and arms high, arm tips fairly narrow; some of the inferomarginal plates have spines; there are stout, **conical**, sharp pointed spines on the five primary radial plates and usually a central spine; there is usually a single series of spines on the radial plates, spines are present on some superomarginal plates, **spines are high and slender** often with a large bare point; pore areas usually well separated, many small pedicellariae in pore areas; two rows of dorsolateral plates; marginal plates covered with fairly even-sized small flat, polygonal granules; many large bivalved pedicellariae on actinal and inferomarginal plates; inferomarginal spines are only on plates near the arm tips; no intermarginal plates; there are 5–8 (sometimes nine) furrow spines and two subambulacral spines in a single row except in very large specimens (R = 145mm), when there may be three rows of subambulacral spines with up to four rows per plate.

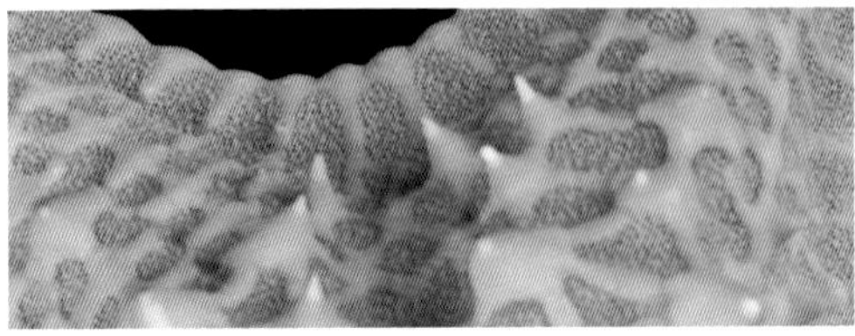

Western Australian Museum (WAMZ18158), abactinal spine characteristics [bottom].

MAXIMUM SIZE
R/r = 145/48 mm (R = 3r).

COLOUR
Variable, sometimes cream or beige with a red or green disc, radial areas and arm tips. Tubercles red, brown or red and cream.

HABITAT, DEPTH AND AUSTRALIAN DISTRIBUTION
Sand and rubble, 1–54 m. In Australia this species is known from a single specimen, from sand and rubble at 35 m in the Houtman Abrolhos, WA.

TYPE LOCALITY
New Caledonia.

FURTHER DISTRIBUTION
Indonesia, the Philippines and Guam, east to New Caledonia and Samoa.

REFERENCES
Clark, A.M. and Rowe, 1971: 56.

Döderlein, 1936: 332–336; pl. 24 figs 1–11; pl. 25 figs 1–7; pl. 29 fig. 10.

REMARKS
Although the single specimen was found deeper than 30 m this species could be expected in shallow water.

GRACILIS SEASTAR
Pentaceraster gracilis
(Lütken, 1871)

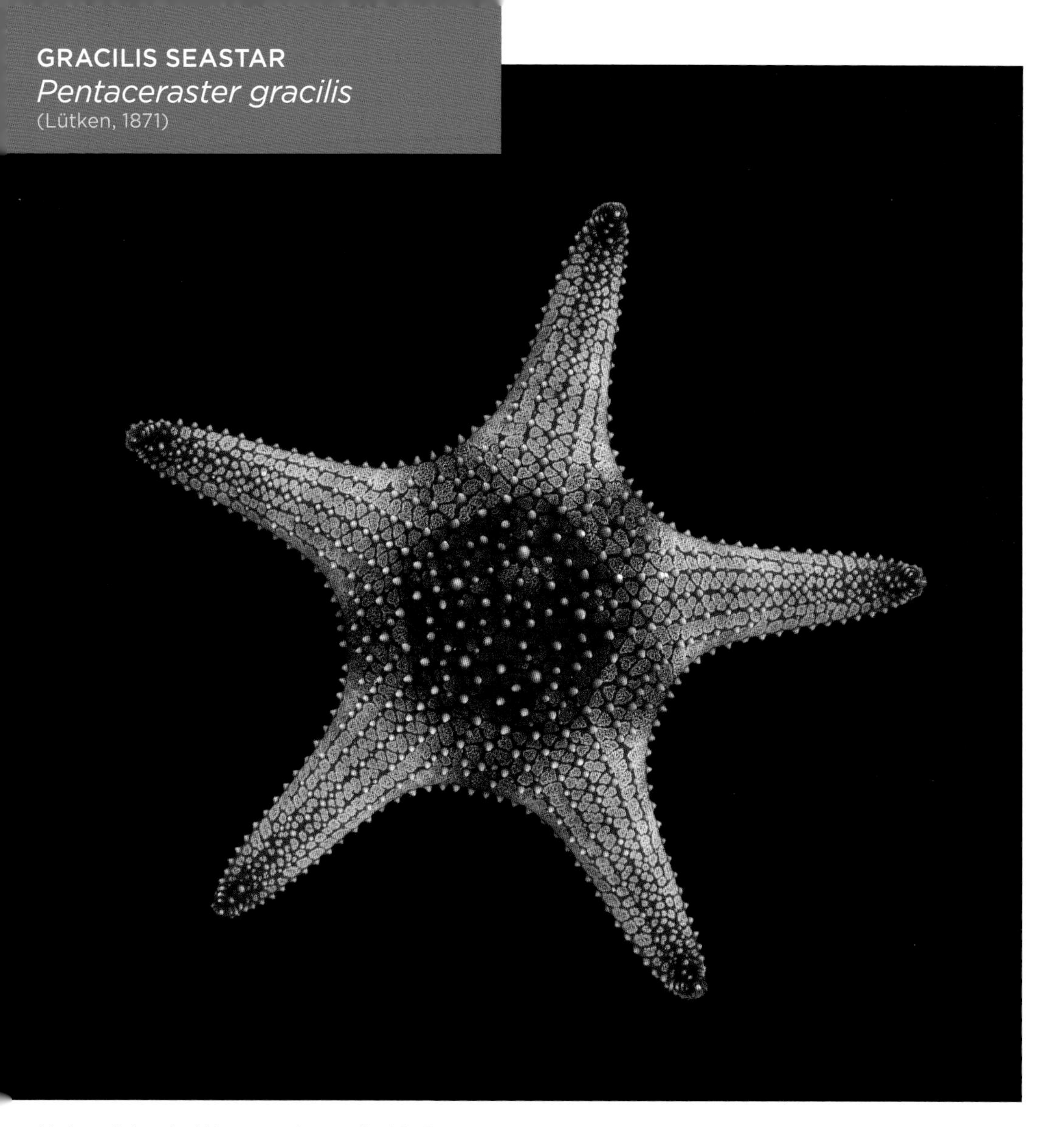

Muiron Islands, Western Australia (C. Bryce).

DESCRIPTION

A massive species with a very large evenly convex disc and relatively slender arms; **the superomarginal plates arch away from the inferomarginals and there is a well developed row of intermarginal plates between them in the arm angle** in animals with $R > 80$ mm; nearly all superomarginal and inferomarginal plates bear moderate sized tubercles; abactinal plates nearly all bear similar tubercles, occasionally the interradial areas lack tubercles; there are small tong-shaped pedicellariae on the abactinal surface and large bivalved pedicellariae on the actinal surface; in dried specimens, the skeleton is visible as regular pentagons with triangular papular areas in most specimens; there are 7–9 (up to 10) furrow spines, 2–4 subambulacral spines, and 2–4 small spines outside the subambulacral spines; there are no large spines.

MAXIMUM SIZE

R/r = 335/110 mm (R = 3r).

COLOUR

Grey or beige to pinkish brown with disc centre and arm tips chocolate brown. Tubercles yellow to orange, or finely mottled cream to brown with bright red tubercles.

FOOD

The stomach is everted on echinoids, coralline algae and sponge (Birtles, personal communication).

HABITAT, DEPTH AND AUSTRALIAN DISTRIBUTION

On sand with seagrass, or silty or shelly sand or mud, 10–146 m. From Houtman Abrolhos, Shark Bay, Exmouth Gulf and the North West Shelf of WA to Lady Elliott Island, Queensland. Moderately common in Shark Bay, WA.

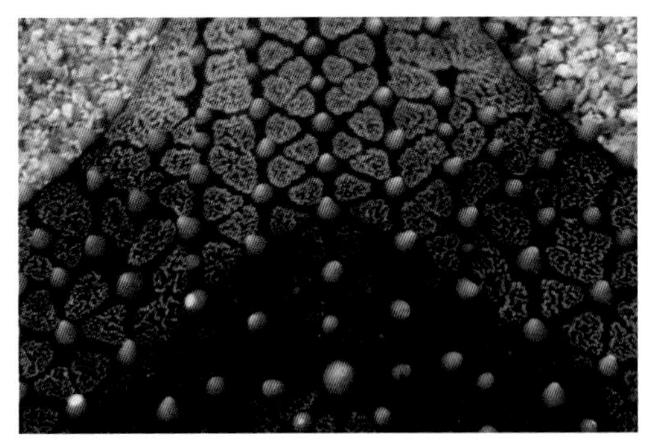

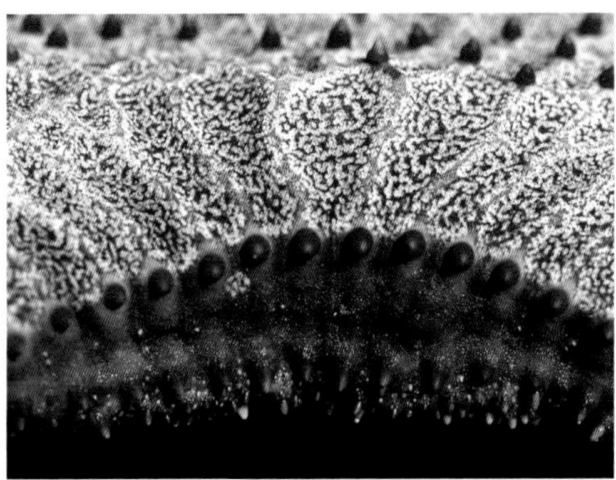

Abactinal spinelet characteristics [top], Shark Bay, Western Australia; intermarginal characteristics [bottom], North West Shelf, Western Australia (L. Marsh).

TYPE LOCALITY

Bowen, Queensland.

FURTHER DISTRIBUTION

Western Indian Ocean, Mozambique, Red Sea, Bay of Bengal and Indonesia.

REFERENCES

Clark, H.L., 1946: 107.

Clark, A.M. and Rowe, 1971: 56.

Marsh and Marshall, 1983: 675.

REMARKS

The smallest specimen identifiable as *Pentaceraster gracilis* has an R = 70 mm. At this size the five primary radial tubercles are prominent, the central disc is raised, and there are few other tubercles except on the inferomarginals. Smaller than this, it is impossible to distinguish whether specimens are *P. regulus* or *P. gracilis*. Clark and Rowe (1971) remark that *P. gracilis* might be simply large specimens of *P. regulus*, but examination of abundant material in the Western Australian Museum indicates that the two species are easily separated except when small.

STOUT-SPINED SEASTAR
Pentaceraster multispinus
(von Martens, 1866)

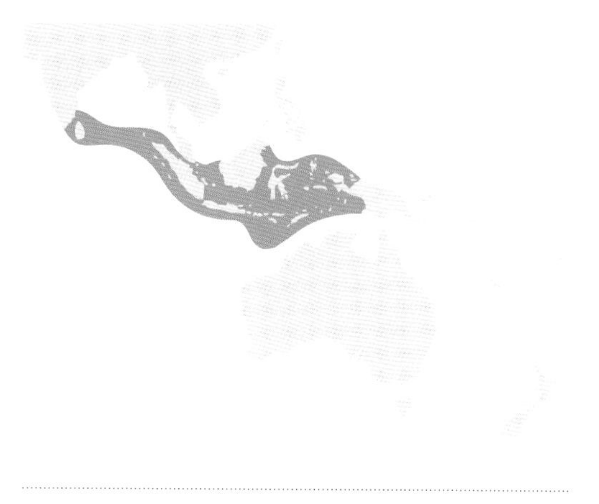

DESCRIPTION

Disc of moderate height; arms broad and short with rounded tips; **spines very plump with a thick base and a large bare tip; apical spines large, other spines slightly smaller**; there is usually a central spine, 5–6 carinal spines and a large dorsolateral spine on either side of the disc; **superomarginal plates have large spines except rarely in the arm angle**; inferomarginals with smaller spines in the arm angle; marginal and abactinal plates covered by fine granules; arms with two rows of dorsolateral plates that usually lack spines but may have large bivalved pedicellariae; there are up to nine furrow spines followed by 2–3 subambulacral spines in two rows; this species is highly variable and some large specimens are hard to distinguish from *P. alveolatus*.

Western Australian Museum (WAMZ18204), abactinal spine characteristics [bottom].

MAXIMUM SIZE

R/r = 155/65 mm (R = 2.2–2.8r).

COLOUR

Not recorded.

HABITAT, DEPTH AND AUSTRALIAN DISTRIBUTION

Intertidal and sandy back reef areas, 0–3 m. In Australia it is only found on Ashmore Reef, Timor Sea.

TYPE LOCALITY

Flores, Indonesia.

FURTHER DISTRIBUTION

Indonesia including Timor, Sri Lanka.

REFERENCES

Clark, A.M. and Rowe, 1971: 56.

Döderlein, 1936: 336–340; pl. 24 fig. 12; pl. 26 figs 1–3.

Marsh and Marshall, 1983: 675.

REGULUS SEASTAR
Pentaceraster regulus
(Müller and Troschel, 1842)

Montebello Islands, Western Australia (J. Keesing).

DESCRIPTION

Body fairly high; five arms taper to blunt tips; there are **five primary radial tubercles of moderate size, conical, with a bare tip**; there is a central tubercle and usually some small tubercles within the primary pentagon; there is a row of carinal tubercles within the primary pentagon continuing on the arms, with small tubercles on some of the dorsolateral plates of disc and arms; **superomarginal plates usually lack spines but the inferomarginals have small spines in the arm angle and larger ones near the arm tip**; papulae are in fairly well defined areas (which may run together), no papulae on the actinal surface; small pedicellariae sometimes on the dorsolateral plates, larger, bivalved pedicellariae on the ventrolateral plates; there are 7–9 furrow spines, 2–4 subambulacral spines and 2–4 enlarged granules outside the subambulacral spines.

MAXIMUM SIZE

R/r = 165/55 mm (R = 2.3–5r).

COLOUR

Variable, often grey, cream and tan, or dark and light grey with brown marginals and actinal surface. In New Caledonia it is a uniform dark red.

FOOD

The substrate biofilm of microorganisms and detritus, also grazes large Foraminifera and bryozoans; preys on heart urchins such as *Breynia, Maretia* and *Metalia* species (Birtles, personal communication).

HABITAT, DEPTH AND AUSTRALIAN DISTRIBUTION

On sand and seagrass flats on coral reefs, mud, sand, shelly sand or gravel, 0–115 m. From the Montebello Islands and North West Shelf, WA, Ashmore Reef, Timor Sea to Double Island Point, Queensland.

TYPE LOCALITY

Pondichery, India.

FURTHER DISTRIBUTION

From India to Thailand, Indonesia and the Philippines to New Caledonia, Fiji and north to the Ryukyu Islands.

REFERENCES

Clark, H.L., 1946: 107 (as *Pentaceraster australis*).
Coleman, 1979: 157 (food).
Döderlein, 1936: 350–352; pl. 26 figs 5–6, 6a.
Marsh and Marshall, 1983: 675.
Rowe and Gates, 1995: 105.

REMARKS

This species is very variable and can be confused with *Pentaceraster alveolatus* which differs in having very few spines in the apical pentagon and rarely any dorsolateral spines, it also has the pore areas well separated and inferomarginal spines long and slender. *Oreaster australis* Lutken, 1871 was synonymised with *Pentaceraster regulus* by Döderlein (1936), however H.L. Clark (1946) believed they were different. Rowe and Gates (1995) agree with Döderlein's decision.

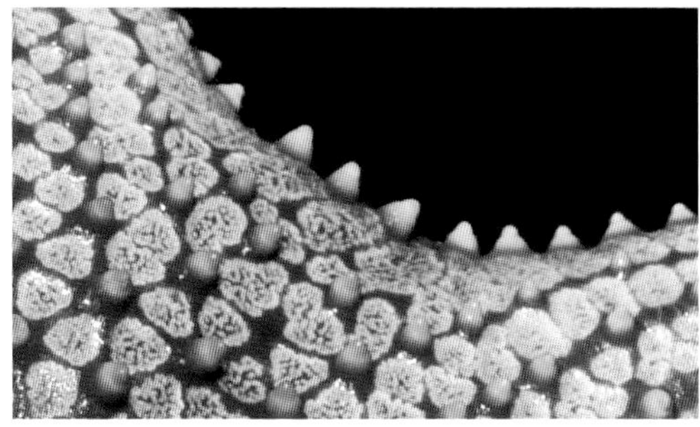

Superomarginal and inferomarginal plate characteristics [above], Montebello Islands, Western Australia (J. Keesing); primary radial tubercle characteristics [left], Muiron Islands, Western Australia (L. Marsh).

SUPERB SEASTAR
Poraster superbus
(Möbius, 1859)

DESCRIPTION

A very large stellate asteroid **with a large disc and relatively slender, tapering arms; the abactinal surface is coarsely granulated, without spines or tubercles except for a carinal row of large conical tubercles** extending to the arm tips; the superomarginal and inferomarginal plates are granule-covered but the inferomarginals also have 1–3 short conical spines; **intermarginal plates are present interradially**; actinal plates have groups of short conical spines or enlarged granules; there are 5–7 furrow spines and 3–4 subambulacral spines on each plate; papulae are in confluent areas on the abactinal surface and occur intermarginally but not on the actinal surface; this is a very distinctive species, unlike any other.

MAXIMUM SIZE

R/r = 240/60 mm (R = 4r).

COLOUR

Cream to beige or pink with the carinal tubercles and outer half of the arms orange-red to red-brown.

FOOD

Stomach observed everted on a sand encased solitary ascidian and it also scavenges, for example on echinoids (Birtles, personal communication).

HABITAT, DEPTH AND AUSTRALIAN DISTRIBUTION

Inner continental shelf on shelly sand, 20–55 m. In Australia it is only known from off Townsville, Queensland.

TYPE LOCALITY

Sumatra, Indonesia.

FURTHER DISTRIBUTION

From east Africa, Bay of Bengal to New Caledonia, where it is locally common, southern China and north to Japan.

REFERENCES

Clark, A.M. and Rowe, 1971: 54.

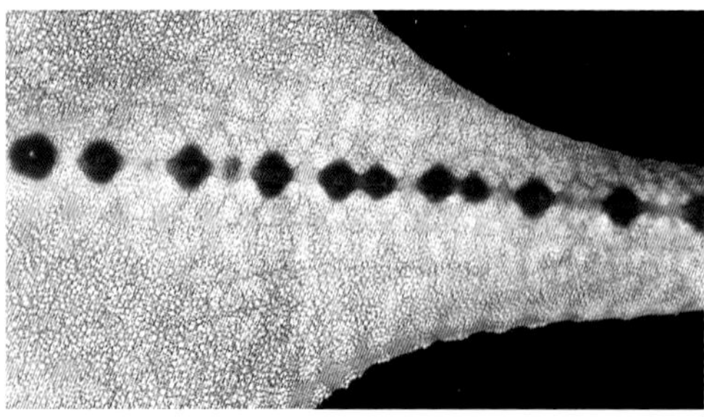

Townsville, Queensland, abactinal plate characteristics [bottom] (L. Marsh).

GENUS *PROTOREASTER* Döderlein, 1916

Species of this genus are large, massive seastars with at least the distal marginal plates and convex parts of the larger abactinal plates covered with a **smooth plastering of unequal, polygonal flattened granules**. The dorsolateral areas of the arms are without any projections except rarely in *Protoreaster lincki*. The superomarginals are spineless except for a few distal spines in *P. lincki*. There are no spines on any inferomarginals. The actinal granulation is smooth and only two rows of adambulacral spines are well developed.

Type species: *Protoreaster nodosus* (Linnaeus, 1758) (Indo-Pacific)

Four species occur worldwide, three of which are found in shallow Australian waters.

GENUS *PSEUDOREASTER* Verrill, 1899

Type species: *Pseudoreaster obtusangulus* (Lamarck, 1816) (Australia)

Monotypic genus. Diagnosis as for the species (page 440).

Protoreaster nodulosus, Dampier Archipelago, Western Australia (C. Bryce).

Protoreaster lincki, Ningaloo Reef, Western Australia (C. Bryce).

KEY TO THE SPECIES OF *PROTOREASTER* IN SHALLOW AUSTRALIAN WATERS

1	A few of the distal superomarginal plates bear laterally projecting, usually conspicuous tapering spines or knobs.	***P. lincki*** page 435
	Marginal plates lacking any conspicuous spines or knobs.	**2**
2	Disc markedly elevated; some carinal plates with conspicuous more or less high, usually pointed elevations, particularly large on the five primary radial plates; papular pore areas confluent; northern and north-eastern Australia.	***P. nodosus*** page 436
	Disc of moderate height; carinal plates with broad, low, rounded cushion-like tubercles, never pointed, many broader than long; papular pore areas separate from each other; north-western Australia.	***P. nodulosus*** page 438

LINCK'S SEASTAR
Protoreaster lincki
(de Blainville, 1830)

DESCRIPTION

Disc markedly elevated, centre of disc and primary radial plates with high pointed tubercles, the connecting ossicles between them dividing the pentagon into five triangular pore spaces; pore areas with large rounded and scattered granules covering the sides of the disc and arms; the marginal and carinal plates are covered with a smooth pavement of polygonal granules of several sizes; **a few distal superomarginal plates with laterally projecting tapering spines or knobs**; a single row of dorsolateral plates also often has spines and rarely there are a few in the interradial area; **there are no inferomarginal spines or tubercles**; actinal plates covered by a smooth pavement of polygonal granules; bivalved pedicellariae are sometimes present on the marginal plates.

MAXIMUM SIZE

R/r = 160/50 mm (R = 2.3–3r).

COLOUR

Grey to white, cream or pale orange with bright red or orange tubercles, reticulations of the skeleton, and arm tips.

FOOD

Substrate biofilm of microorganisms and detritus, likely also scavenges dead animals.

REPRODUCTION

Ebert (1976) observed *Protoreaster lincki* spawning during early May in Seychelles (4°37'S), he also noted that males were smaller than females. In north-western Australia spawning was observed during November (Keesing, personal communication).

HABITAT, DEPTH AND AUSTRALIAN DISTRIBUTION

On sandy reef flats or seagrass beds, or on algal covered rocks on coral reefs, 0–10 m. *P. lincki* has been found at few localities in Australia: Ningaloo Reef, North West Cape and near Onslow, WA, and on Ashmore Reef, Timor Sea.

TYPE LOCALITY

Not recorded.

FURTHER DISTRIBUTION

It is predominantly a western Indian Ocean species, but also in the Red Sea, and from east Africa to Sri Lanka and Java, Indonesia.

REFERENCES

Döderlein, 1936: 328–330; pl. 22 fig. 6; pl. 23 figs 1–7; pl. 28 fig. 12.

Ebert, 1976: 71–77.

Marsh and Marshall, 1983: 675.

Sloan, 1980: 69 (food).

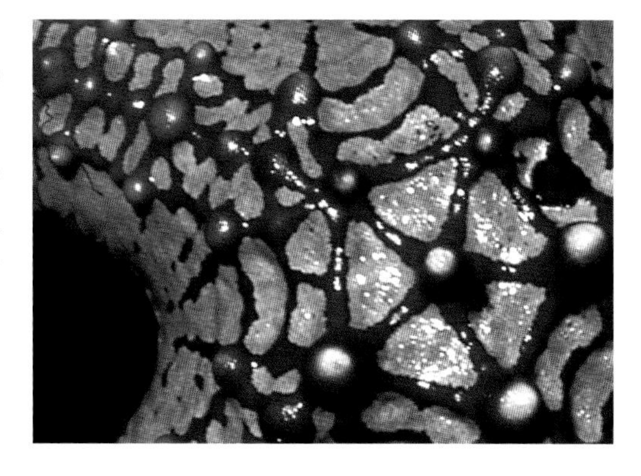

Abactinal plate and tubercle characteristics, Ashmore Reef, Timor Sea (L. Marsh).

RHINOCEROS SEASTAR
Protoreaster nodosus
(Linnaeus, 1758)

DESCRIPTION

Disc markedly elevated; **some of the carinal plates with high, rounded or more usually conical, sometimes sharply pointed tubercles**, particularly large on the five primary radial plates; pore areas confluent; no tubercles on superomarginal or inferomarginal plates; the marginal and carinal plates are covered with a smooth pavement of flat polygonal granules of several sizes; this species is abundant in places.

MAXIMUM SIZE

R/r = 200/80 mm (R = 2–3r).

COLOUR

Variable, often cream with brown patches around black tubercles, or pinkish cream with red disc and patches surrounding black tubercles, sometimes all cream.

FOOD

Algae, substrate biofilm of detritus and microorganisms. Scavenges dead animals.

REPRODUCTION

In Palau reproduction takes place from May to July. Eggs (200 µm) give rise to a planktotrophic bipinnaria larvae that become brachiolaria larvae before settling after 14 days. Metamorphosis has not been observed. The larvae swim near the seabed.

HABITAT, DEPTH AND AUSTRALIAN DISTRIBUTION

On mud, sand and seagrass flats, and sandy coral reef flats, occasionally on outer reef slopes, 0–30 m. From Ashmore Reef, Timor Sea, NT, the Great Barrier Reef and Queensland coast south to Mackay. Not found in WA where it is replaced by *P. nodulosus*.

TYPE LOCALITY

'Indian Ocean'.

FURTHER DISTRIBUTION

Distributed from east Africa to southern China (Hainan Island), Japan and eastwards to Guam and New Caledonia.

REFERENCES

Clark, A.M. and Rowe, 1971: 54; pl. 6 fig. 3.

Döderlein, 1936: 324–328; pl. 22 figs 7–8; pl. 23 figs 8–12.

Yamaguchi, 1975: 20 (food).

Yamaguchi, 1977b: 283–296 (reproduction).

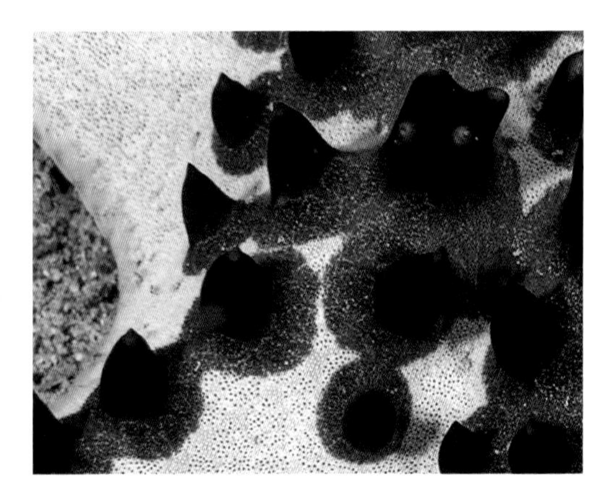

Abactinal tubercles and pore area characteristics, North Sulawesi, Indonesia (S. Morrison).

Ashmore Reef, Timor Sea (L. Marsh).

North Sulawesi, Indonesia (S. Morrison).

NODULOSE SEASTAR
Protoreaster nodulosus
(Perrier, 1875)

DESCRIPTION

Disc of moderate height; **five primary radial plates with very large, rounded, cushion-like tubercles**; carinal plates with similar but smaller tubercles extending to the arm end; the marginal and carinal plates are covered with a smooth pavement of polygonal granules of mixed sizes; **there are no spines or pointed tubercles on any plates**; pore areas are well separated from each other; the primary radial tubercles are evident in juveniles from R/r = 13/8 mm and carinal tubercles are gradually added during growth; this easily recognised species is confined to WA.

MAXIMUM SIZE

R/r = 170/75 mm (R = 2–2.9r).

COLOUR

Very variable, often olive green but can be white, pink, mauve, purple, dark brown and violet, grey-green, red and green, or various combinations of orange, tan, black, brown and grey.

FOOD

Although apparently rigid and inert, juveniles have been observed bending in a circle to open and feed on mussels in aquaria, they probably also graze on substrate biofilm.

REPRODUCTION

Spawning was observed in north-western Australia during November. Juveniles are common in Shark Bay, WA.

HABITAT, DEPTH AND AUSTRALIAN DISTRIBUTION

On sand flats, with or without algae, seagrass, rocks and sponges, 0–64 m. From the Houtman Abrolhos and Shark Bay to Cape Voltaire on the Kimberley coast, WA. Occurs on the coast, inner shelf and nearshore islands, not found on the offshore reefs. Western Australian endemic.

TYPE LOCALITY

'Australia'.

FURTHER DISTRIBUTION

Australian endemic.

REFERENCES

Clark, H.L., 1938: 130–31.

Döderlein, 1936: 323–324; pl. 22 figs 1–5.

Abactinal plate and tubercle characteristics, Ningaloo Reef, Western Australia (R. Lasley).

Colour variations, Dampier Archipelago, Western Australia [top];
Shark Bay, Western Australia [bottom] (C. Bryce).

OBTUSE SEASTAR
Pseudoreaster obtusangulus
(Lamarck, 1816)

Western Australian Museum (WAMZ99860).

DESCRIPTION

Similar in appearance to *Gymnanthenea globigera*; disc large, arms blunt-ended; **larger specimens have a reticulate skeleton of stellate plates covered by bare shiny skin; in the centre of each plate is a small or large tubercle and/or a small bivalved pedicellaria; usually the carinal tubercles are enlarged**, convex, up to 6 mm diameter, 4 mm high but sometimes only one or two on each arm are enlarged; the outer row of about seven distal dorsolateral plates

Exmouth Gulf, Western Australia (N. Coleman).

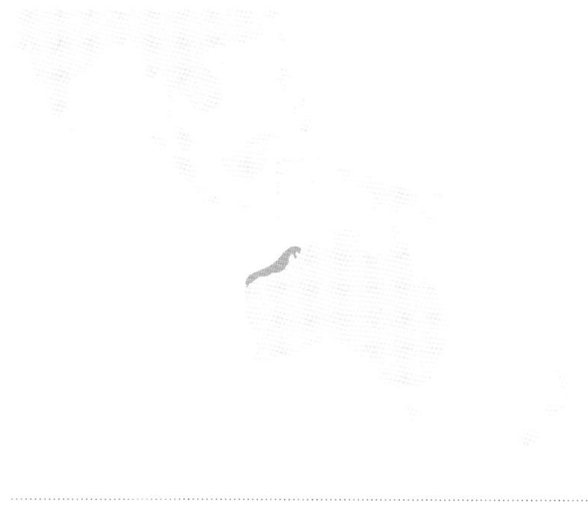

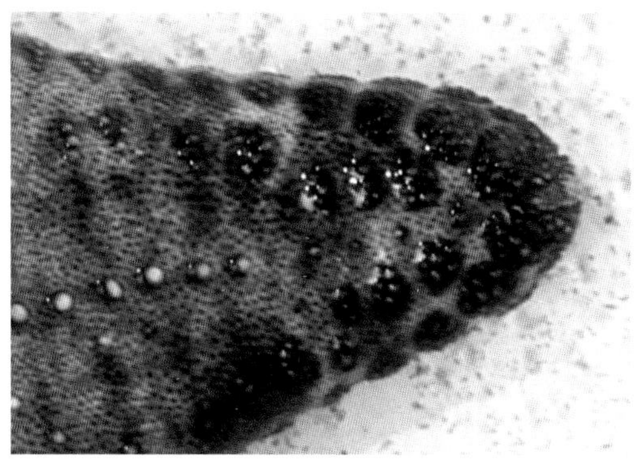

Dampier Archipelago, Western Australia [top] (C. Bryce); abactinal plate characteristics [above], Dampier Archipelago, Western Australia (S. Morrison).

are convex with up to 20 coarse granules; sometimes a granulated plate replaces a tubercle or vice versa; superomarginals and inferomarginals are covered by even-sized coarse granules except for a narrow bare margin; one bivalved pedicellaria is on most inferomarginals; there are four furrow spines and two unequal subambulacral spines; actinal plates each have a large bivalved pedicellaria.

MAXIMUM SIZE

R/r = 115/56 mm (R = 2r),
R usually < 90 mm.

COLOUR

Variable, for example bright red with black superomarginal granules and brown or red carinal tubercles; variegated with dull greys, brown, reds and purple; grey and red; mottled browns and greens; brown with white spots; or greenish orange.

HABITAT, DEPTH AND AUSTRALIAN DISTRIBUTION

Usually intertidal, on muddy sand flats or sandy reefs, sometimes on algal covered rock platforms, 0–10 m. Endemic to the coast and nearshore islands of north-western Australia, from Exmouth Gulf and Barrow Island to Augustus Island off the Kimberley coast. North-western Australian endemic.

TYPE LOCALITY

'North-western Australia'.

FURTHER DISTRIBUTION

Australian endemic.

REFERENCES

Clark, H.L., 1938: 104–105; pl. 6.
Clark, A.M. and Rowe, 1971: 52; pl. 6 fig. 1.

VELATIDA

Euretaster insignis, Cockburn Sound, Western Australia (D. Lane).

Species of this order are stellate, five armed or multi-armed seastars, often with a large disc. The marginal, abactinal and actinal plates are often small, and may be paxilliform. The marginal plates are inconspicuous and the actinal plates may be absent. The actinal skeleton is normally open and irregularly reticulate with paxillae on the plates. The madreporite is small, except in the Pterasteridae. Papulae occur on the abactinal and marginal surfaces. Species have prominent oral plates which may be keeled, and furrow and subambulacral spines may be webbed. Pedicellariae are absent. Suckered tube feet occur in two or four rows.

(Modified from Byrne et al., 2017.)

PTERASTERIDAE

Pterasteridae seastars are distinguished by having a stellate to pentagonal shape with five arms. The abactinal skeleton consists of cruciform or lobed plates bearing groups of spinelets that support a membrane distinct from the abactinal surface, forming a brood cavity from which the young emerge through a central valved aperture (osculum). Lateral spines on the adambulacrals either support the actinal web or merge in the actinal surface. Actinal intermediate plates are absent.

The family is cosmopolitan with eight known genera of which two occur in Australian waters but only one (with one species) occurs in depths of less than 30 m.

Euretaster insignis, Cockburn Sound, Western Australia (A. Brearley).

GENUS *EURETASTER* Fisher, 1940

Type species: *Euretaster insignis* (Sladen, 1882) (Indo-Pacific)

One species is found in shallow Australian waters. Diagnosis as for the species (page 444).

STRIKING SEASTAR
Euretaster insignis
(Sladen, 1882)

Minden Reef, Western Australia (C. Wood).

DESCRIPTION

Body stellate with five short, thick, slightly tapering arms with rounded tips; interradial arcs rounded; abactinal surface convex, margins rounded, actinal surface flat; **the supradorsal membrane is perforated by papular pores and is conspicuously reticulated, the membrane is supported by paxillae spinelets arranged in regular lines, joined by fibres forming large uniform rhomboidal meshes** filled in with regularly reticulate tissue; the interspaces of these meshes are pierced by minute spiracula, up to 100 in each mesh; the opposite angles of the

Dampier Archipelago, Western Australia (Z. Richards).

rhomboidal areas are joined by robust fibres forming a right angled cross in the centre; a conspicuous spinelet protrudes at the angle of the meshes; **in the centre of the abactinal surface is a small osculum** closed by pseudo-valves with slightly prominent spinelets; **furrows are narrow and sunken, fringed by the ventrolateral spines that stand at right angles to the furrows joined by a delicate web**; the ventrolateral spines are robust and longer than the outermost adambulacral spine; the outermost adambulacral spine is joined to the adjacent ventrolateral spine by a continuation of the web; the oral plates bear about five webbed spines on the furrow margin and a single, robust, cylindrical suboral spine standing perpendicular, not webbed.

Furrow characterics [top], Cockburn Sound, Western Australia (S. Morrison); osuculum, supradorsal membrane and papular area characteristics [bottom], Komodo Island, Indonesia (R. Steene).

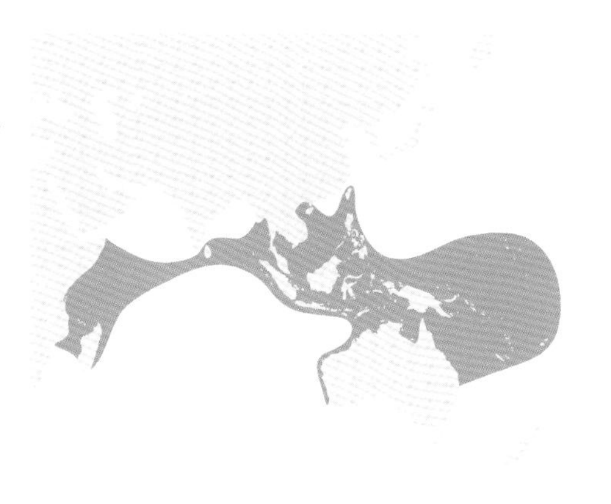

MAXIMUM SIZE

R/r = 110/45 mm (R = 2.4r).

COLOUR

Usually mottled red and yellow or completely scarlet. The type specimen had a white supradorsal membrane with dark purple to black spinelets.

REPRODUCTION

It is not confirmed whether *E. insignis* broods its young as do some other Pterasterids.

HABITAT, DEPTH AND AUSTRALIAN DISTRIBUTION

On mixed rock and sand or mud, 0–132 m. From Dunsborough south-western Australia, WA, around northern Australia to Moreton Bay, Queensland.

TYPE LOCALITY

Arafura Sea.

FURTHER DISTRIBUTION

Widely distributed in the Indo-West Pacific from east Africa and Madagascar to the central Pacific, north to Taiwan.

REFERENCES

Clark, H.L., 1946: 152 (as *Retaster insignis*).

SPINULOSIDA

Plectaster decanus, Jervis Bay, New South Wales (J. Turnbull).

Spinulosids have a small disc and long, usually cylindrical arms. They do not have pedicellariae and the skeleton is reticulate. Only one family, the Echinasteridae, is recognised in this order.

ECHINASTERIDAE

Echinasteridae is the sole family in the order Spinulosida and is distingushed by containing seastars with a small disc and long slender arms. The abactinal skeleton is reticulate or loosely tessellate overlaid by thick or thin skin with spines single or in groups. There are no pedicellariae, the ampullae are single, and the furrow and adambulacral spines are usually in a vertical series. The marginal plates are prominent and the papulae are isolated (only on the abactinal surface in *Metrodira*).

The family is cosmopolitan with eight genera, three of which (comprising 10 species) are found in depths of less than 30 m in Australian waters.

KEY TO THE GENERA OF ECHINASTERIDAE IN SHALLOW AUSTRALIAN WATERS

1	Furrow spines 3–4 in a comb-like row in line with the furrow; abactinal surface with a mosaic pattern formed by ridges of minute blunt spinelets.	***Plectaster*** page 448
	Furrow spines in a vertical series at right angles to the furrow.	**2**
2	Furrow spines in a vertical series of four that divide the furrow into separate circular compartments for tube feet; abactinal skeleton tessellate in small specimens tending to become reticulate in large ones.	***Metrodira*** page 448
	Furrow and adambulacral spines in a vertical series without specialistion; abactinal skeleton reticulate bearing single or grouped spines.	***Echinaster*** page 448

Echinaster callosus, with episymbiont comb jelly, *Coeloplana astericola*, Mindoro, the Philippines (N. Probst).

GENUS *ECHINASTER* Müller and Troschel, 1840

Seastars of this genus have a small disc and long arms that are cylindrical or tapered and slightly wider at the base. The abactinal skeleton is reticulate bearing spines singly in lines or groups. The furrow and adambulacral spines are in a vertical series, without specialisation. The papulae are abactinal and sometimes intermarginal.

Type species: *Echinaster seposita* (Retzius, 1783) (North Atlantic, Mediterranean Sea)
The genus is found in temperate and tropical waters of the Indo-Pacific and Atlantic Oceans. Eight species occur in Australian waters in depths of less than 30 m.

GENUS *METRODIRA* Gray, 1840

Type species: *Metrodira subulata* Gray, 1840 (Indo-West Pacific)
Monotypic genus. Diagnosis as for the species (page 462).

GENUS *PLECTASTER* Sladen, 1889

Type species: *Plectaster decanus* (Müller and Troschel, 1843) (Australia)
Monotypic genus. Diagnosis as for the species (page 464).

Echinaster superbus, Dampier Archipelago, Western Australia (C. Bryce).

KEY TO THE SPECIES OF *ECHINASTER* IN SHALLOW AUSTRALIAN WATERS

1	Abactinal spinelets large, conspicuous, up to 4–5 mm long, each enclosed in a well spaced balloon-like vesicle, singular on the primary plates only; arms cylindrical.	***E. callosus*** page 452
	Abactinal spinelets smaller, < 2 mm long, numerous, singular or in groups on primary plates, sometimes occurring on secondary abactinal plates; arms cylindrical or widened at base.	**2**
2	Abactinal spinelets singular, on primary plates; skeletal reticulum with relatively small papular areas; arms cylindrical or widened at the base.	**3**
	Abactinal spinelets in small (2–4) to large (5–60) groups on primary plates, sometimes spinelets occur on secondary plates; relatively large papular areas; arms cylindrical.	**6**
3	Arms cylindrical; disc small.	**4**
	Arms widened at base; disc relatively large.	***E. stereosomus*** page 458
4	Arms 5–7, autotomous; usually more than one madreporite; subambulacral spines well developed.	***E. luzonicus*** page 456
	Arms five, not autotomous; single madreporite; subambulacral spines not well developed.	**5**
5	Abactinal and actinal spinelets include stout, chisel-shaped or club-shaped forms.	***E. superbus*** page 459
	Abactinal and actinal spinelets uniform in size and shape, bluntly rounded at tips or truncate, some may be pitted at tip.	***E. varicolor*** page 460
6	Abactinal spinelets in large discrete groups on primary plates (5–60); papulae extend to inferomarginal line.	***E. glomeratus*** page 455
	Abactinal spinelets in small groups on primary plates (2–4), also singly on secondary plates; papulae restricted to abactinal surface above the superomarginal line.	**7**
7	Papular pore areas up to about 15 mm diameter, each with 11–40 (up to about 60) papulae.	***E. arcystatus*** page 450
	Papular pore areas up to 6.5 mm diameter, each with 6–8 (up to about 14) papulae.	***E. colemani*** page 454

(Rowe and Albertson, 1987.)

NETTED SEASTAR
Echinaster arcystatus
H.L. Clark, 1914

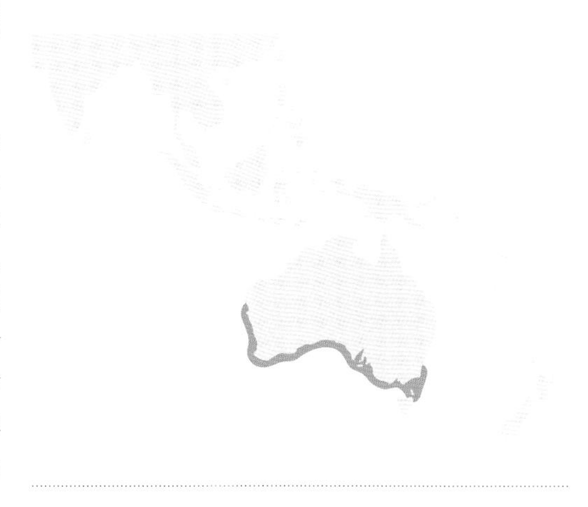

DESCRIPTION

A large non-autotomous species of *Echinaster*; disc small; arms rounded, tapering to a blunt tip; the abactinal skeleton forms a distinct network with meshes up to 15 mm across, with 10–60 papular pores; **the skeletal ridges carry numerous, spaced, bluntly pointed spinelets to 1.5 mm high**; the body is covered by a thick skin but the network pattern of spines is visible in live specimens; no spines in actinal interradial area.

MAXIMUM SIZE

R/r = 200/30 mm (R = 4.5–9r).

COLOUR

Dull yellow to red-brown. Commonly light brown with cream skeletal ridges and light red papulae.

FOOD

Encrusting sponges, molluscs and detritus.

HABITAT, DEPTH AND AUSTRALIAN DISTRIBUTION

On or among algal covered rocks or on sand, 0–46 m. From Montague Island, NSW, around southern and south-western Australia to Shark Bay, WA. Southern Australian endemic.

TYPE LOCALITY

Between Fremantle and Geraldton, WA.

FURTHER DISTRIBUTION

Australian endemic.

REFERENCES

Clark, H.L., 1914: 148–150; pl. 21.
Clark, H.L., 1946: 146.
Rowe and Albertson, 1987: 197, 199, 201; fig. 3.
Shepherd, 1968: 750 (food).

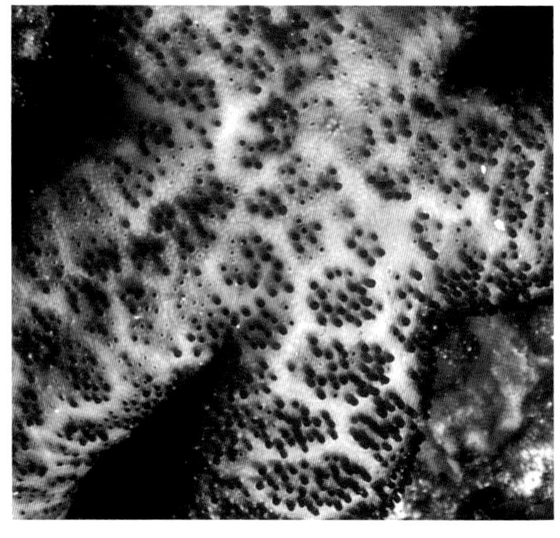

Papular area characteristics, Geographe Bay, Western Australia [above] (S. Morrison), Twilight Cove, Western Australia [right] (G. Edgar).

Rottnest Island, Western Australia (B. Wilson).

Rottnest Island, Western Australia (C. Bryce).

CALLUSED SEASTAR
Echinaster callosus
von Marenzeller, 1895

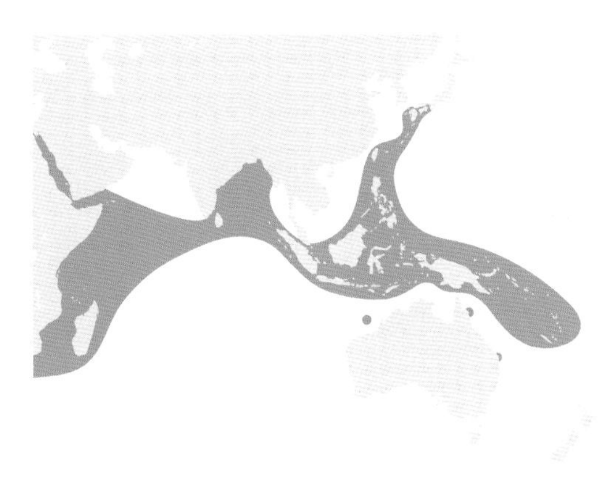

DESCRIPTION
Disc small; five long, cylindrical arms tapering to a narrow tip; in living animals **the skin is very thick and pustulose obscuring the long (2–5 mm) well spaced spines** on the primary plates of the skeletal network; the intervening papular areas are large with 10–20 papulae.

MAXIMUM SIZE
R/r = 215/21.5 mm (R = 9–10r).

COLOUR
Red, purple, brown or green with cream arm pustules and arms banded distally with cream. Papulae yellow or green.

HABITAT, DEPTH AND AUSTRALIAN DISTRIBUTION
On coral reef slopes, 2–35 m; very uncommon. In Australia it is only known from Mermaid Reef, WA, Lizard Island, Great Barrier Reef and Double Island Point (~26°S), Queensland.

TYPE LOCALITY
Solomon Islands.

FURTHER DISTRIBUTION
Found from east Africa and the Red Sea, through South-East Asia, Indonesia and the Philippines to southern Japan, east to New Caledonia.

REFERENCES

Clark, A.M. and Rowe, 1971: 72.

Fisher, 1919: 428–429; pl. 112 fig. 2, pl. 122 figs 4–5; pl. 132 fig. 5a–e.

Rowe and Albertson, 1987: 197.

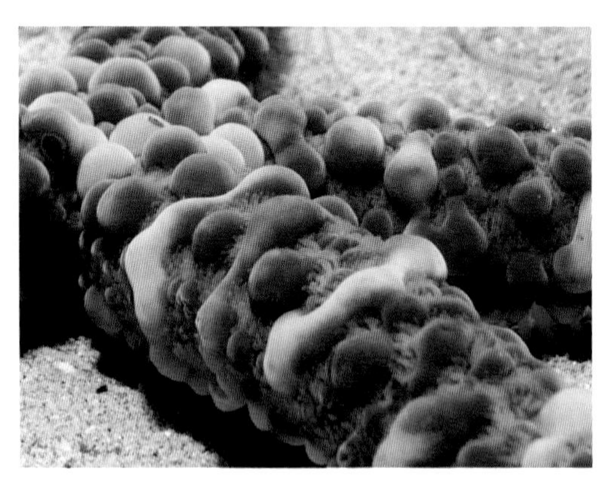

Varying abactinal vesicle characteristics, Mindoro, the Philippines [top] (N. Probst), Bintan Island, Indonesia [bottom] (C. Bryce).

REMARKS
Echinaster callosus can be readily identified by the conspicuous vesicles on the abactinal surface.

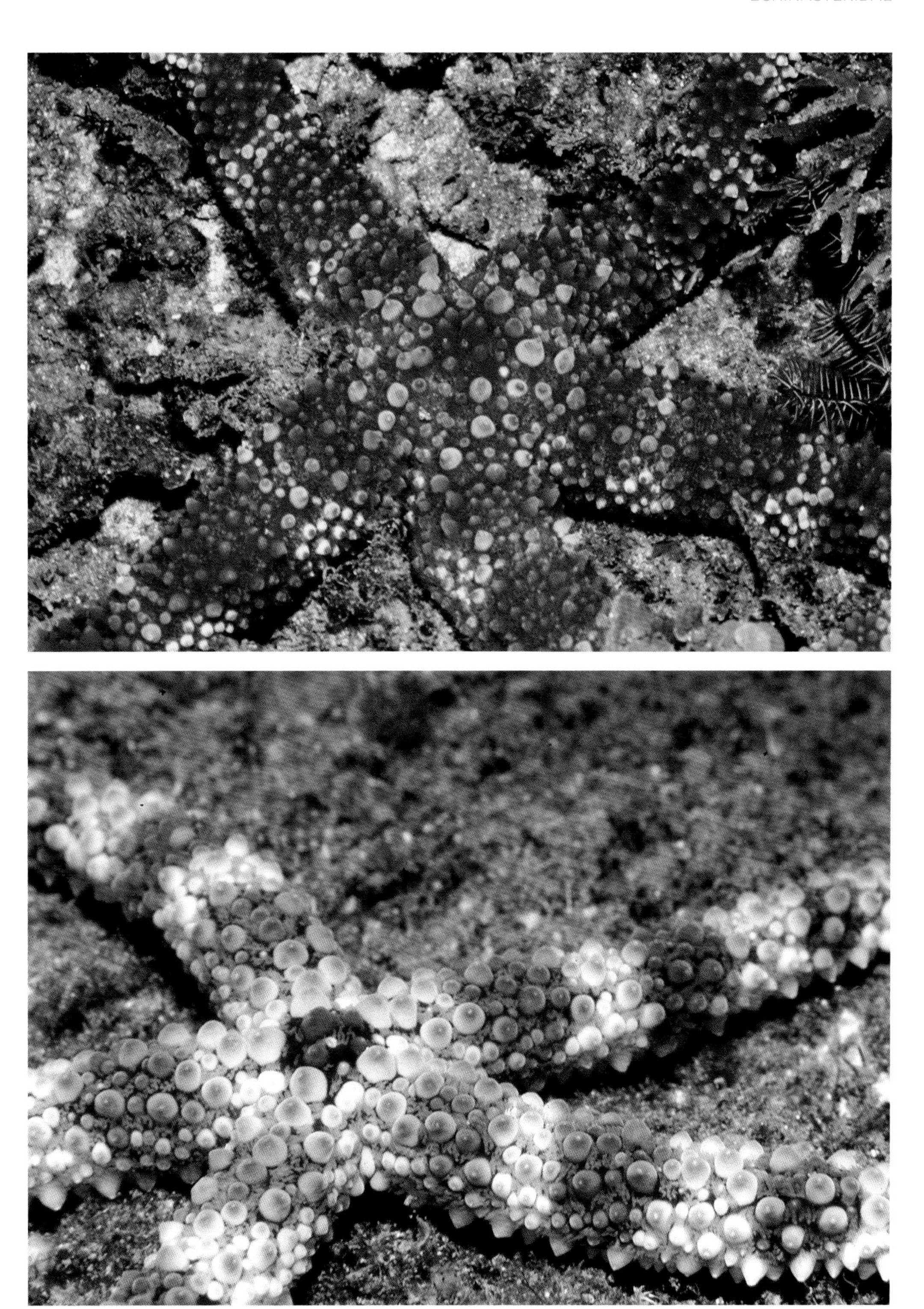

Milne Bay, Papua New Guinea [top] (R. Steene); Bintan Island, Indonesia [bottom] (C. Bryce).

COLEMAN'S SEASTAR
Echinaster colemani
Rowe and Albertson, 1987

DESCRIPTION

Five arms, rounded in cross section, tapering evenly to a blunt tip, slightly constricted at the base; the body is covered by a thick skin; the abactinal skeleton forms an open reticulum; the papular areas have 6–8 papulae (extremes 3–14), no papulae below superomarginals; **reticulum junctions have groups of 2–4 small bluntly pointed spinelets,** 1–4 occur singly on the secondary plates; superomarginal plates are four-lobed and have 2–3 spinelets decreasing to one distally; inferomarginal plates similar, but larger, than superomarginal plates; **intermarginal plates extend to about a quarter the arm length,** many with a single spinelet; a row of actinal plates also extends to about a quarter the arm length, each with a single spinelet; adambulacral plates each have a single small furrow spine followed by 2–3 subambulacral spines in a series at right angles to the furrow.

MAXIMUM SIZE

R/r = 137.5/18.5 mm (R = 4.48–8.4r).

COLOUR

Velvety brown with purple papulae.

HABITAT, DEPTH AND AUSTRALIAN DISTRIBUTION

Found on rock substrates, sometimes with sponges and coral, 5–77 m. From Moreton Bay, Queensland, south to Ulladulla NSW, and at Norfolk Island, Tasman Sea. Eastern Australian endemic.

TYPE LOCALITY

Bate Bay, off Cronulla, NSW.

FURTHER DISTRIBUTION

Australian endemic.

REFERENCES

Rowe and Albertson, 1987: 195–202; figs 1a–b, 2.

REMARKS

Echinaster colemani is most closely related to *E. arcystatus* and *E. glomeratus*.

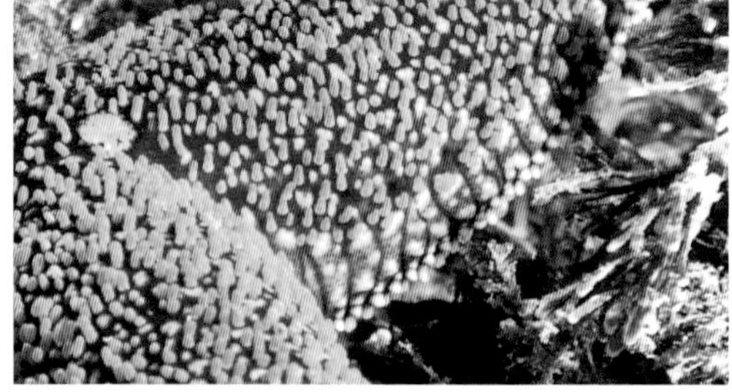

Port Stephens, New South Wales [left] (N. Coleman); intermarginal characteristics, Jervis Bay, New South Wales [above] (J. Turnbull).

GLOBULAR SEASTAR
Echinaster glomeratus
H.L. Clark, 1916

DESCRIPTION
Non autotomous; large disc, moderately elevated as are the bases of the broad, evenly tapering arms; abactinal skeleton forming a wide meshed network with large papular areas with 12–15 papulae; **primary abactinal plates at the junction of the meshes bear clusters of up to 20 stout, sharp spinelets, 0.5–1 mm high**; the actinal interradial plates each have 2–3 spines; the body is covered by a thick skin but the clusters of spinelets are visible in live specimens.

MAXIMUM SIZE
R/r = 135/21 mm (R = 5–6.6r).

COLOUR
Yellow, orange to dark red.

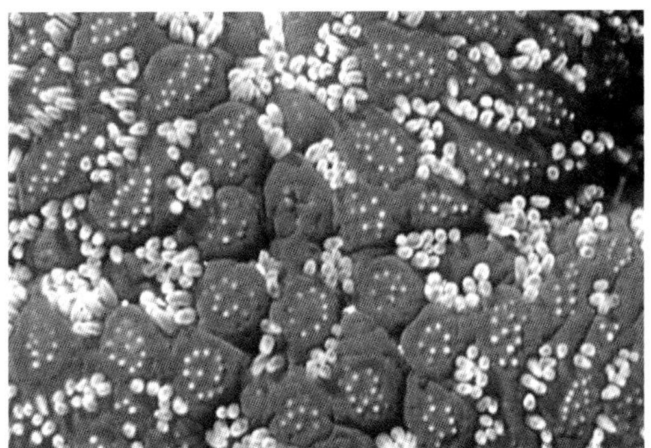

Rottnest Island, Western Australia [top], abactinal plate characteristics [bottom] (S. Morrison).

FOOD
Encrusting sponges and ascidians.

HABITAT, DEPTH AND AUSTRALIAN DISTRIBUTION
In intertidal pools or on algal covered rock, uncommon, 0–64 m. From Cape Jervis, SA to the Houtman Abrolhos (~28°S), WA. It is rarely found north of Cape Naturaliste. Southern Australian endemic.

TYPE LOCALITY
Off Cape Marsden, Kangaroo Island, SA.

FURTHER DISTRIBUTION
Australian endemic.

REFERENCES
Clark, H.L., 1916: 62–64; fig. 8; pls 22–23.

Clark, H.L., 1946: 146.

Coleman, 2007: 23 (food).

Rowe and Albertson, 1987: 197, 199; fig. 4.

LUZON SEASTAR
Echinaster luzonicus
(Gray, 1840)

DESCRIPTION
Disc small; 5–6 (occasionally seven) slender and cylindrical arms; **autotomous** with single separated arms regenerating to produce comet forms; **skeleton a close mesh with numerous short, blunt spines (~1 mm)** and small papular areas with ~3–10 pores; often two madreporites.

MAXIMUM SIZE
R/r = 125/30.5 mm (R = 4.1–9.1r).

COLOUR
Red to dark green, nearly black, occasionally brown.

FOOD
Substrate biofilm and encrusting sponges (Birtles, personal communication).

REPRODUCTION
Asexual reproduction producing comets by autotomy is common.

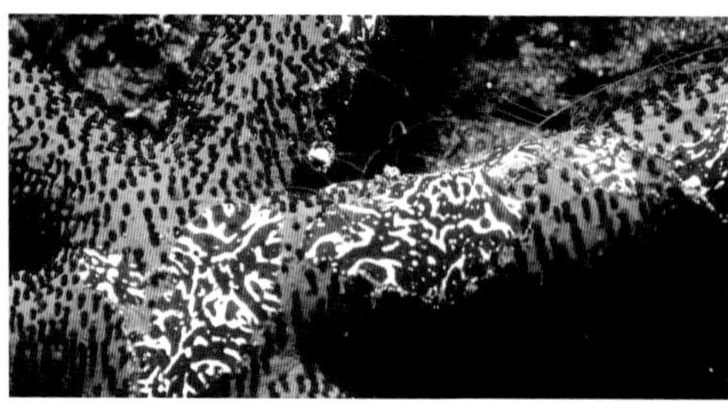

Palau [top] (L. Colins); with episymbiont comb jellies, *Coeloplana astericola* [bottom], Bali, Indonesia (C. Bryce).

HABITAT, DEPTH AND AUSTRALIAN DISTRIBUTION
On coral reefs and rocky shores in the tropics, intertidal to 60 m; very common. From Ningaloo Reef (23°30'S), the inshore and offshore islands of north-western Australia, WA, and the Great Barrier Reef, Queensland, south to Solitary Islands (~30°S), NSW .

TYPE LOCALITY
Luzon, the Philippines.

FURTHER DISTRIBUTION
From the Maldives and Bay of Bengal to the eastern Indian Ocean (Cocos (Keeling) Islands), through Indonesia, the Philippines and Palau, New Caledonia, north to Japan and east to Palmyra Island (160°W) in the Pacific.

REFERENCES
Clark, H.L. 1946: 147–148.

Clark and Rowe, 1971: 72.

Marsh, 1977: 278.

Rowe and Albertson, 1987: 197.

REMARKS
The comb jelly *Coeloplana astericola* is often observed living as an episymbiont on *Echinaster luzonicus* and *E. callosus.*

Ashmore Reef, Timor Sea [top]; Scott Reef, Western Australia [bottom] (C. Bryce).

THORNED SEASTAR
Echinaster stereosomus
Fisher, 1913

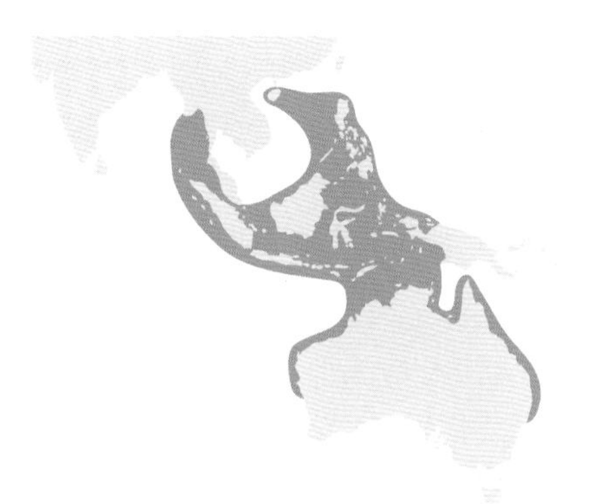

DESCRIPTION
Disc relatively large; five broad based arms, tapering evenly to a narrow tip; **spines short, conical, sharp, in rows**; skeletal meshwork close set with small papular areas with 1–6 pores; the body is covered by skin, leaving only the tips of the spines exposed; there is a single madreporite; not autotomous.

MAXIMUM SIZE
R/r = 110/17 mm (R = 6.5r).

COLOUR
Yellow-brown to pink, sometimes with darker arm tips.

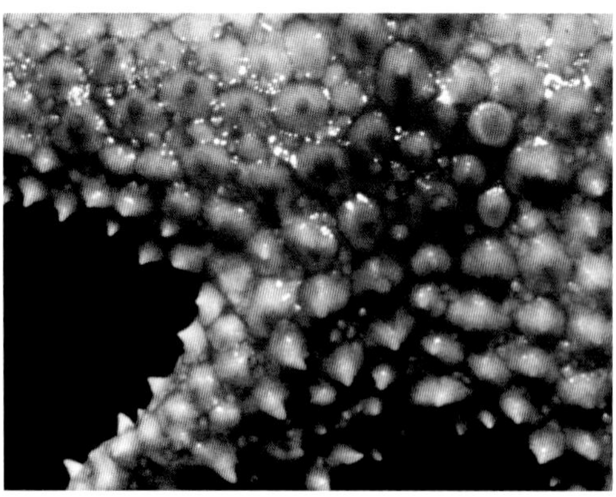

Western Australian Museum (WAMZ99859), abactinal plate characteristics [bottom].

HABITAT, DEPTH AND AUSTRALIAN DISTRIBUTION
On sand, often with shells and rubble, 5–146 m (most in 50–100 m). In Shark Bay and on the inner continental shelf off WA, from Kalbarri (27°40'S), around north-western and northern Australia and on the Queensland Shelf to Dangar Point, NSW.

TYPE LOCALITY
Off Cabugan Grande Island, the Philippines.

FURTHER DISTRIBUTION
Indonesia, the Philippines, Hainan Island (southern China).

REFERENCES
Clark, H.L., 1946: 146 (as *Echinaster acanthodes*).
Fisher, W.K. 1919: 430–432; pl. 122 figs 2–3; pl. 132 figs 6a–e.
Jangoux, M., 1978: 298; pl. 3.
Rowe and Albertson, 1987: 197.

REMARKS
Echinaster stereosomus is sometimes host to a *Dendrogaster* crustacean endoparasite that lives in the coelomic cavity of the arms where it may make a closed gall-like pouch.

SPLENDID SEASTAR
Echinaster superbus
H.L. Clark, 1916

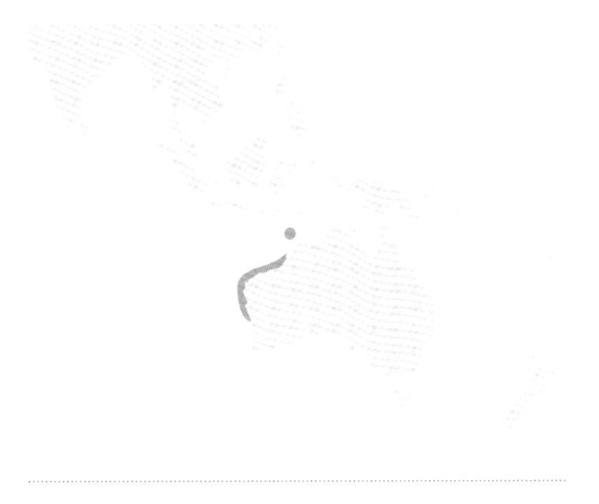

DESCRIPTION
A very large species with a small disc, not elevated; five arms nearly round in section, tapering evenly to a blunt tip; abactinal skeleton forming a rather fine mesh with small papular areas, with ~5–30 pores; **abactinal spines not in groups, numerous, 1–2 mm high by 0.5–1.75 mm thick, most have a club-shaped or blunt chisel-shaped tip sometimes with 3–4 flanges**; club-shaped spines occur on some of the actinal interradial plates; the body is covered with a thick skin through which the spine tips project; not autotomous.

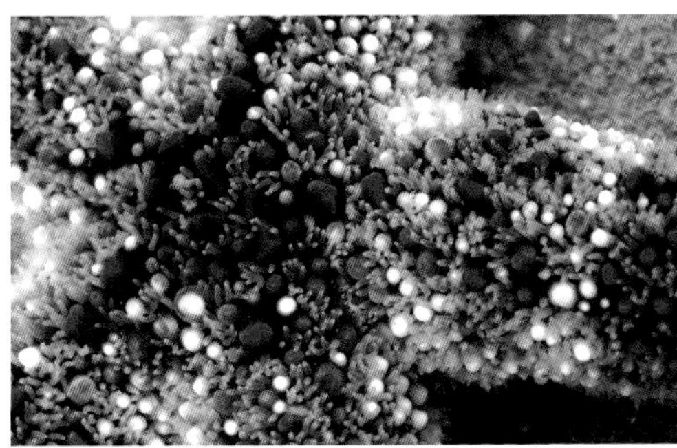

Houtman Abrolhos, Western Australia [top] (C. Bryce); abactinal plate characteristics [bottom], Dampier Archipelago, Western Australia (Z. Richards).

MAXIMUM SIZE
R/r = 145/21 mm (R = 5.75–7r).

COLOUR
Light yellowish green with darker banded areas in centre of disc and radiating along the arms. In the darker bands the spines are old rose colour, elsewhere they are cream white. Papulae olive green.

HABITAT, DEPTH AND AUSTRALIAN DISTRIBUTION
On intertidal rock platforms at Barrow Island, probably on rocky substrates in deeper water, 0–62 m; rare. Only known from north-western Australia between the Houtman Abrolhos and Cape Leveque north of Broome, WA, Ashmore Reef. North-western Australian endemic.

TYPE LOCALITY
Broome, WA.

FURTHER DISTRIBUTION
Australian endemic.

REFERENCES

Clark, H.L., 1916: 64–65; figs 9–10; pls 24–25.

Clark, H.L., 1946: 147.

Clark, A.M. and Rowe, 1971: 72.

Rowe and Albertson, 1987: 197.

VARICOLOURED SEASTAR
Echinaster varicolor
H.L. Clark, 1938

DESCRIPTION
A very large species with a small disc; five arms, fairly stout, cylindrical, tapering distally to a blunt tip; skeletal mesh is open with fairly large papular areas with up to 20 pores; **cylindrical, blunt-tipped spinelets (1–1.5 per mm) 1–2 mm long on the skeletal meshes; the spines are stouter and more widely spaced than on *E. luzonicus*** and some may be pitted at the tips; not autotomous.

MAXIMUM SIZE
R/r = 200/25 mm (R = 8r).

COLOUR
Variable, commonly orange to red but often spotted, for example, pale orange with bright orange spots, dark red-brown with very dark spots, grey with navy blue spots. H.L. Clark reported uniformly violet individuals from the Broome area.

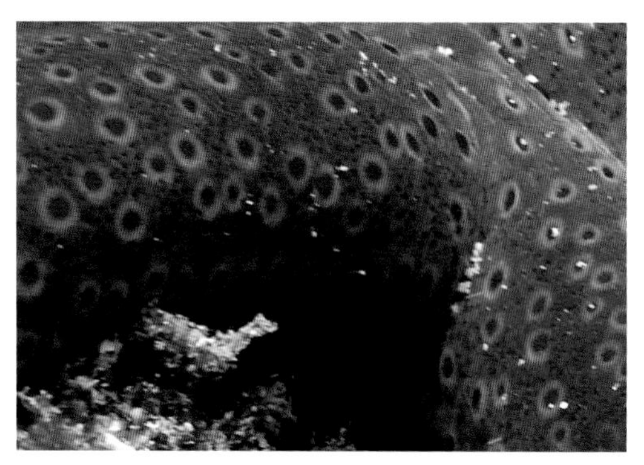

Papular area characteristics [top], Minden Reef, Western Australia (C. Wood), Western Australian Museum [bottom] (WAMZ99854).

HABITAT, DEPTH AND AUSTRALIAN DISTRIBUTION
On sand, seagrass, rock or coral in sheltered waters, moderately common in suitable habitats, 0–52 m. From Esperance, WA to Beagle Gulf, NT.

TYPE LOCALITY
South-west of Broome, WA.

FURTHER DISTRIBUTION
Indonesia, New Caledonia.

REFERENCES
Clark, H.L., 1938: 184; pl. 11 fig. 1.
Clark, H.L., 1946: 148.
Clark, A.M. and Rowe, 1971: 73.
Jangoux, 1986: 152–153.
Rowe and Albertson, 1987: 197.

REMARKS
Specimens from Geographe Bay and the south coast of WA are much smaller than those from the Fremantle area and northwards.

Shark Bay, Western Australia (C. Bryce) [top];
King George Sound, Western Australia (S. Morrison).

PAINTED SEASTAR

Metrodira subulata

Gray, 1840

Western Australian Museum (WAMZ20216).

Dampier Archipelago, Western Australia (N. Coleman).

DESCRIPTION

Small disc with slender, tapering arms; **skeletal plates tessellate, partly overlapping, covered by a thin skin, most plates bear small spines; marginal plates large, forming a side wall to the arm; intermarginal plates present proximally**; papulae are only on the abactinal surface, 1–3 in each area; the most striking recognition character is **the arrangement of the furrow spines, which divide the ambulacral furrow into a separate circular compartment for**

each tube foot; the furrow spines are in a vertical series of four, with three or more spines on the furrow margin.

MAXIMUM SIZE

R/r = 85/12 mm (R = 7r).

COLOUR

Cream to grey or pink with 3–5 brown or blue bands across the arms.

HABITAT, DEPTH AND AUSTRALIAN DISTRIBUTION

On muddy sand, 0–200 m. From off Cape Leeuwin, WA to Moreton Bay, Queensland.

TYPE LOCALITY

Migupou, the Philippines.

FURTHER DISTRIBUTION

Sri Lanka to Indonesia and the Philippines, Gulf of Thailand, Hainan Island to Taiwan Straits.

REFERENCES

Blake, 1980: 163–182.
Clark, H.L., 1938: 187; pl. 11 fig. 3.
Clark, H.L., 1946: 111.
Clark, A.M. and Rowe, 1971: 57; pl. 10 figs 5–6.

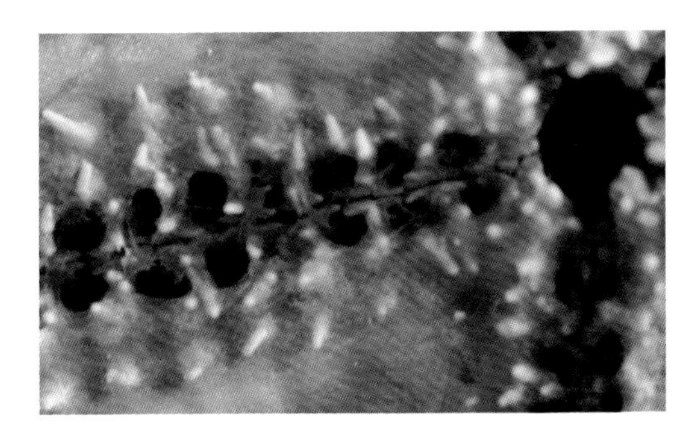

Furrow [top] and abactinal plate characteristics [bottom], Western Australian Museum (WAMZ20216).

REMARKS

Formerly in the monotypic family Metrodiridae, Blake (1980) regarded *Metrodira* as not sufficiently distinct from Echinasteridae to warrant retaining it in a separate family, however the abactinal and marginal plates are unlike any other seastar.

MOSAIC SEASTAR
Plectaster decanus
(Müller and Troschel, 1843)

DESCRIPTION
Superficially similar in appearance to *Echinaster arcystatus* but the **large papular areas are outlined by ridges covered with close set groups of granuliform spinelets** that are not obscured by skin; **there are 3–4 furrow spines in a comb-like row and three subambulacral spines**; the arrangement of furrow spines is unlike any other echinasterid, as they are in line with the ambulacral furrow.

MAXIMUM SIZE
R/r = 125/25 mm (R = 5r).

COLOUR
Variable, for example purple papular areas outlined by red ridges; light brown or orange papular areas outlined with dark olive green; papular areas red and purple with bright yellow skeletal ridges; or papular areas white, papulae and skeletal ridges red. Tube feet lilac coloured or white.

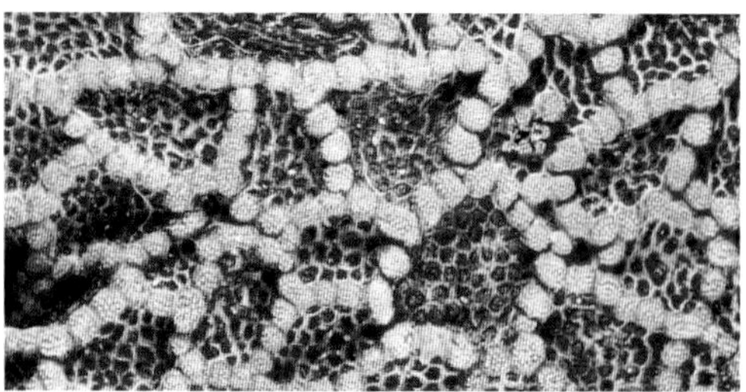

Papular area characteristics [top], Bare Island, New South Wales (J. Turnbull), South Australian Museum [bottom] (SAMAK4137).

FOOD
Encrusting sponges and ascidians.

HABITAT, DEPTH AND AUSTRALIAN DISTRIBUTION
Intertidal, in pools, amongst rocks or under reef ledges on open coast reefs, 0–200 m. From Byron Bay, NSW to Yanchep, WA, including Tasmania. Southern Australian endemic.

TYPE LOCALITY
'South-western Australia'.

FURTHER DISTRIBUTION
Australian endemic.

REFERENCES
Clark, H.L., 1946: 149.

Rowe and Albertson, 1988: 84; fig. 1.

Shepherd, 1968: 751 (food).

REMARKS
Rowe and Albertson (1988) noted that the furrow spines are in a comb-shaped row, unlike any other echinasterid, and suggested that it does not belong in the Echinasteridae.

Jervis Bay, New South Wales (B. Hutchins).

King George Sound, Western Australia (P. Southwood).

FORCIPULATIDA

Coscinasterias muricata, Dusky Sound, New Zealand (R. Robinson).

Forcipulatids generally have a small, well defined disc, and a small and inconspicuous madreporite near the disc centre. They have a thick body and long, flexible arms that are usually round or elliptical in cross section. The arms may be constricted at the base, as they may shed them as a response to disturbance. The seastars have a distinct, reticulate, open meshed skeleton. The abactinal plates are not paxillate and the marginal plates are inconspicuous. Spines or tubercles and papulae are scattered over the seastar surface, the latter extend to the ambulacral row. These seastars have complex pedicellariae with either crossed or straight valves, often both are present. The ambulacral plates are numerous, short, compressed and rarely overlap, and the oral plates are small. The seastars have a strong skeletal framework around the jaw which assists their predatory feeding strategy. The tube feet are in four rows in shallow water species.

(Modified from Byrne et al., 2017.)

ASTERIIDAE

Asteriid seastars often have more than five arms and some are fissiparous. The disc may be sharply delimited from the arms and the tube feet are in 2–4 rows. The skeleton is reticulate but at least the carinal and marginal rows are usually distinct, the abactinal and marginal spines are wreathed with small crossed pedicellariae in some genera. The actinal plates are in one or more rows or are absent. The adambulacral plates are short with spines in a single transverse series and the spines are with or without pedicellariae. Straight and crossed pedicellariae are present on different parts of the body.

This is a large cosmopolitan family of 39 genera which is most diverse in the colder waters of the northern and southern hemispheres. In Australian waters there are seven genera and nine species with five genera and six species in depths of less than 30 m (including an introduced species of the genus *Asterias*).

KEY TO THE GENERA OF ASTERIIDAE IN SHALLOW AUSTRALIAN WATERS

1	Crossed pedicellariae in wreaths around abactinal and superomarginal spines; complete or partial pedicellariae wreaths are present on at least one series of inferomarginal spines.	2
	Actinal plates in one series; superomarginals form the ventrolateral margin; inferomarginals strictly actinal in position; clusters of straight pedicellariae on adambulacral spines.	***Asterias*** page 468
2	Arms fissiparous; one adambulacral furrow spine, without attached pedicellariae.	3
	Two adambulacral furrow spines.	4
3	Arms 7–14; multiple madreporites; alternate carinal plates without lateral lobes; one adambulacral furrow spine.	***Coscinasterias*** page 478
	Arms 6–9; all carinal plates four-lobed.	***Stolasterias*** page 478
4	Five arms; a web of skin links the outer inferomarginal spines along arm.	***Sclerasterias*** page 478
	Arms 5–7, not fissiparous; no web of skin linking the outer inferomarginal spines.	***Astrostole*** page 472

GENUS *ASTERIAS* Linnaeus, 1758

The abactinal skeleton of *Asterias* seastars is an irregular net with meshes of various sizes. The plates are sometimes imbricated by their lobes, and the abactinal plates are not in obvious longitudinal series but occasionally in transverse series, while the carinal series are usually distinguishable and often irregular. The **adambulacral spines have scattered straight pedicellariae**. There is a single series of actinal plates, either spineless or with a single small spine, and a single series of actinal papular areas alternate with the actinal plates. The inferomarginal plates are strictly actinal in position and the superomarginal plates define the margin.

Type species: *Asterias rubens* Linnaeus, 1758 (North Atlantic)

One introduced species, *A. amurensis* (page 469), is found in shallow Australian waters.

Asterias amurensis, Primorsky Krai, Russia (A. Nekrasov).

NORTHERN PACIFIC SEASTAR

Asterias amurensis

Lütken, 1871

South Australian Museum (SAMAK4210).

DESCRIPTION

A very variable species with five stout arms tapering evenly to a sharp tip; lateral margin of arms sharply differentiated from the actinal surface by superomarginals; the abactinal surface is densely covered with very short spines, singly or in groups of 2–4, irregularly arranged, they may form a wavy line along the midline of the arms; spines are conical or cylindrical-conical with 1–3 points, they are more dense on the disc than on the arms; papulae are numerous in the skeletal reticulations abactinally, intermarginally and on the actinal surface; superomarginal plates have 6–8 flat, grooved, blunt spines 2.5–3 mm long; inferomarginal plates have 2–3 spines 3–4.7 mm long; there is an actinal row of spines separated from the marginals and subambulacrals by a bare strip either side; furrow spines single, on alternate plates, are compressed, and bluntly pointed followed by two longer, blunt spines, grooved on their outer side, with three on alternate plates; pedicellariae are abundant; **small ovate to lanceolate straight pedicellariae are scattered on the abactinal surface, superomarginal and inferomarginals and intermarginal plates, bases of spines, and on adambulacral and mouth plates, there are 5–10 on the furrow spines; small crossed pedicellariae are on both surfaces, in a circle around actinal spines and in clusters on marginal and adambulacral plates, but not on furrow spines, which have straight pedicellariae; the presence of pedicellariae on the furrow spines distinguishes it at once from *Uniophora*,** large specimens of which are superficially similar; tube feet are in two rows in specimens with R < 10 mm, in four rows in adults.

MAXIMUM SIZE

R/r = 250/73 mm (R = 3.4–5r).

COLOUR

Variable, yellow, orange, light brown often mottled with purple or with purple arm tips, sometimes all purple.

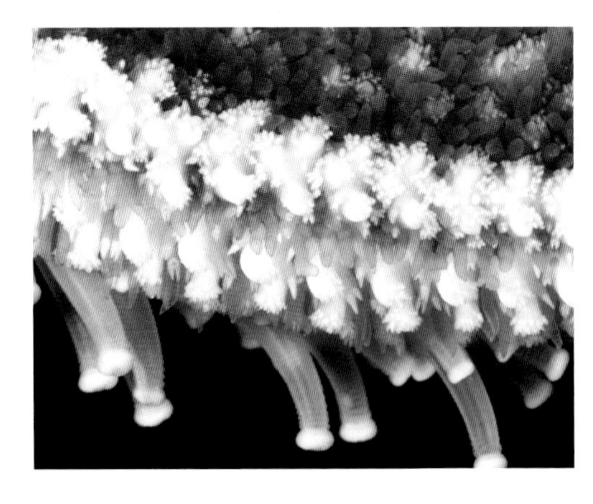

Intermarginal and pedicellariae characteristics, Primorsky Krai, Russia (A. Nekrasov).

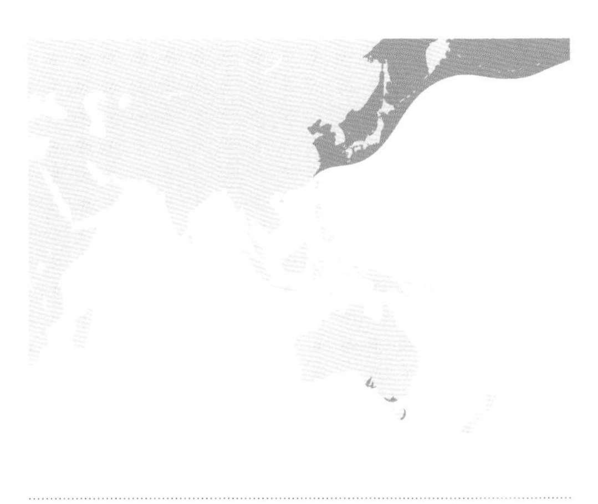

FOOD

A voracious predator feeding on polychaetes, crustaceans, echinoderms, ascidians and a variety of gastropods and large bivalves; it also scavenges dead vertebrates and invertebrates. It is a significant predator on scallops, oysters and mussels.

REPRODUCTION

Gonopores open on the abactinal surface. An adult female may release 5–25 million eggs during winter and spring (July to October), eggs are small (110–150 µm), fertilised externally and have a planktonic larval life of 50–120 days. After metamorphosis, sexual maturity may be reached in as little as one year at R = ~50 mm.

HABITAT, DEPTH AND AUSTRALIAN DISTRIBUTION

On mud, sand, gravel and stones in sheltered waters, 0–200 m. In Australia it is so far known from 0–35 m. Optimum temperature range is 8–22°C, but it can survive and reproduce up to 25°C. First reported from eastern Tasmania, also reported from Port Phillip, Victoria. There have also been recent observational records from SA, around Whyalla in the Spencer Gulf and both sides of Gulf St Vincent (Atlas of Living Australia, 2019).

TYPE LOCALITY

Amur county, northern part of the Gulf of Tartary (Tatarskiy Proliy) near Sakhalin Island (~52°N), Russia.

FURTHER DISTRIBUTION

Japan to the Bering Sea and Arctic Ocean to Alaska in temperatures of 7–10°C.

REFERENCES

Buttermore et al., 1994: 21–25.
Byrne et al., 1997: 673–685 (reproduction and distribution).
Edgar, 1997: 348–349.
Fisher, 1930: 5–23; pls. 1–5; pl. 6 figs 2–8, 10, 11; pl. 7.
Hayashi, 1936: 5–20; pls 1–2.
Zeidler, 1992: 28–29.

REMARKS

Asterias amurensis was first collected from the Derwent River estuary in 1986 but was not correctly identified until 1992, by which time it was abundant. In 1996 the population in the estuary was estimated to be 28 million, with a density of up to nine seastars per m^2. Larvae were introduced in ships' ballast water from Japan (confirmed by genetic studies) in the early 1980s leading to a population explosion within 10 years. It has the potential to cause local extinction of many native species as it is now the top carnivore in the ecosystem with no known predators.

Derwent River estuary, Tasmania (J. Turnbull).

GENUS *ASTROSTOLE* Fisher, 1923

Species of this genus have 5–7 arms, which are not fissiparous. The adambulacral plates have two spines. There is one series of actinal plates bearing spines, while alternate superomarginals are spineless. The superomarginals have a conspicuous beaded area. Pedicellariae are only on the outer of the two inferomarginal spines and the inferomarginal web is absent or rudimentary. Straight pedicellariae are smooth or denticulate and crossed pedicellariae are with or without a slight enlargement of the terminal lateral teeth.

Type species: *Astrostole scabra* (Hutton, 1872) (New Zealand, Australia)

Two species are found in Australia in depths of less than 30 m; *A. rodolphi* (page 473) and *A. scabra* (page 476).

Astrostole rodolphi, Lord Howe Island, Tasman Sea (N. Coleman).

RODOLPH'S SEASTAR
Astrostole rodolphi
(Perrier, 1875)

Australian Museum (AMJ151).

DESCRIPTION

Disc small; seven (occasionally five) arms tapering to an acute or blunt tip, not fissiparous; there is a carinal row of sharply pointed spines up to 5 mm long, one on every plate except near the arm tip, and an incomplete row on the dorsolateral areas of each arm, each spine has a wreath of crossed pedicellariae around its base; proximally most superomarginal plates have one or two spines distally, every third plate carries a large spine wreathed with crossed pedicellariae; inferomarginals large, forming a conspicuous margin to the arm, each plate with two spines, the upper spine is 4–5 mm long, bluntly pointed, the lower is ~3.5 mm long, with a square-cut tip; the upper inferomarginal spine has a cluster of straight pedicellariae on its upper surface; actinal plates small, each with a flattened, square-cut spine; two furrow spines, the inner shorter and narrower than the outer; lanceolate pedicellariae are few, scattered on the abactinal surface; a few large straight pedicellariae (1 mm long) are present on the adambulacral plates near the mouth, and small ones in the actinal areas; papulae small, up to 10 in larger skeletal meshes, most meshes have 3–5, a single series of actinal papulae, 1–3 in each papular area.

MAXIMUM SIZE

R/r = 120/14 mm (R = ~8r, ~9.5br).

COLOUR

Variable, commonly mottled red, yellow, brown, blue to nearly black, often brown and greenish blue. Large marginal spines are bright blue, arm tips often blue (dark purple in specimens from Kermadec Island).

FOOD

A predator and scavenger.

HABITAT, DEPTH AND AUSTRALIAN DISTRIBUTION

On rocky shores of exposed coasts, 0–250 m. From Point Danger, Queensland 28°10'S, 153°32'E, to Bondi, NSW and Norfolk and Lord Howe Islands, Elizabeth and Middleton Reefs, Tasman Sea.

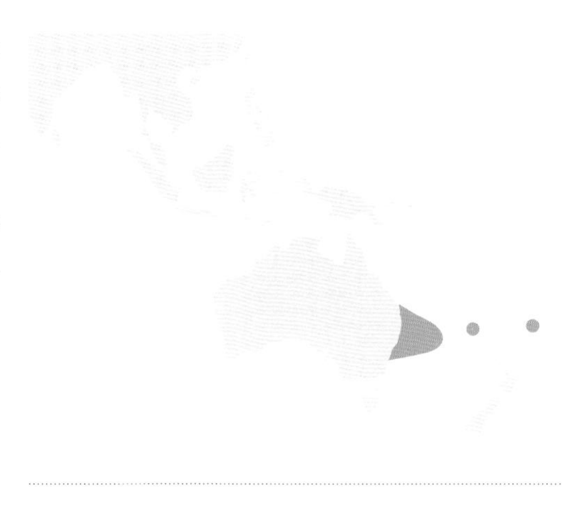

TYPE LOCALITY

Kermadec Island, New Zealand.

FURTHER DISTRIBUTION

Kermadec Island, New Zealand.

REFERENCES

Clark, A.M., 1950: 808; pls 11–12 (as *Astrostole multispina*).

Clark, H.L., 1938: 191; pl. 8 fig. 1 (as *Astrostole insularis*).

Clark, H.L., 1946: 155 (as *Astrostole insularis*).

Clark, A.M. and Mah, 2001: 267.

Mah et al., 2009: 396.

Rowe, 1989: 290 (synonymy).

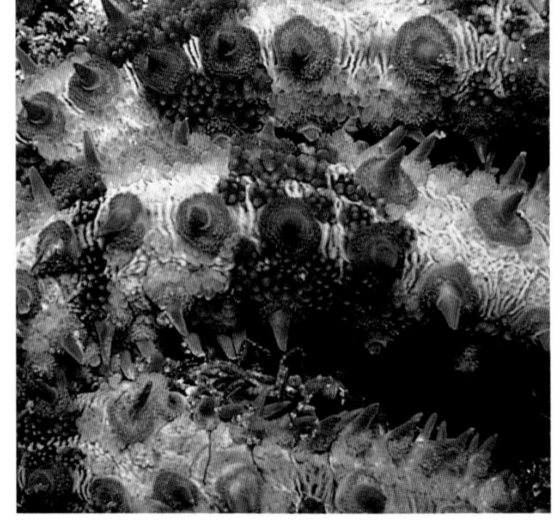

Middleton Reef, Tasman Sea [above] (G. Edgar); abactinal plate and pedicellariae characteristics [right], Lord Howe Island, Tasman Sea (N. Coleman).

Lord Howe Island, Tasman Sea (J. Turnbull).

Lord Howe Island, Tasman Sea (A. Green).

ROUGH-BACK SEASTAR

Astrostole scabra

(Hutton, 1872)

Kaikoura, New Zealand (P. Ryan).

DESCRIPTION

Disc small; **seven arms** tapering to a blunt tip, arms rounded abactinally, flat actinally; **disc plates loosely arranged in several circles, at the arm base there are seven longitudinal series of abactinal plates reducing to five about halfway down the arm**; carinal and superomarginal plates lobed or cruciform, superomarginals with a beaded area; two series of slightly lobed dorsolateral plates to about halfway down the arm, then one; inferomarginal plates form the ventrolateral margin of the arm; carinal plates may have a single spine, rarely two or none; beyond the first quarter of the arm alternate carinals usually lack a spine; proximally most upper dorsolaterals have one spine; over most of each arm every second superomarginal has a spine, inferomarginals have two flattened truncate spines; a single series of actinal plates extends to about halfway down the arm, most have a single spine near the lower inferomarginal spine; adambulacral plates with a pair of flattened truncate furrow spines; papulae small, up to seven in larger skeletal meshes, actinally one per papular area; crossed pedicellariae small, in wreaths around abactinal and superomarginal spines and on the outer face of outer inferomarginal spines, absent from inner spines and actinal spines; straight pedicellariae but variable in shape and size; small lanceolate pedicellariae are present on the furrow margin and on oral spines; larger lanceolate or spatulate pedicellariae are present on the actinal surface; spatulate forms are scattered sparsely over the abactinal surface with small crossed pedicellariae.

Bruny Island, Tasmania [top] (N. Coleman); wreaths of pedicellariae around the abactinal spines [bottom], Tasman Bay, New Zealand (A. Barnett).

MAXIMUM SIZE

R < 365 mm (R = 9.4r),
R/r usually = 150/16 mm.

COLOUR

Grey-blue, blotched with dark grey-brown. Tube feet bright orange.

FOOD

Gastropods, chitons, bivalves and crustaceans, 70 prey species have been recorded. Also a scavenger and sometimes taken in rock lobster pots where it is attracted to the bait.

REPRODUCTION

Spawning takes place during late winter and spring (August and September), development is planktotrophic through bipinnaria and brachiolaria stages. Sexual maturity is attained when R = 110 mm.

HABITAT, DEPTH AND AUSTRALIAN DISTRIBUTION

Under boulders in intertdial rock pools and subtidal rock and rubble, 0–146 m. Coast and continental shelf of NSW, Victoria and Tasmania, rare in NSW, common on the east coast of Tasmania.

TYPE LOCALITY

New Zealand.

FURTHER DISTRIBUTION

New Zealand.

REFERENCES

Dartnall, 1969a: 54–55; figs 1–2.
Hutton, 1872: 5.
Mah et al., 2009: 396.
Town, 1979: 385–395; figs 1–3.
Town, 1980: 111–132; figs 1–14 (reproduction).
Town, 1981: 69–80 (food).

REMARKS

Dartnall (1969a) included *Astrostole scabra* among species believed to have been introduced to Tasmania from New Zealand.

GENUS *COSCINASTERIAS* Verrill, 1867

Species of this genus are fissiparous, reported predominantly in juveniles, with 7–12 arms. Alternate carinal plates are oval, without lateral lobes. Pedicellariae are only on the outer of the two inferomarginal spines. There are one or more series of spiniferous actinal plates. Crossed pedicellariae with an enlarged tooth are on the outer side of the terminal lip, large straight pedicellariae have toothed jaws. The adambulacral plates have a single spine.

Type species: *Coscinasterias muricata* Verrill, 1867 (New Zealand, Australia)

Only the type species (page 480) is found in shallow Australian waters.

GENUS *SCLERASTERIAS* Perrier, 1891

These seastars lack crossed pedicellaria on the inner inferomarginal spine and have an inferomarginal web in the restriction of the dorsolateral plates to, at most, a single adradial series. There is one terminal lateral enlarged tooth on the outer face of each jaw of the crossed pedicellariae (instead of one enlarged tooth on each side of each jaw). The arms are pentagonal in section and the dorsolateral skeleton is regular, consisting in the adult of two regular series of transverse ossicles overlapping about midway between the carinal and superomarginal plates. Superimposed on these overlapping ends, the three- or four-lobed dorsolateral plates form a single series and the dorsolateral spines are in a single series, although usually poorly developed. The superomarginal plates are large with a broad descending lobe and their surfaces have an area of crystal bodies. Alternate superomarginal plates are spineless. Beyond the base of the arm alternate carinal plates are usually spineless. The outer series of inferomarginal spines are united by a retractile web. The mouth angle has one or two pairs of contiguous postoral adambulacral plates with two spinelets. There are no longitudinal intermediate ossicles in the carinal or superomarginal series. Spines are large, usually solitary, spike-like. The abactinal and superomarginal spines are encircled by a wreath of crossed pedicellariae, and the outer inferomarginal spine is encircled by a half wreath. There is a series of small actinal plates that are sometimes spiniferous proximally. Straight pedicellaria are ovate to lanceolate and acute, not spatulate or dentate. Each crossed pedicellaria has one enlarged tooth on the outside of the expanded lip of each jaw and that tooth is not significantly larger than the other terminal teeth. The young of three species are known to have a six-armed fissiparous stage.

Type species: *Sclerasterias guernei* Perrier, 1891 (North Atlantic)

One species, *S. dubia* (page 483), is recorded from shallow Australian waters.

GENUS *STOLASTERIAS* Sladen, 1889

These seastars are fissiparous as adults as well as juveniles. The genus differs from *Coscinasterias* in having all carinal plates four-lobed.

Type species: *Stolasterias tenuispina* (Lamarck, 1816)
(North and South Atlantic, Mediterranean Sea)

One species, *S. calamaria* (page 484), occurs in Australian waters in depths of less than 30 m.

Coscinasterias muricata, Port Phillip, Victoria (M. Norman).

Stolasterias calamaria, False Bay, South Africa (G. Jones).

SOUTHERN ELEVEN-ARMED SEASTAR
Coscinasterias muricata
Verrill, 1867

Western Australian Museum (WAMZ99857).

DESCRIPTION

Disc small; arms 9–11, slender, rounded above, flat below, narrowed at the base, tapering to the tip; four rows of tube feet; skeletal plates form an open mesh, those of the abactinal surface each bear a strong, pointed spine wreathed by minute crossed pedicellariae; **the carinal row consists of four-lobed, spine bearing plates alternating with oval plates (this character distinguishes it from *Stolasterias calamaria*,** with which it was formerly considered synonymous); there are two series of oval dorsolateral plates; superomarginal plates are lobed, strongly imbricating with one or two spines, and with a small beaded area; inferomarginals are oval, with two slightly flattened spines up to 5 mm long; there are up to eight papulae in the larger skeletal meshes, actinally 1–3 in each area; scattered over the abactinal surface and between the spines of the actinal surface are numerous larger, straight pedicellariae; ambulacral furrows are broad, adambulacrals have a single furrow spine; actinal plates are inconspicuous, spineless, to the last third of the arm; arms are readily autotomised from the disc, madreporites are usually multiple as a result of fissiparity.

MAXIMUM SIZE

R/r = 250/35 mm (R = 7.1r).

COLOUR

Variable, for example brown and cream, blue pedicellariae and pinkish brown papulae.

Arms usually banded or blotched. Wreaths of pedicellariae and papulae may be white, base of spines blue. Tube feet white.

FOOD

Preys mainly on bivalve and gastropod molluscs (including abalone), echinoids, crustaceans, occasionally corals and scavenges dead animals. Cannibalism has been observed.

REPRODUCTION

Spawning takes place mainly between November and January in New Zealand; the small eggs (140–160 µm) are red to orange and develop into a feeding bipinnaria larva that becomes a brachiolaria with a sucker and adhesive arms after 18 days, metamorphosis into a juvenile seastar starts after 27 days from fertilisation, and the full transformation into a mobile juvenile seastar takes another six days; the juvenile commences feeding a few days later at a diameter of 950 µm. Asexual reproduction by fission is more common in populations in which food is less abundant or of poorer quality. These seastars put less energy into sexual reproduction than those that have plentiful food (e.g. mussels) available.

Carinal plate characteristics [top], Western Australian Museum (WAMZ99857); Marmion Lagoon, Western Australia [bottom] (C. Bryce).

HABITAT, DEPTH AND AUSTRALIAN DISTRIBUTION

On algal covered reefs, and rock, sand or mud in moderately protected places, 0–140 m. Port Denison, Queensland, to Tasmania, west to the Houtman Abrolhos, WA and from Norfolk and Lord Howe Islands, Tasman Sea.

TYPE LOCALITY

Auckland, New Zealand

FURTHER DISTRIBUTION

New Zealand.

REFERENCES

Barker, 1978: 32–46; figs 1, 3 (as *Coscinasterias calamaria*)

Crump and Barker, 1985: 109–127 (reproduction and food, as *Coscinasterias calamaria*).

Edgar, 1997: 347–348 (as *Coscinasterias calamaria*).

Mah et al., 2009: 396.

Rowe, 1989: 290 (as *Coscinasterias calamaria*).

Shepherd, 1968: 751–752 (food).

Lucky Bay, Western Australia (B. Hutchins).

REMARKS

Fisher (1928) separated as subgenera *Coscinasterias* (with alternating four-lobed and elliptical carinal plates) and *Stolasterias* with all carinal plates four-lobed. Rowe (1989) and Rowe (in Rowe and Gates, 1995) raised the subgenera to full genera with *Coscinasterias muricata* as type species of *Coscinasterias* and *Stolasterias calamaria* as a species of *Stolasterias* of which *S. tenuispina* (Lamarck) is the type species.

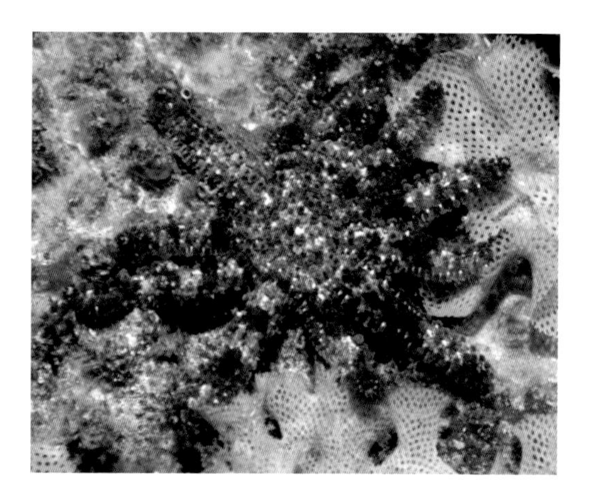

Juvenile, Houtman Abrolhos, Western Australia (C. Bryce).

DUBIUS SEASTAR
Sclerasterias dubia
(H.L. Clark, 1909)

DESCRIPTION
Disc small; **five arms** constricted at the base and tapering to a moderately sharp tip; interradial arcs acute; the abactinal surface is covered by thick skin and has conspicuous spines (about 20) 3–5 mm high on the disc and in five rows on the arms; **each of the abactinal spines is surrounded by a wreath of minute crossed pedicellariae**; the reticulate abactinal skeleton is stout with small papular areas each with up to six pores; large straight pedicellariae (2 mm long) are scattered on the disc and arm bases; there are two cylindrical furrow spines, one outside the other; the actinal surface of the arm has three rows of spines in oblique rows of 2–3, with the marginal spines the longest (~5 mm); the marginal spines are partly webbed forming a marginal fringe, each is deeply furrowed on the actinal side and a cluster of pedicellariae covers half of the upper side; the spines below the marginal fringe lack pedicellariae and membranes, and are shorter and more slender.

MAXIMUM SIZE
R/r = 138/16 mm (R = 5r).

COLOUR
Not recorded.

Holotype, Australian Museum (AMJ11432).

HABITAT, DEPTH AND AUSTRALIAN DISTRIBUTION
On fine sand or rock, 27–500 m. Off Crowdy Head, NSW, to Tasmania and south coast of SA. Southern Australian endemic.

TYPE LOCALITY
Off Botany Bay and off Coogee, NSW.

FURTHER DISTRIBUTION
Australian endemic.

REFERENCES
Clark, H.L., 1909a: 532; pl. 49 figs 3–4; pl. 50 figs 1–2 (as *Coscinasterias dubia*).

Clark, H.L., 1946: 156 (as *Australiaster dubia*).

Fisher, 1923: 253 (as *Australiaster dubia*).

Mah and Foltz, 2011b: 649–650, 653, 655 (as *Sclerasterias dubia*).

REMARKS
Mah and Foltz (2011b) used phylogenetic and molecular studies to determine that *Australiaster* (where *Sclerasterias dubia* was previously located) was not a monotypic genus and synonymised *Australiaster* with *Sclerasterias*. This species is most closely related to *S. mollis* from New Zealand.

ELEVEN-ARMED SEASTAR

Stolasterias calamaria

(Gray, 1840)

Western Australian Museum (WAMZ99858).

DESCRIPTION

In general appearance this species resembles *Coscinasterias muricata* but differs principally in the carinal plates, which are all four-lobed so there is no alternation of dissimilar plates as in *C. muricata*; on the west coast of Australia the distribution of the two species overlaps at the Houtman Abrolhos, otherwise *Stolasterias calamaria* is generally found in tropical waters, *C. muricata* in temperate areas.

MAXIMUM SIZE

R/r = 60/8.5 mm (R = 7r).

COLOUR

Wine red with darker patches.

FOOD

Carnivorous, probably similar to *C. muricata*.

HABITAT, DEPTH AND AUSTRALIAN DISTRIBUTION

On sand or rock substrates, 0–18 m. In Australia it only occurs at the Houtman Abrolhos, Shark Bay and Pt Maud, WA and Heron Island, Queensland.

TYPE LOCALITY

Mauritius.

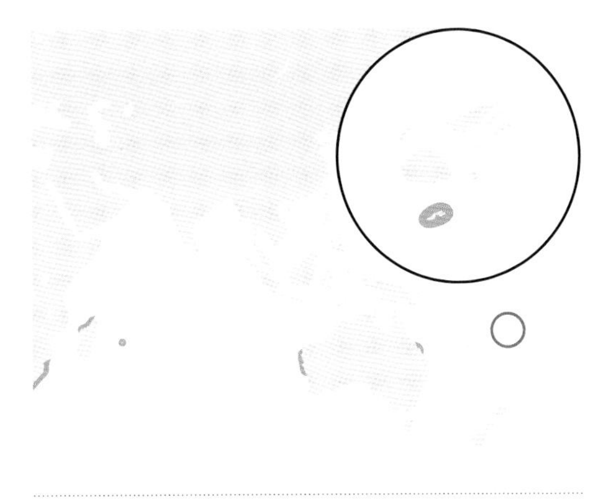

REMARKS

Coscinasterias muricata was described by Verrill (1867), but for many years was included in the synonymy of *C. calamaria* Gray, 1840. Rowe (1989) revived Verrill's original name, retaining Gray's name for the species that occurs principally in the tropical Indian Ocean. Rowe (in Rowe and Gates, 1995) referred *C. calamaria* to Sladen's genus *Stolasterias*, which clarified the difference between *calamaria* (all carinal plates four-lobed) and *muricata* (with alternating four and oval-lobed carinals). Waters and Roy (2003) found *C. muricata* to be a basal species, highly distinct from all other members of the genus. This presents the problem of determining which species is intended when interpreting historical records. Material from southern Australia would almost certainly be *C. muricata* while tropical specimens are likely to be *C. calamaria*.

Mah (2019) accepts *Coscinasterias (Stolasterias) calamaria* as an 'alternate representation'. Clark and Courtman-Stock (1976) note differences between South African and Australian species but did not formally separate them.

FURTHER DISTRIBUTION

South-eastern Africa, Madagascar, Mauritius and Kandavu, Fiji.

REFERENCES

Clark, A.M. and Downey, 1992: 426–427; fig. 63f–g.

Clark, A.M. and Courtman-Stock, 1976: 92–93 (as *Coscinasterias calamaria*).

Clark, A.M. and Rowe, 1971: 40–41 (distribution), 71; pl. 12 figs 5–6.

Mah, 2019: (alternate representation).

Rowe and Gates, 1995: 30 (synonymy).

Waters and Roy 2003: 190 (genetics).

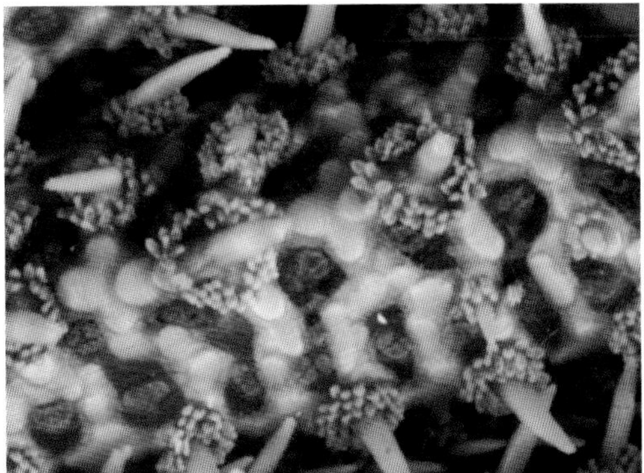

Juvenile, False Bay, South Africa [left] (G. Jones); carinal plate characteristics [above], Western Australian Museum (WAMZ99858).

STICHASTERIDAE

Seastars with either straight or crossed pedicellariae, or both together. They normally have five arms protected by a strong skeleton formed of longitudinal rows of contiguous, triangular, imbricated plates. They have four rows of tube feet or fewer at the base of the arms. The mouth is ambulacral.

Stichasteridae were long considered to belong to the Asteriidae but Mah and Foltz (2011b) published a molecular phylogenetic revision of the Forcipulatacea, and reinstated the family Stichasteridae Perrier (1885) and Slader (1889), and added *Smilasterias* to the family. *Uniophora* was provisionally included on morphological grounds. Stichasterids have minute crossed pedicellariae scattered over the body surface, but these never occur in wreaths around the spines, as in the Asteriidae. Stichasterids have rows of papulae along the arms and blunt, short spines. The endoskeleton is more strongly developed than in Asteriids, and is not as reticulate. Reproduction is both sexual and asexual.

There are three genera of Stichasteridae in Australian waters, each with three species, and all nine species occur in depths of less than 30 m.

KEY TO THE GENERA OF STICHASTERIDAE IN SHALLOW AUSTRALIAN WATERS

1	Arms five, non fissiparous; one madreporite.	2
	Arms 5–8, often of two sizes, may be fissiparous; two or more madreporites; carinal and superomarginal plates broader than other plates; dorsolateral area narrow with one row of plates.	***Allostichaster*** page 487
2	Abactinal skeleton with small carinal plates linked to superomarginals by small dorsolateral plates that may be transversely elongate and form a broad dorsolateral area.	***Smilasterias*** page 493
	Abactinal skeleton formed of large transverse oblong plates, each bearing a large unequal sized subglobular articulated spine, that form longitudinal series.	***Uniophora*** page 498

GENUS *ALLOSTICHASTER* Verrill, 1914

Seastars of this genus have scattered, crossed pedicellariae on the abactinal surface that are not in wreaths around the spines. The skeleton is usually closeknit, with the plates in definite long series. The carinals and superomarginals are broader than the other plates while the adradial plates are narrow, in a single straight or zigzag series. The inferomarginal plates form the edge of the arm. There is usually one series of actinal plates and the adambulacral plates have one furrow and one subambulacral spine. The superomarginal plates have a beaded area. The seastars are usually fissiparous with 5–8 arms and several madreporites.

Type species: *Allostichaster polyplax* (Müller and Troschel, 1844) (New Zealand, Australia) Three species are known from Australian waters in depths of less than 30 m.

Allostichaster polyplax, Deal Island, Tasmania (G. Edgar).

KEY TO THE SPECIES OF *ALLOSTICHASTER* IN SHALLOW AUSTRALIAN WATERS

1	Arms five, non fissiparous.	***A. regularis*** page 492
	Arms more than five, fissiparous.	**2**
2	Arms broad, not tapering, blunt, inferomarginal spines usually single, narrow basally, short broad blade distally, tiny.	***A. palmula*** page 488
	Arms narrow, tapering, pointed, inferomarginal spines usually paired, elongate, club-shaped.	***A. polyplax*** page 490

PALM SEASTAR

Allostichaster palmula

Benavides-Serrata and O'Loughlin, 2007

Paratype, Museums Victoria (NMVF132700).

DESCRIPTION

A very small, fissiparous species usually with six arms, sometimes five; the arms are wide, not tapering and have rounded tips; arms are domed abactinally, flat actinally, margin acute, defined by inferomarginal plates; abactinal plates thick, imbricate; disc plates irregular in form and arrangement, imbricate; carinal plates four-lobed, narrowly imbricating longitudinally, imbricating with superomarginals laterally; rare small dorsolateral plates; superomarginal plates largest, transversely elongate, proximal lobes narrowly imbricating longitudinally and laterally with carinals and inferomarginals; papular areas distinct, smaller than plates, two longitudinal series between carinals and superomarginals; inferomarginal plates longitudinally elongate with single spines, no actinal plates, adambulacral plates transversely narrow, 2–3 contiguous with each inferomarginal plate and 2–3 spinelets on each disc and carinal or superomarginal plate; spinelets widened distally or columnar, coarsely spinous distally; spinelets distributed over abactinal surface of arms and disc; crossed pedicellariae on abactinal and marginal surface; on inferomarginal plates **usually single spines with a proximal stem and a distal broad flat, ribbed blade with a**

serrated margin; furrow spines one per plate with one subequal subambulacral spine; tube feet in four rows.

MAXIMUM SIZE

R = 5 mm (R = 5r).

COLOUR

Brown with white arm tips. Actinal surface white, inferomarginal spines brown.

FOOD

Observed feeding on encrusting bryozoan (*Membranipora*) on a brown algal frond.

REPRODUCTION

Asexually by fissiparity. Sexual reproduction is unknown.

HABITAT, DEPTH AND AUSTRALIAN DISTRIBUTION

On sediment, 0–12 m. Mallacoota to Port Phillip, Victoria. Victorian endemic.

FURTHER DISTRIBUTION

Australian endemic.

TYPE LOCALITY

San Remo, Victoria.

REFERENCE

Benavides-Serrato et al., 2007: 71, 78; figs 1–4.

Abactinal plate and inferomarginal characteristics, Port Phillip, Victoria (L. Altoff).

REMARKS

Allostichaster palmula differs from *A. polyplax* in having broad, blunt-ended arms and not having globose abactinal spinelets. The former has four-lobed carinal plates (triangular in *A. polyplax*), and predominantly single spines on the inferomarginal plates. The inferomarginal spines narrow basally with short, broad, flat ribs and a palm leaf-like blade, while *A. polyplax* has elongate club-shaped spines. The adambulacral plates have subequal furrow and subambulacral spines, while in *A. polyplax* the subambulacral spine is broader than the furrow spine.

FOUR AND FOUR SEASTAR
Allostichaster polyplax
(Müller and Troschel, 1844)

DESCRIPTION

A small **fissiparous** seastar with scattered abactinal crossed pedicellariae that are not in wreaths around spines, and tube feet in four rows; disc small, slightly domed, **eight arms, four usually larger than the rest; two madreporites in opposite interradial areas**; skeleton closeknit, plates in definite long series; inferomarginal plates form a ventrolateral border to the arm; abactinal plates have 1–5 short, stubby spines with a slightly expanded tip; cleaned superomarginals have a beaded surface; inferomarginal plates have two longer and heavier spines flattened and truncate at the tips; actinal plates in a single series extending to about half the arm length, often with a spine similar to, but shorter, than the inferomarginal spines; adambulacral plates each bear two spines, oral plates small with three spines, papular areas small with up to 1–3 papulae in each, no papulae on the actinal surface; **straight and crossed pedicellariae small, scattered over abactinal surface and a few along the furrow margin**; a single straight pedicellaria usually occurs on the actinal surface between the inferomarginal and actinal spines.

Abactinal and inferomarginal plate characteristics, South Australian Museum (SAMAK4186).

MAXIMUM SIZE

R/r = 40/6 mm (R = 6.6r).

COLOUR

Variable, usually mottled brown, greenish, cream, orange and red.

FOOD

Small gastropod and bivalve molluscs.

REPRODUCTION

Asexual reproduction predominates. Opposing halves grip the substrate and pull against each other until the seastar tears in two after 1–24 hours.

HABITAT, DEPTH AND AUSTRALIAN DISTRIBUTION

Under intertidal boulders on rock platforms, among mussels and seagrass or on mud, 0–238 m. South Solitary Island, NSW to Houtman Abrolhos, WA, including Tasmania.

TYPE LOCALITY

Tasmania.

FURTHER DISTRIBUTION

New Zealand.

REFERENCES

Clark, H.L., 1946: 157.
Emson, 1978: 321–329.
Emson and Wilkie, 1980: 155–250.
Shepherd, 1968: 752 (food).

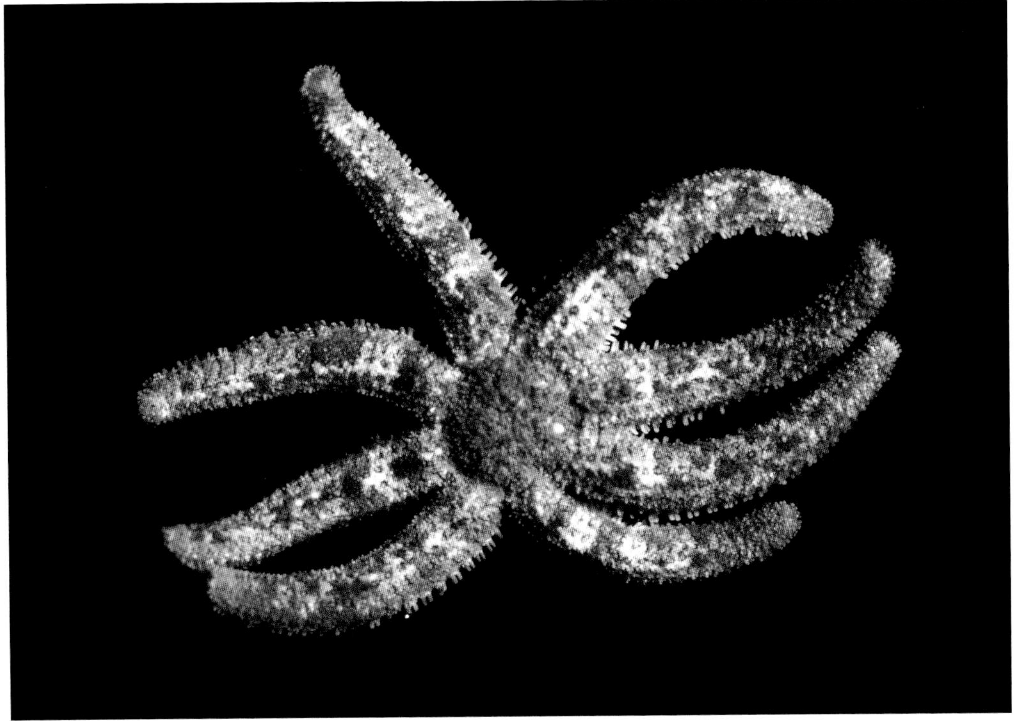

Cape Paterson, Victoria [top] (D. Paul);
Geographe Bay, Western Australia [bottom] (L. Marsh).

REGULAR SEASTAR
Allostichaster regularis
H.L. Clark, 1928

DESCRIPTION

A small seastar with **five fairly regular arms, tapering to a blunt tip; not readily autotomous**; skeleton open, reticulate, but only 1–3 pores in each area, none on the actinal surface; **abactinal plates carry capitate spinelets and a few scattered small crossed pedicellariae**; superomarginals have a beaded surface and about five small, somewhat capitate spinelets; carinal plates regular, corresponding to superomarginals; on the basal part of the arm there is a single series of dorsolateral plates, distally the carinal plates reach the superomarginals; inferomarginals form a distinct angular margin to each arm; actinal surface flat; each inferomarginal carries four spines and ~6 pedicellariae; no actinal plates.

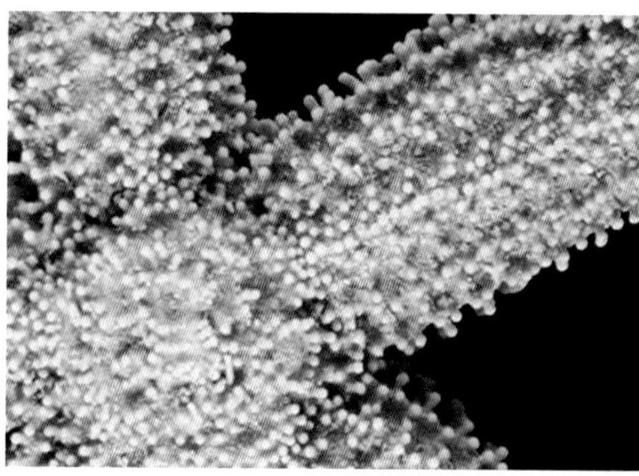

Holotype, South Australian Museum (SAMAK169), abactinal and superomarginal plate characteristics [bottom].

MAXIMUM SIZE

R/r = 35/7 mm (R = 5r).

COLOUR

Variable, usually mottled brown, greenish, cream, orange and red.

FOOD

Small gastropod and bivalve molluscs.

HABITAT, DEPTH AND AUSTRALIAN DISTRIBUTION

Intertidal under rocks, and subtidally, 0–174 m. From off Moreton Bay, Queensland and north-eastern Tasmania to Spencer Gulf, SA. South-eastern Australian endemic.

TYPE LOCALITY

Gulf St Vincent, SA.

FURTHER DISTRIBUTION

Australian endemic.

REFERENCES

Clark, H.L.,1928: 400–401; fig. 115.
Zeidler and Shepherd, 1982: 417.

GENUS *SMILASTERIAS* Sladen, 1889

Species of this genus are quite small with R < 65 mm. They have five subcylindrical arms and one madroporite, and are not fissiparous. The abactinal skeleton is finely reticulate and the carinal plates are small and often irregular. The carinal plates form a fine median longitudinal ridge that is linked to the superomarginal plates by 7–16 small dorsolateral plates, which are commonly transversely elongate. This creates transverse ribbing in an irregular series of longitudinal linkages. Both series of marginal plates are small but distinct. There are no actinal papulae. The inferomarginal plates have an oblique comb of 2–3 flattened spines while the adambulacral plates have 2–4. The abactinal spinelets are numerous and spaced or grouped on plates, they are slightly tapering to clavate and vary from stout to thin. Crossed and straight pedicellariae are present and are not clustered around or on spines or spinelets.

Type species: *Smilasterias scalprifera* (Sladen, 1889) (Subantarctic, South Atlantic)

Seven species are recognised, most are from Subantarctic waters but three species are found in Tasmania. Two of these species extend into Victoria and one, *Smilasterias irregularis*, is also found in NSW and SA.

KEY TO THE SPECIES OF *SMILASTERIAS* IN SHALLOW AUSTRALIAN WATERS

1	A short series of thin actinal plates present on arms; pedicellariae present on upper abactinal surface; oral spines lacking pedicellariae; arms long, with only slight proximal swelling; R > 4br.	**2**
	Series of actinal plates lacking; upper abactinal surface lacking pedicellariae; a few pedicellariae on oral spines; arms short, swollen proximally; R < 4br.	***S. tasmaniae*** page 497
2	Superomarginal plates contiguous or imbricating longitudinally, transverse papular areas very rarely continuous between them; superomarginal plates with mostly three spinelets, one on a prominent proximal lobe; mostly one irregular series of longitudinal dorsolateral plate linkages along mid-arm; readily autotomous; live colour mostly reddish brown and cream.	***S. irregularis*** page 494
	Superomarginal plates often separated longitudinally, with transverse papular areas continuous between them; superomarginal plates mostly with two spinelets, aligned transversely; mostly two irregular series of longitudinal dorsolateral plate linkages along mid-arm; not readily autotomous; live colour very dark grey over pale cream.	***S. multipara*** page 495

IRREGULAR SEASTAR
Smilasterias irregularis
H.L. Clark, 1928

DESCRIPTION

Five arms each constricted basally, slightly swollen proximally, tapering to a rounded tip and readily autotomized; the abactinal skeleton is closely reticulated, at least distally the carinal row is distinct; superomarginal plates lobed, with three spinelets per plate imbricating or at least touching, separating papular areas above and below; mostly a single series of longitudinal dorsolateral plate linkages along the mid-arm; up to four actinal plates in a single series; straight pedicellaria may occur (three per plate) actinally, marginally and abactinally; up to six papulae per area.

Kangaroo Island, South Australia [top] (N. Coleman); abactinal plate characteristics [bottom], South Australian Museum (SAMAK1784).

MAXIMUM SIZE

R/r = 65/6.5 mm (R = 9–10r).

COLOUR

Predominantly mottled red or reddish brown and cream.

FOOD

Small gastropods and chitons.

HABITAT, DEPTH AND AUSTRALIAN DISTRIBUTION

Under boulders or on subtidal rocky substrate, 0–30 m. Found from Shellharbour, NSW to Nuyts Archipelago, SA and north coast of Tasmania. South-eastern Australian endemic.

TYPE LOCALITY

Spencer Gulf or Gulf St Vincent, SA.

FURTHER DISTRIBUTION

Australian endemic.

REFERENCES

Edgar, 1997: 350.

O'Loughlin and O'Hara, 1990: 309–311, 317–318; figs 1–2; pl. 1h.

Shepherd, 1968: 752–753 (food).

BROODING SEASTAR
Smilasterias multipara
O'Loughlin and O'Hara, 1990

Table Cape, Tasmania (L. Marsh).

DESCRIPTION

Five arms, subcylindrical, constricted basally, then slightly swollen before tapering to a rounded tip, not readily autotomous; the abactinal skeleton is very finely reticulate; narrow superomarginal plates often separated longitudinally with papular areas continuous between them, two spinelets on each superomarginal; mostly two irregular series of longitudinal dorsolateral plate linkages along the mid-arm; up to six actinal plates in a single series; straight pedicellariae, few on actinal surface; 1–5 papulae per papular area.

MAXIMUM SIZE

R/r = 38/5 mm (R = 7.5r).

COLOUR

Pale cream banded with varying amounts of dark brownish grey or greenish grey. Actinal surface white to cream with dark grey or red-brown mosaic at ends of arms.

REPRODUCTION

Sexes are separate, eggs are fertilised externally then transferred along the ambulacral furrow to the mouth and brooded in pockets in the cardiac stomach, with up to 300 juveniles (R ≤ 1 mm) in one animal, that have 5–10 pairs of tube feet when released. Brooding females have been found in October and November.

HABITAT, DEPTH AND AUSTRALIAN DISTRIBUTION

Found on exposed ocean platforms on the underside of boulders, intertidal and shallow subtidal, 0–3 m; abundant in some areas. In Victoria, west of Wilsons Promontory, on the north and east coasts of Tasmania and the Bass Strait islands. Southern endemic.

TYPE LOCALITY

Flinders, Victoria.

FURTHER DISTRIBUTION

Australian endemic.

REFERENCES

Edgar, 1997: 350–351.

O'Loughlin and O'Hara, 1990: 309–315; figs 1–3; pl. 1a–b.

REMARKS

This species was named *multus* (many) *parere* (to bear) in reference to its brooding habit.

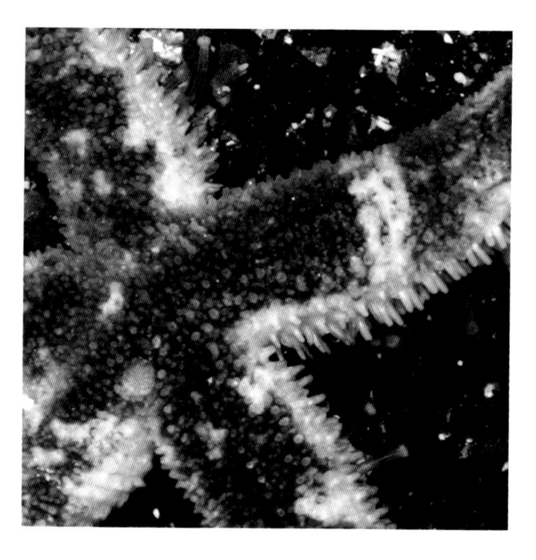

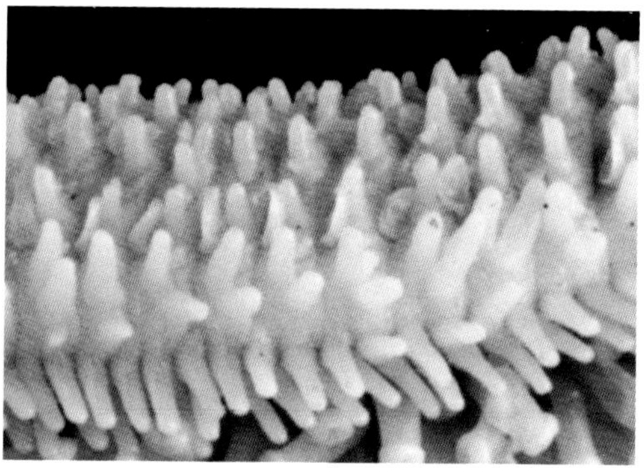

Superomarginal characteristics [above], paratype, Western Australian Museum (WAMZ1721), Port Philip Bay, Victoria [left] (M. Norman).

BRUNY ISLAND SEASTAR
Smilasterias tasmaniae
O'Loughlin and O'Hara, 1990

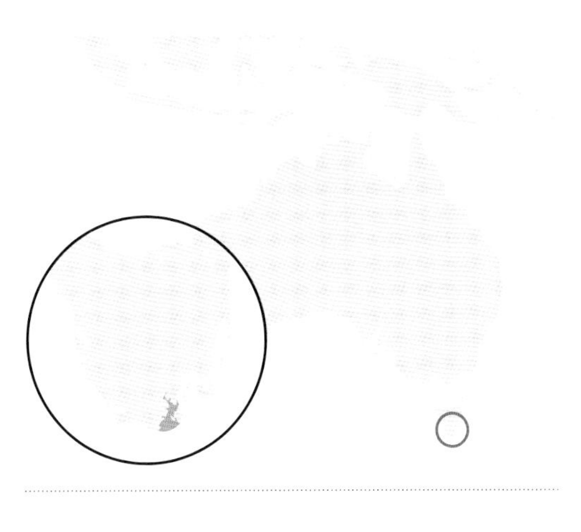

DESCRIPTION
Five subcylindrical arms, constricted basally, slightly swollen proximally, narrowing distally to a rounded tip, not readily autotomous; the abactinal skeleton is reticulate and compact, with longitudinal and transverse series of plates not very evident; dorsolateral plates with up to three irregular series of longitudinal linkages; superomarginal plates with 2–3 spinelets; no actinal plates; a few straight pedicellariae on the oral spines; papulae single, no actinal papulae.

MAXIMUM SIZE
R/r = 20/5 mm (R = 4r).

COLOUR
Tan or brown.

HABITAT, DEPTH AND AUSTRALIAN DISTRIBUTION
Found under rocks at low tide and on rock with brown kelp, 0–8 m. Only found around Bruny Island (~43°S), Tasmania. South-eastern Tasmanian endemic.

TYPE LOCALITY
Lighthouse Bay, Bruny Island, Tasmania.

FURTHER DISTRIBUTION
Australian endemic.

REFERENCES
Edgar, 1997: 350–351.

O'Loughlin and O'Hara, 1990: 309–311, 315–316; fig. 1; pl. 1c–e.

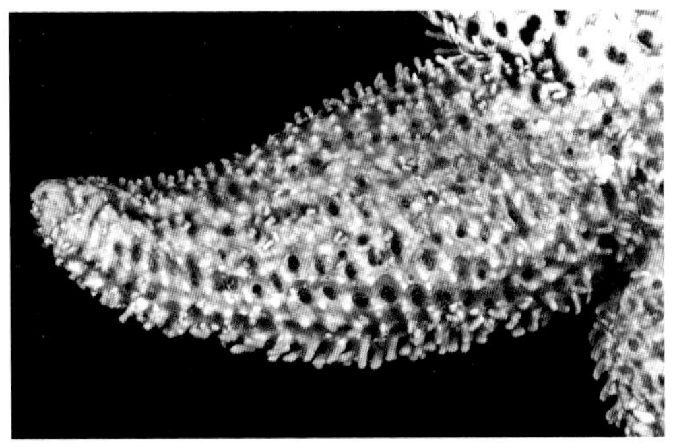

Bruny Island, Tasmania [left] (J. Esling); abactinal plate characteristics [above], holotype, Australian Museum (AMJ11395).

GENUS *UNIOPHORA* Gray, 1840

Seastars that have five arms and four rows of tube feet, at least basally. The carinal plates are large in straight, zigzag or irregular pattern. The abactinal skeleton has large irregular spaces in two dorsolateral series each side of the carinals, with one well defined but irregular series of dorsolateral plates. The abactinal spines are coarse, globose, capitate, or in *Uniophora nuda*, small and pointed. The superomarginals have one or more stout capitate spines while the inferomarginals have 1–3 spines. The actinal plates are usually in 2–3 series. The adambulacrals have two spines that never have pedicellariae. Straight pedicellariae may be on oral spines and on the furrow margin of adambulacral plates while small crossed pedicellariae are found all over the abactinal surface.

Type species: *Uniophora granifera* (Lamarck, 1816) (Australia)

Uniophora is an endemic southern Australian genus with species found from northern NSW to Lancelin (~31°S) in WA. The species are highly variable, with nine species being recognised by H.L. Clark (1946). A critical examination of a large number of specimens by Shepherd (1967b) reduced the South Australian species to two, *Uniophora granifera* and *U. nuda*. *U. granifera* is the most widespread (from NSW to Tasmania and SA) while *U. nuda* is endemic to SA and *U. dyscrita* to WA. Genetic studies may clarify the distinction between the three species.

Large specimens of *Uniophora* superficially resemble *Asterias amurensis* but the presence of pedicellariae on the furrow spines of *Asterias* easily distinguishes the two genera.

(Shepherd, 1967b.)

Uniophora granifera (left) and *U. nuda* (right), Chain of Bays, South Australia (A. Green).

KEY TO THE SPECIES OF *UNIOPHORA* IN SHALLOW AUSTRALIAN WATERS

1	Abactinal spines coarse, globose, capitate.	**2**
	Abactinal spines small and pointed.	***U. nuda*** page 504
2	Distribution west of the Great Australian Bight (Esperance to Lancelin, WA).	***U. dyscrita*** page 500
	Distribution east of the Great Australian Bight (NSW to Spencer Gulf, SA).	***U. granifera*** page 501

Uniophora granifera, Jervis Bay, New South Wales (A. Green).

VEXING SEASTAR
Uniophora dyscrita
H.L. Clark, 1923

DESCRIPTION
The carinal series of plates is usually zigzag in arrangement and the carinal spines are capitate or bluntly pointed; dorsolateral spines are irregularly arranged, often in groups, smaller than carinals; abactinal plates form an open network, the meshes are in four rows at the arm base; superomarginal spines are capitate or flattened; ventrolateral spines are present; papulae are in groups of 4–12 abactinally and intermarginally; pedicellariae are absent from furrow spines.

MAXIMUM SIZE
R/r = 70/17.5 mm (R = 4r).

COLOUR
Variable, for example white with pink markings; grey with pink spines; reddish or greenish brown; olive green with pink or orange-brown spines; or mottled pink and dark brown with pink spines.

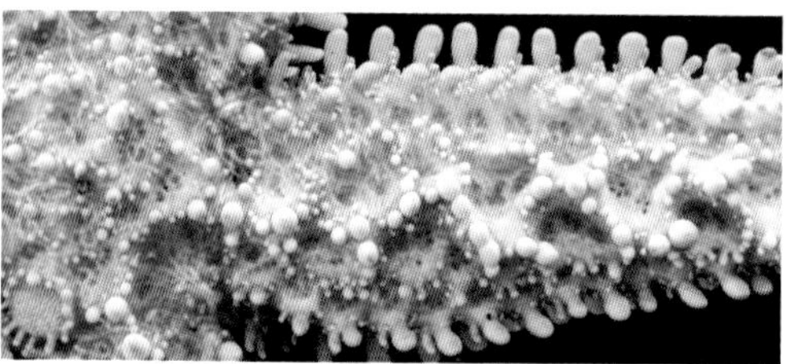

Western Australian Museum (WAMZ99864), abactinal plate characteristics [bottom].

FOOD
Carnivorous, probably feeds on bivalves and encrusting organisms on jetty piles.

HABITAT, DEPTH AND AUSTRALIAN DISTRIBUTION
Among rocks, sand, seagrass and mussel beds, and on jetty piles, rare except in Geographe Bay and Cockburn Sound, 2–185 m. From Esperance to Lancelin (31°S), WA. South-western Australian endemic.

TYPE LOCALITY
Garden Island, near Fremantle, WA.

FURTHER DISTRIBUTION
Australian endemic.

REFERENCES
Shepherd, 1967b: 7–8; table 2; fig. 3; pl. 1.

REMARKS
Large specimens have a superficial resemblance to *Asterias amurensis* but *Uniophora dyscrita* lacks pedicellariae on the furrow spines. Further studies may show that *U. dyscrita* and *U. granifera* are not separated by the Great Australian Bight and the two species may intergrade.

BEADED SEASTAR
Uniophora granifera
(Lamarck, 1816)

Australian Museum (AMJ6884).

DESCRIPTION

The carinal series of plates are straight, zigzag distally or zigzag, sometimes irregularly with globose capitate or bluntly pointed spines, single or in groups on carinal plates; dorsolateral spines in 1–2 regular or irregular series each side of carinals; superomarginal spines more or less capitate, often in groups of 2–3; ventrolateral spines are present; pedicellariae are absent from furrow spines; there is great variability within this species attested by the six specific names it has received; Shepherd (1967b) recognised a 'granifera' form and a 'multispina' form but regarded them as variations of the one species.

Port Jackson, New South Wales (N. Coleman).

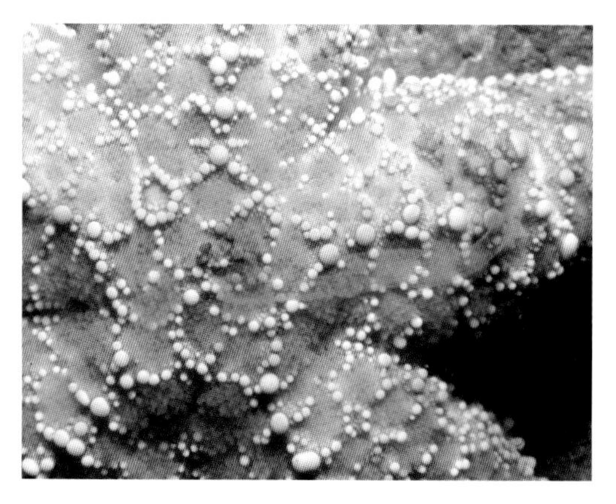

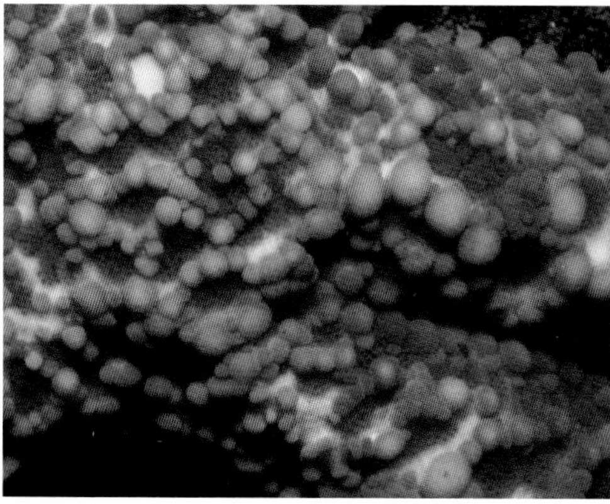

Abactinal plate characteristics, Port Jackson, New South Wales [top], Flinders Bay, Victoria [bottom] (N. Coleman).

MAXIMUM SIZE

R < 102 mm.

COLOUR

Cream to pink, red to orange, occasionally mottled. Papular areas sometimes bright blue.

FOOD

Small ascidians, bivalve and gastropod molluscs.

HABITAT, DEPTH AND AUSTRALIAN DISTRIBUTION

Rocky or sandy bottom, 0–143 m. From North-west Solitary Island, NSW to Spencer Gulf, SA, including Tasmania. South-eastern Australian endemic.

TYPE LOCALITY

'Australian seas'.

FURTHER DISTRIBUTION

Australian endemic.

REFERENCES

Clark, H.L. 1946: 159 ((as *Uniophora granifera, U. sinusoida* (Perrier, 1875), *U. multispina* Clark, 1928, *U. obesa* Clark, 1928, *U. fungifera* (Perrier, 1875), *U. uniserialis* Clark, 1928)).

Shepherd, 1967b: 3–6; tables 1–2; figs 1–3.

Shepherd, 1968: 753 (food).

REMARKS

H.L. Clark (1928) described four new species of *Uniophora* and recognised five others, one described by Lamarck (1816) and three by Perrier (1875), all from SA except *U. dyscrita* Clark 1923, from WA. Shepherd (1967b) revised the genus and determined from morphology and field studies that six species should be synonymised with *U. granifera* with the possibility that *U. dyscrita* might be shown to fall within the range of *U. granifera.*

Feeding on a bed of mussels (*Mytilus*), Port Phillip, Victoria (A. Green).

Cape Paterson, Victoria (A. Green).

NAKED SEASTAR
Uniophora nuda
(Perrier, 1875)

DESCRIPTION
Similar to *Uniophora granifera* but the dorsal surface is covered with a thick skin and is almost devoid of spines, except for a few small pointed spines near the arm tips; carinal plates are in a straight or zigzag series; dorsolateral spines are absent, superomarginal and ventrolateral spines are rare or absent.

MAXIMUM SIZE
R/r = 35/9 mm (R = 3.9r).

COLOUR
Beige to reddish brown.

FOOD
Hammer oysters (*Malleus*) and other bivalves and ascidians.

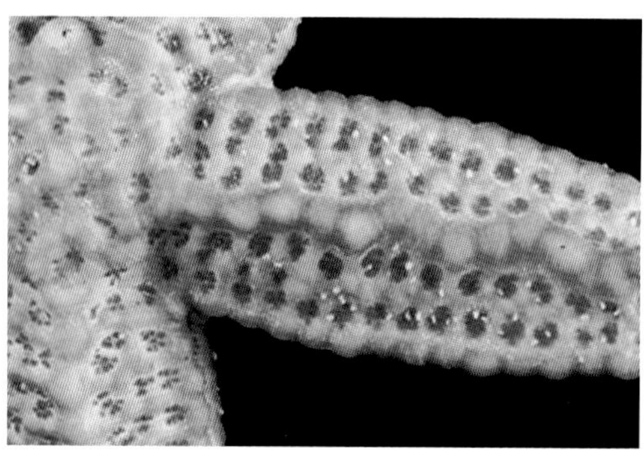

South Australian Museum (SAMAK599), abactinal plate characteristics [bottom].

HABITAT, DEPTH AND AUSTRALIAN DISTRIBUTION
On seagrass beds, and sandy and shelly bottom, 0–60 m. Relatively rare and known only from the lower regions of Spencer Gulf and Gulf St Vincent and the north coast of Kangaroo Island, SA. South Australian endemic.

TYPE LOCALITY
Port Lincoln, SA.

FURTHER DISTRIBUTION
Australian endemic.

REFERENCES
Clark, H.L., 1946: 160–161 (as *Uniophora nuda* and *U. gymnonota* H.L. Clark 1928).

Shepherd, 1967b: 6–7.

Shepherd, 1968: 753–754 (food).

REMARKS
It is possible that further study will show that *Uniophora nuda* intergrades with *U. granifera*.

Gulf St Vincent, South Australia [top] (D. Muirhead);
Chain of Bays, South Australia [bottom] (A. Green).

GLOSSARY

Abactinal (or **aboral**) **surface**: the upper surface of the seastar on which the madreoporite is found (Figure 1A).

Abactinal (or **aboral**) **plates** (or **ossicles**): (Figure 3C–E) can take the form of paxillae (column-like, e.g. *Astropecten* and *Luidia*), tabulae (table-like, e.g. *Nectria*), or flat or convex plates that may be covered by bare skin (e.g. *Leiaster*), the skin may be covered by granules (e.g. *Fromia*) or spinules (e.g. *Nepanthia*). Plates may be close fitting like a pavement (e.g. *Tosia* and *Stellaster*), or overlapping like roof tiles (imbricate, e.g. *Patiriella*), or may be narrow, skin-covered and widely spaced, leaving large papular areas or spaces between (e.g. *Echinaster, Coscinasterias*).

Acicular: needle-like.

Actinal (or **oral**) **surface**: the lower surface of the seastar on which the mouth is found centrally, and the ambulacral groove radially in the arms (Figure 1B). The actinal interradial area is the area between the arms on the actinal surface (Figure 3A).

Ab-: away from, for example the abactinal surface (upper surface away from the mouth).

Ad-: near to, for example the adambulacral surface (adjacent to the ambulacral furrow or groove).

Adradial plates: the row or rows of plates between the carinal and superomarginal plates (Figure 3C–D).

Ambitus: the outer margin of a seastar.

Ambulacral groove: the arm furrow (Figures 1B, 3B) on the actinal surface of the arms that contain the tube feet. The furrow sides are formed by the adambulacral plates, and the arm furrow is roofed by the ambulacral plates (Figure 3C).

Ampulla: the bulb-like portion of the tube foot internal to the ambulacral plate; it may be single or double (Figures 2, 5).

Anus: the external opening of the digestive tract in most seastars, it is central on the abactinal surface of the disc (Figures 1A, 5).

Arc: applied to the curved margin of the disc between two arms (Figure 1A).

Arm (or **ray**): a movable or rigid ambulacral projection distal to the disc, that carries a radial branch of the water vascular system and the nervous system (Figure 1A).

Asteroid: seastar, a member of the echinoderm class Asteroidea with a central often pentagonal or stellate disc and typically five arms (Figure 1A), but these may be 4 to ~50 in number. The underside of the arms has a median groove or ambulacral furrow (Figures 1B, 3A–B) into which hydraulically operated tube feet or podia protrude for locomotion (Figures 2, 5). Seastars have an internal skeleton of separate calcareous ossicles on plates, bound by connective tissue and usually bearing externally projecting spines, tubercles or knobs. The arms contain gonads and a pair of digestive caecae (Figure 5).

Autotomous: capable of shedding an arm, to escape a predator or as part of a process of asexual reproduction e.g. *Linckia multifora* (Figure 6).

Benthic: living at the bottom of the sea.

Bilateral symmetry: the condition when an organism can be divided into similar halves.

Bipinnaria: a free swimming larval stage of some asteroids having blunt larval appendages that support two ciliated bands (Figure 7). It may develop into a brachiolaria larva or undergo metamorphosis in the plankton (*Luidia* and *Astropecten*).

Brachiolaria: the pre-settlement stage of seastar larvae (Figure 7). Characterised by having three short blunt brachiolar arms tipped with adhesive cells and between them a sucker used for attachment to the substrate when ready for metamorphosis.

br: arm breadth, measured at the base of the arm (Figure 8).

Brooding: the reproductive mode in which the embryos are protected on, in, or beneath the parent and emerge as tiny crawl away juveniles.

Calcite: the mineral form of calcium carbonate that makes up the echinoderm skeleton.

Carinal: refers to plates on the midline of the abactinal surface (Figure 3C–D).

Chevron: inverted V shape.

Coelom: a fluid filled cavity within a seastar's body, surrounding the digestive, reproductive and water vascular systems (Figure 5A). Outgrowths of the coelom protrude through the body wall as breathing organs (papulae) (Figure 5A).

Commensal: an organism that lives in association with another organism, and which usually benefits from the partnership without harming its host.

Congeners: the species belonging to a single genus.

Crystal bodies: the tiny glassy beads on the surface of skeletal plates in some asteroid species. Also called glassy convexities, or glassy tubercles.

Dermis: the layer of cells beneath the epidermal covering of the body wall. The skeleton develops in the dermis, embedded in connective tissue.

Digitiform: finger-like.

Disc: the round or pentagonal central body region containing the body's organs (Figure 1A).

Distal: in a direction away from the centre of the body (i.e. toward the tip of the arm).

Dorsolateral (or **adradial**) **plates**: lateral abactinal plates, often in the interradial area.

Embryo: an early developmental stage that is enclosed in a fertilisation membrane or protected by the body of the parent.

Endemic: a term used in biogeography to refer to species confined to a limited area that may range from a few hundred metres of coast to a whole continent.

Epiproctal cone: a small blind tube projecting from the centre of the abactinal surface in some Astropectinidae, in place of an anus.

Fascioles: the channels between rows of plates, commonly lined with spinules.

Fission: asexual reproduction by splitting of the body into two parts, each of which regenerates into a complete seastar. Fissiparity refers to the process of fission.

Forficiform: shaped like earwig pincers.

Furrow spines: the spines on the inner furrow margin of the adambulacral plates, usually projecting over the furrow, backed by the subambulacral spines in most species (Figure 3B).

Gamete: reproductive product, egg or sperm.

Gonad: the reproductive organ in the arms of seastars (Figure 5A). Can be male, female or sometimes hermaphroditic.

Gonopores: the openings for the release of gametes. There may be a single pore at the base of the arm, one on each side of the arm, or arranged serially along each arm.

Gonochoric: used for organisms having separate sexes, as opposed to hermaphroditic.

Granules: small, often spherical mounds on the surface of plates (Figure 5A).

Hermaphroditic: used for organisms having both male and female reproductive structures. Hermaphroditic individuals may express both sexes simultaneously, alternately or sequentially.

Holotype: the individual specimen on which the description of a new species is based (type specimen).

Infauna: animals living within the sediment on the seafloor, as opposed to epifauna (animals living on the surface of the seafloor, or on other animals).

Inferomarginal: a row of plates that define the ventral edge of the asteroid body (Figure 3A, C, E).

Integument: the outer organic layer of a seastar, analogous to skin.

Interbrachial septum: a muscular or calcareous strut between the mouth and the interradial margin.

Interradial: area between the arms on either surface of the seastar (Figure 3A, D).

Intermarginal: between the superomarginal and inferomarginal plates. Can refer to plates or pores (Figure 3E).

Intragonodal: within the gonad.

Larva: a post embryonic stage of development after hatching from the fertilisation membrane. Bipinnaria and brachiolaria are stages of asteroid larvae.

Lecithotrophy: a mode of reproduction in which free swimming larvae develop using yolk deposits in a large egg, which may be supplemented by the intake of nutrients dissolved in seawater.

Lectotype: a syntype designated as the single name-bearing type specimen subsequent to the establishment of a nominal species or subspecies.

Madreporite: the ridged interradial sieve plate on the upper side of the disc forming the external opening of the water vascular system (Figures 1A, 2, 5).

Neotype: the single specimen designated as the name-bearing type of a nominal species or subspecies when there is a need to define the nominal taxon objectively and no name-bearing type is believed to exist.

Oral (mouth) plates: ossicles that converge at the mouth and support the mouth region.

Ossicle: any calcified element of the skeleton; larger ones are termed plates (Figure 3).

Papulae: small finger-like transparent respiratory processes that project from the coelom through pores in the body wall, mainly on the upper side (Figure 5A). Their arrangement is sometimes characteristic of a genus or species.

Papulate areas: the parts of the abactinal surface where papulae occur.

Papular space: a restricted area with papular pores.

Paratypes: specimens of the same species as the holotype, described at the same time and preferably from the same locality.

Paxillae: columnar plates, only occurring on the abactinal surface, that bear an apical cluster of spinelets or granules (Figure 9). Characteristic of Astropectinidae, Luidiidae, Chaetasteridae and some Asterinidae.

Pedicellariae: small stalked, or unstalked, organs on the surface of the body or embedded in a skeletal plate that are used in defence and feeding. Bivalved pedicellariae have two low, broad valves lying in a depression in a plate (Figure 4A). Sugar tong-shaped pedicellariae look like tongs and are hinged (Figure 4B). Split-granule pedicellariae are very small and consist of a granule split in two (Figure 4C). Fasciculate pedicellariae have 2–5 straight or slightly curved, upright, movable spines (Figure 4D).

Pelagic: living in the surface waters of the sea.

Planktotrophy: the mode of nutrition of free swimming larvae that feed on particulate matter in the plankton.

Proximal: toward the centre of the body.

R (major radius): a body dimension measured from the centre of the disc to the arm tip (Figure 8).

r (minor radius): a body dimension measured from the centre of the disc to the edge of the disc in the middle of an interradius (Figure 8).

Radial symmetry: a pattern of symmetry in which identical segments of the body are arranged around a central axis. Asteroids generally have a five part (pentamerous) near-radial symmetry.

Ring canal: a canal encircling the mouth (Figure 2). It is part of the water vascular system.

Sacciform: sac-like.

Skeleton: the hard, calcite ossicles of the seastar body.

Spines: movable structures that are short or long, usually slender and attenuated, attached to an ossicle (plate) by muscle. Small structures fixed to the surface of ossicles or plates and include: granules (minute and nearly equidimensional); spinelets that may be acicular (slender, stiff and pointed, needle-like), or have various numbers of apical projections, for example bifid (2), trifid (3) or multifid (many). Sacciform spinelets have bag-like or pouch-like structures at their base; scales form an overlapping snake-like covering. (Spinelets are differentiated from spines by size with the former being smaller than the latter).

Squamous: (of scales). Overlapping like roof tiles.

Stone canal: a tube, usually reinforced with ossicles, leading from the madreporite to the water vascular ring canal (Figure 2).

Subacicular: not quite or nearly acicular.

Subsacciform: not quite or nearly sacciform.

Superactinal plates: linking abactinal and actinal plates to support the interradial margin. They are embedded in connective tissue that must be removed with bleach to make them visible (Figure 3C).

Superambulacral plates: small internal plates above the junction of the ambulacral and adambulacral plates (Figure 3C).

Superomarginals: a row of plates defining the abactinal edge of the body (Figure 3C–E). They overlie a row of inferomarginals.

Syntypes: two or more specimens selected from the available specimens to serve as types if the holotype is lost. Each specimen of a type series from which neither a holotype nor a lectotype has been designated. The syntypes collectively constitute the name-bearing type.

Terminal (or **ocular**) **plate**: an unpaired plate at the arm tip (Figure 3D) bearing a sensory tentacle with a pigmented light sensitive optic cushion, projecting from the ventral surface of the base of the tentacle.

Transactinals: very small internal skeletal ossicles that help support the actinal inner arm. Connective tissue has to be removed with bleach to make them visible.

Tube feet: fluid filled, finger-like extensions of the water vascular system that protrude between skeletal plates in the ambulacral groove (Figures 1B, 2, 5). They are principally used in locomotion; they are tipped with suckers in all seastars except the Paxillosida.

Tubercle: a rounded prominence on the skeleton.

Ventrolateral (or **actinolateral**) **plates**: lateral actinal plates, between the adambulacral and inferomarginal plates.

Viviparous: live bearing. In echinoderms this is defined as brooding embryos in the gonad or genital tract, releasing them as crawl away juveniles.

REFERENCES

Antokhina, T.I. and Britayev, T.A. (2012). Sea stars and their macrosymbionts in the Bay of Nhatrang, Southern Vietnam. *Paleontological Journal* **46**(8): 894–908.

Atlas of Living Australia (2019). [online] available at: www.ala.org.au

Australian National Tide Tables (2015). Australian Hydrographic Publication 11. Australian Government Department of Defence.

Baker, A.N. and Marsh, L.M. (1976). The rediscovery of *Halityle regularis* Fisher (Echinodermata: Asteroidea). *Records of the Western Australian Museum* **4**(2): 107–116.

Barker, M.F. (1978). Descriptions of the larvae of *Stichaster australis* (Verrill) and *Coscinasterias calamaria* (Gray) (Echinodermata: Asteroidea) from New Zealand, obtained from laboratory culture. *Biological Bulletin* **154**: 32–46.

Bell, F.J. (1884). Echinodermata. *In: Report on the zoological collections made in the Indo-Pacific Ocean during the voyage of H.M.S. Alert 1881–2*. British Museum (Natural History), London. pp. 117–177, 509–512; pls 8–17, 45.

Bell, F.J. (1894). On the echinoderms collected during the voyage of the H.M.S. *Penguin* and by H.M.S. *Egeria*, when surveying Macclesfield Bank. *Proceedings of the Zoological Society of London,* 1894: 391–413, pls 23–27.

Benavides-Serrato, M. O'Loughlin, P.M. and Rowley, C. (2007). A new fissiparous micro-asteriid from southern Australia (Echinodermata: Asteroidea: Asteriidae). *Memoirs of Museum Victoria* **64**: 71–78, figs 1–4.

Benham, W. B. (1911). Stellerids and echinids from the Kermadec Islands. *Transactions of the New Zealand Institute* **43**: 140–163.

Blake, D.B. (1978). The taxonomic position of the modern sea-star *Cistina* Gray, 1840. *Proceedings of the Biological Society of Washington* **91**(1): 234–241.

Blake, D.B. (1979). The affinities and origins of the crown-of-thorns sea star *Acanthaster* Gervais. *Journal of Natural History* **13**: 303–314.

Blake, D.B. (1980). On the affinities of three small sea star families. *Journal of Natural History* **14**: 163–182.

Blake, D.B. (1987). A classification and phylogeny of post-Palaeozoic sea stars (Asteroidea: Echinodermata). *Journal of Natural History* **21**: 481–528.

Blake, D.B. (1988). Paxillosidans are not primitive asteroids: a hypothesis based on functional considerations. *In:* R.D. Burke, P.V. Mladenov, P. Lambert and R.L. Parsley (eds), *Echinoderm Biology*. Balkema, Rotterdam. pp. 309–314.

Bunt, J.S. 1987. The Australian Marine Environment. *In*: G.R. Dyne and D.W. Walton (eds), *Fauna of Australia (vol. 1A): General Articles*. Australian Government Publishing Service, Canberra. pp. 17–42.

Buttermore, R.E. Turner, E. and Morrice, M.G. (1994). The introduced northern Pacific seastar *Asterias amurensis* in Tasmania. *Memoirs of the Queensland Museum* **36**(1): 21–25.

Byrne, M. (1991). Developmental diversity in the starfish genus *Patiriella* (Asteroidea: Asterinidae). *In:* T. Yanagisawa, L. Yasumasu, C. Oguro, N. Suzuki and T. Motokawa (eds), *Biology of Echinodermata*. Balkema, Rotterdam. pp. 499–508.

Byrne, M. (1992). Reproduction of sympatric populations of *Patiriella gunnii, P. calcar* and *P. exigua* in New South Wales, asterinid seastars with direct development. *Marine Biology* **114**: 297–316.

Byrne, M. (1995). Changes in larval morphology in the evolution of bentic development by *Patiriella exigua* (Asteroidea), a comparison with the larvae of *Patiriella* species with planktonic development. *Biological Bulletin* **188**: 293–305.

Byrne, M. (1996). Viviparity and intragonadal cannibalism in the diminative seastars *Patiriella vivipara* and *P. parvivipara* (family Asterinidae). *Marine Biology* **125**: 551–567.

Byrne, M. (2006). Life history diversity and evolution in the Asterinidae. *Integrative and Comparative Biology* **46**(3): 243–254.

Byrne, M. and Anderson, M.J. (1994). Hybridization of sympatric *Patiriella* species (Echinodermata: Asteroidea) in New South Wales. *Evolution* **48**(3): 564–576.

Byrne, M. and Barker, M.F. (1991). Embryogenesis and larval development of the asteroid *Patiriella regularis* viewed by light and scanning electron microscopy. *Biological Bulletin* **180**: 332–345.

Byrne, M. and Cerra, A. (1996). Evolution of intragonadal development in the diminative asterinid seastars *Patiriella vivipara* and *P. parvivipara* with an overview of the development in the Asterinidae. *Biological Bulletin* **191**: 17–26.

Byrne, M., Hart, M.W., Cerra, A. and Cisternas, P. (2003). Reproduction and larval morphology of broadcasting and viviparous species in the *Cryptasterina* species complex. *Biological Bulletin* **205**: 285–294.

Byrne, M. Morrice, M.G. and Wolf, B. (1997). Introduction of the northern Pacific asteroid (*Asterias amurensis*) to Tasmania: reproduction and current distribution. *Marine Biology* **127**: 673–685.

Byrne, M. and O'Hara, T. D. (2017). *Australian Echinoderms: Biology, Ecology and Evolution.* CSIRO Publishing, Melbourne and Australian Biological Resources Study, Canberra.

Byrne, M. Rowe, F.W.E. Marsh, L.M. and Mah, C.L. (2017). Asteroidea. *In:* M. Byrne and T. D. O'Hara (eds), *Australian Echinoderms: Biology, Ecology and Evolution.* CSIRO Publishing, Melbourne, and Australian Biological Resources Study, Canberra. pp. 231–293.

Chao, S.M. (1999). Revision of Taiwan starfish (Echinodermata: Asteroidea), with description of ten new records. *Zoological Studies* **38**(4): 405–415.

Chao, S.M. (2001). Seven newly recorded starfish from Taiwan (Echinodermata: Asteroidea). *Bulletin National Museum of Natural Science* **13**: 1–11.

Chen, C.P. and Run, J.Q. (1991). Induction of larval metamorphosis of *Archaster typicus* (Echinodermata: Asteroidea). *Bulletin of the Institute of Zoolology, Academia Sinica* **30**(3): 257–260.

Chia, F.S. (1976). Reproductive biology of an intraovarian brooding starfish *Patiriella vivipara* Dartnall, 1969. *American Zoologist* **16**: 181.

Chiu, S.T. Lam, V.W.W. and Shin, P.K.S. (1990a). Further observations on the feeding biology of *Luidia* spp. in Hong Kong. *In:* B. Morton (ed), *Proceeding of the Second International Marine Biology Workshop: the Marine Flora and Fauna of Hong Kong and Southern China, 1986.* Hong Kong University Press. pp. 907–933.

Chiu, S.T. Lam, V.W.W. and Shin, P.K.S. (1990b). Notes on the diet of the asteroid *Astropecten polyacanthus* (Müller and Troschel, 1842) in Hong Kong. *In:* B. Morton (ed.), *Proceeding of the Second International Marine Biology Workshop: the Marine Flora and Fauna of Hong Kong and Southern China, 1986.* Hong Kong University Press. pp. 1033–1038.

Clark, A.M. (1950). A new species of sea-star from Norfolk Island. *Annals and Magazine of Natural History* **3**(77): 808–811.

Clark, A.M. (1951). On some echinoderms in the British Museum (Natural History). *Annals and Magazine of Natural History* **4**: 1256–1268.

Clark, A.M. (1953a). Notes on asteroids in the British Museum (Natural History). III. *Luidia*. *Bulletin of the British Museum (Natural History) (Zoology)* **1**: 380–396, pls 39–41.

Clark, A.M. (1953b). Notes on asteroids in the British Museum (Natural History). IV. *Tosia* and *Pentagonaster*. *Bulletin of the British Museum (Natural History) (Zoology)* **1**: 396–412, pls 42–46.

Clark, A.M. (1962). Asteroidea. *Reports of the British, Australian and New Zealand Antarctic Research Expedition 1929–1931, Series B* **9**(1): 1–104.

Clark, A.M. (1966). Port Phillip Survey 1957–1963, Echinodermata. *Memoirs of the National Museum of Victoria* **27**: 289–384.

Clark, A.M. (1967). Notes on asteroids in the British Museum (Natural History). V. *Nardoa* and some other Ophidiasterids. *Bulletin of the British Museum (Natural History) (Zoology)* **15**: 169–198.

Clark, A. M. (1970). The name of the starfish, *Anthenea acuta* (Perrier), preoccupied. *Proceedings of the Linnean Society of New South Wales* **95**(2): 157.

Clark, A.M. (1984). Notes on Atlantic and other Asteroidea. 4. Families Poraniidae and Asteropidae. *Bulletin of the British Museum (Natural History) (Zoology)* **47**(1): 19–51.

Clark, A.M. (1989). An index of names of recent Asteroidea — Part 1: Paxillosida and Notomyotida. *In*: M. Jangoux and J.M. Lawrence (eds), *Echinoderm Studies 3*. Balkema, Rotterdam. pp 225–347.

Clark, A.M. (1993). An index of names of recent Asteroidea — Part 2: Valvatida. *In*: M. Jangoux and J.M. Lawrence (eds), *Echinoderm Studies 4*. Balkema, Rotterdam. pp. 187–366.

Clark, A.M. and Courtman-Stock, J. (1976). *The echinoderms of southern Africa*. British Museum (Natural History) London, Publication 776.

Clark, A.M. and Downey, M.E. (1992). *Starfishes of the Atlantic*. Natural History Museum and Chapman and Hall, London.

Clark, A.M. and Mah, C. (2001). An index of names of recent Asteroide — Part 4: Forcipulatida and Brisingida. *In*: M. Jangoux and J.M. Lawrence (eds), *Echinoderm Studies 6*. Balkema, Rotterdam. pp. 229–347.

Clark, A.M. and Rowe, F.W.E. (1971). *Monograph of shallow-water Indo-West Pacific Echinoderms*. British Museum (Natural History), London. i–vii, 1–238, pls 1–31.

Clark, H.E.S. and McKnight, D.G. (2000). The Marine Fauna of New Zealand: Echinodermata: Asteroidea (Sea-stars). Order Paxillosida and Order Notomyotida. *NIWA Biodiversity Memoir* **116**: 1–196.

Clark, H.L. (1909a). Scientific results of the trawling expedition of HMCS *Thetis,* off the coast of New South Wales in February and March, 1898, Echinodermata. *Memoirs of the Australian Museum* **4**(11): 519–564, pls 47–58.

Clark, H.L. (1909b). Notes on some Australian and Indo-Pacific echinoderms. *Bulletin of the Museum of Comparative Zoology* **52**(7): 107–137, pl. 1.

Clark, H.L. (1914). The echinoderms of the Western Australian Museum. *Records of the Western Australian Museum* **1**(3): 132–173, pls 17–26.

Clark, H.L. (1916). Report on the sea lilies, starfishes, brittle stars and sea urchins obtained by the F.I.S. "Endeavour" on the coasts of Queensland, New South Wales, Tasmania, Victoria, South Australia and Western Australia. *In: Biological Results of the Fishing Experiments carried on by the F.I.S 'Endeavour' 1901–14, Volume 4.* Department of Trade and Customs, Sydney. pp. 1–123, pls 1–44.

Clark, H.L. (1921). The echinoderm fauna of Torres Strait; its composition and its origin. *Papers from the Department of Marine Biology of the Carnegie Institution of Washington* **10**: 1–223, pls 1–38.

Clark, H.L. (1923). Some echinoderms from West Australia. Percy Sladen Trust Expedition to the Abrolhos Islands. *Journal of the Linnean Society of London* (Zoology) **35**: 229–251.

Clark, H.L. (1928). The sea lilies, sea stars, brittle stars and sea urchins of the South Australian Museum. *Records of the South Australian Museum* **3**: 361–482, figs 108–142.

Clark, H.L. (1938). Echinoderms from Australia. An account of collections made in 1929 and 1932. *Memoirs of the Museum of Comparative Zoology, Harvard College* **55**: 1–596, pls 1–28.

Clark, H.L. (1946). The echinoderm fauna of Australia. Its composition and its origin. *Carnegie Instution of Washington Publication* **556**: 1–567.

Coleman, H.L. (1911). Scientific results of the trawling expedition of H.M.C.S. "Thetis," off the coast of New South Wales, in February and March, 1898. Supplement to Echinodermata. *Memoirs of the Australian Museum* **4**(14): 699–701, pl. 83.

Coleman, N. (1979). *A Field Guide to Australian Marine Life.* Rigby, Adelaide.

Coleman, N. (1994). *Sea stars of Australasia and their relations.* Neville Coleman's Underwater Geographic. Springwood, Qld.

Coleman, N. (2007). *Sea stars — Echinoderms of the Asia/Indo-Pacific.* Neville Coleman's Underwater Geographic. Springwood, Qld.

Crawford, T.J. and Crawford, B.J. (2007). *Linckia multifora* (Echinodermata: Asteroidea) in Rarotonga, Cook Islands: Reproductive mechanisms and ecophenotypes. *Pacific Science* **61**(3): 371–381, pl. 1.

Cresswell, G.R. and Golding, T.J. (1980). Observations of a south flowing current in the southeastern Indian Ocean. *Deep-Sea Research* **27**(A): 449–466.

Crump, R.G. (1971). Annual Reproductive cycles in three geographically separated populations of *Patiriella regularis* (Verrill), a common New Zealand asteroid. *Journal of Experimental Marine Biology and Ecology* 7: 137–162.

Crump, R.G. and Barker, M.F. (1985). Sexual and asexual reproduction in geographically separated populations of the fissiparous asteroid *Coscinasterias calamaria* (Gray). *Journal of Experimental Marine Biology and Ecology* **88**: 109–127.

D'Adamo, N., Fandry, C., Buchan, S. and Domingues, C. (2009). Northern Sources of the Leeuwin Current and the "Holloway Current" on the North West Shelf. *Journal of the Royal Society of Western Australia* **92**: 53–66.

Dartnall, A.J. (1967). New Zealand marine animals for Channel waters. *Tasmanian Fisheries Research* **1**(3): 4–5.

Dartnall, A.J. (1969a). New Zealand sea stars in Tasmania. *Papers and Proceedings of the Royal Society of Tasmania* **103**: 53–55.

Dartnall, A.J. (1969b). A viviparous species of *Patiriella* (Asteroidea, Asterinidae) from Tasmania. *Proceedings of the Linnean Society of New South Wales* **93**: 294–296, pl. 29.

Dartnall, A.J. (1970a). Some species of *Asterina* from Flinders, Victoria. *The Victorian Naturalist* **87**(1): 1–4, figs 1–3.

Dartnall, A.J. (1970b). The Asterinid sea stars of Tasmania. *Papers and Proceedings of the Royal Society of Tasmania* **104**: 73–77, pl. 1.

Dartnall, A.J. (1970c). A new species of *Marginaster* (Asteroidea: Poraniidae) from Tasmania. *Proceedings of the Linnean Society of New South Wales* **94**(3): 207–211, pl. 13.

Dartnall, A.J. (1971). Australian sea stars of the genus *Patiriella* (Asteroidea, Asterinidae). *Proceedings of the Linnean Society of New South Wales* **96**(1): 39–49, pls 3–4.

Dartnall, A.J. (1980). Tasmanian Echinoderms. *Fauna of Tasmania Handbook* No. 3. Fauna of Tasmania Committee. Hobart.

Dartnall, A.J. Byrne, M. Collins, J. and Hart, M.W. (2003). A new viviparous species of asterinid (Echinodermata, Asteroidea, Asterinidae) and a new genus to accommodate the species of pan-tropical exiguoid sea stars. *Zootaxa* **359**: 1–14.

Dartnall, A.J. Stevens, H. Byrne, M. (2013). How to lose a population: The effect of Cycloone Larry on a population of *Cryptasterina pentaagona* at Mission Beach, North Queensland. *In:* C. Johnson (ed.), *Echinoderms in a changing World. Proceedings of the Thirteenth International Echinoderm Conference, University of Tasmania, Hobart, Tasmania, Australia, 5–9 January 2009.* CRC Press, Boca Raton. pp. 181–184

Döderlein, L. (1889). Echinodermen von Ceylon. Bericht über die von den Herren Dres, Sarasin gesammelten Asteroidea, Ophiuroidea und Echinoidea. *Zoologische Jahrbücher* **3**: 822–846, pls 31–33.

Döderlein, L. (1896). Bericht über die von Herren Prof. Semon Bei Amboina und Thursday Island gesammelten Asteroidea. *Denkschriften der Medicinisch-Naturwissenschaftlichen Gesellschaft zu Jena* **8**: 301–322, pls 18–22.

Döderlein, L. (1915). Die Arten der Asteroiden-Gattung *Anthenea* Gray. *Jahrbücher des Nassauischen Vereins für Naturkunde*: **68**: 21–55, pls 1–11.

Döderlein, L. (1916). Uber die Gattung *Oreaster* und Verwandte. *Zoologische Jahrbücher. Abteilung für Systematik, Geographie und Biologie der Tiere* **40**: 409–440.

Döderlein, L. (1917). Die Asteriden der Siboga Expedition. I. Die Gattung *Astropecten* und ihre Stammes geswchichte. *Siboga Expedition.* **46**(A): 1–191, pls 1–17.

Döderlein, L. (1920). Die Asteriden der Siboga Expedition. II. Die Gattung *Luidia* und ihre Stammesgeschichte. *Siboga Expedition* **46**(B): 193–291, pls 18–20.

Döderlein, L. (1926). Über Asteriden aus dem Museum von Stockholm. *Kungl. Svenska Vetenskapsakademiens Handlingar, Tredje Serien* **2**(6): 1–22, pls 1–4.

Döderlein, L. (1935). Die Asteriden der Siboga Expedition. III. Oreasteridae. *Siboga Expedition* **46**(C): 71–110, pls 20–27.

Döderlein, L. (1936). Die Asteriden der Siboga Expedition. III. Die Unterfamilie Oreasterinae. *Siboga Expedition* **46**(C): 295–369, pls 21–32.

Ebert, T.A. (1976). Natural History notes on two Indian Ocean starfishes in Seychelles: *Protoreaster lincki* (de Blainville) and *Pentaceraster horridus* (Gray). *Journal of the Marine Biological Association of India* **18**(1): 71–77.

Edgar, G.J. (1997). *Australian Marine Life, the plants and animals of temperate waters*. Reed Books, Kew, Victoria.

Edgar, G. (2019). Tropical Marine Life of Australia. Plants and Animals of the central Indo-Pacific. New Holland Publishers Pty. Ltd, Australia.

Edmondson, C.H. (1935). Autotomy and regeneration in Hawaiian starfishes. *Occasional Papers of the Bernice P. Bishop Museum* **11**(8): 1–20.

Emson, R.H. (1978). Some aspects of fission in *Allostichaster polyplax*. *In:* D.S. McLusky and A.J. Berry (eds), *Physiology and behaviour of marine organisms*. Proceedings of the 12th European Marine Biology Symposium. Pergamon Press, Oxford. pp. 321–329.

Emson R.H. and Wilkie, I.C. (1980). Fission and autotomy in echinoderms. *Oceanography and Marine Biology Annual Review* **18**: 155–250.

Ely, C.A. (1942). Shallow water Asteroidea and Ophiuroidea of Hawaii. *Bernice P. Bishop Museum Bulletin* **176**: 1–63, pls 1–13.

Fabricius K.E., Okaji, K. and De'ath, G. (2010). Three lines of evidence to link outbreaks of the crown-of-thorns seastar *Acanthaster planci* to the release of larval food limitation. *Coral Reefs* **29**: 593–605.

Fell, H.B. (1959). Starfishes of New Zealand. *Tuatara* 7: 127–142.

Fisher, W.K. (1906). The starfishes of the Hawaiian Islands. *Bulletin of the United States Fish Commission* **23**(3): 989–1130, pls 1–49.

Fisher, W.K. (1913). Four new genera and fifty-eight new species of starfishes from the Philippine Islands, Celebes, and the Moluccas. *Proceedings of the United States National Museum* **43**: 599–648.

Fisher, W.K. (1919). Starfishes of the Philippine seas and adjacent waters. *Bulletin of the United States National Museum* **100**(3): 1–712.

Fisher, W.K. (1928). Asteroidea of the North Pacific and adjacent waters. Part 2. Forcipulata (part). *Bulletin of the United States National Museum* **76**(2): 1–245.

Fisher, W.K. (1930). Asteroidea of the North Pacific and adjacent waters. Part 3. Forcipulata (concluded). *Bulletin of the United States National Museum* **76**(3): 1–356.

Fisher, W.K. (1941). A new genus of sea stars (*Plazaster*) from Japan, with a note on the genus *Parasterina*. *Proceedings of the United States National Museum* **90**: 447–456, pls 66–70.

Gale, A.S. (1987). Phylogeny and classification of the Asteroidea (Echinodermata). *Zoological Journal of the Linnaean Society* **87**: 107–132.

Gale, A.S. (2011). The phylogeny of post-Paleozoic Asteroidea (Neoasteroidea, Echinodermata). *Special Papers in Paleontology* **38**: 1–112.

Geocentric Datum of Australia 2020 (2020). [online] available at: www.ga.gov.au/scientific-topics/positioning-navigation/geodesy/datums-projections/gda2020.

Gibbs, P.E. Clark, A.M. Clark, C.M. (1976). Echinoderms from the northern region of the Great Barrier Reef, Australia. *Bulletin of the British Museum of Natural History (Zoology)* **30**(4): 101–144, pl. 1.

Godfrey, J.S. and Ridgway, K.R. (1985). The large-scale environment of the poleward-flowing Leeuwin Current, Western Australia. Longshore steric height gradients, wind stresses and geostrophic flow. *Journal of Physical Oceanography* **15**: 481–495.

Grice, A.J. and Lethbridge, R.C. (1988). Reproductive studies on *Patiriella gunnii* (Asteroidea: Asterinidae) in South-western Australia. *Australian Journal of Marine and Freshwater Research* **39**: 399–407.

Gray, J.E. (1840). A synopsis of the genera and species of the class Hypostoma (*Asterias* Linnaeus). *Annals and Magazine of Natural History. Series 1* **6**: 175–184, 275–290.

Gray, J.E. (1847). Descriptions of some New Genera and Species of Asteriadae. *Proceedings of the Zoological Society of London* **15**: 72–83.

Gray, J.E. (1866). Synopsis of the Species of Starfish in the British Museum (with figures of some of the new species). John Van Voorst, London. pp. 1–17, pls 1–16.

Guille, A. Laboute, P. Menou, J.-L. (1986). Guide des étoiles de mer, oursins et autres échinodermes du lagon de Nouvelle Calédonie. Editions de l'Orstom.

Hart, M.W. Byrne, M. and Johnston, S.L. (2003). Cryptic species and modes of development in *Patiriella pseudoexigua. Journal of the Marine Biological Association of the United Kingdom* **83**: 1109–1116.

Haszprunar, G. and Spies, M. (2014). An integrative approach to the taxonomy of the crown-of-thorns starfish species group (Asteroidea: Acanthaster): A review of names and comparison to recent molecular data. *Zootaxa* **3841**(2): 271–284.

Haszprunar, G., Vogler, C. and Wörheide, G. (2017). Persistent gaps of knowledge for naming and distinguishing multiple species of Crown-of-Thorns-Seastar in the *Acanthaster planci* species complex. *Diversity* **9**(2): 1–10.

Hayashi, R. (1936). Variations of the Sea-star, *Asterias amurensis* Lütken, due to growth stages. *Journal of the Faculty of Sciences, Hokkaido Imperial University, Series VI, Zoology* **5**(1): 5–20, pls 1–2.

Hayashi, R. (1938). Sea-stars of the Caroline Islands. *Studies from the Palao Tropical Biological Station* **3**: 417–446, pls 2–4.

Hutton, F.W. (1872). *Catalogue of the Echinodermata of New Zealand, with diagnoses of the species.* Wellington, Colonial Museum and Geological Survey Department.

Jangoux, M. (1972). Note anatomique sur *Archaster angulatus* Müller et Troschel. *Revue Zoologie et Botanie Africaine* **86**: 163–172, figs 1–9.

Jangoux, M. (1973). Le genre *Neoferdina* Livingstone (Echinodermata, Asteroidea: Ophidiasteridae). *Revue Zoologie et Botanie Africaine* **87**(4): 775–794, pl. 4.

Jangoux, M. (1974). Sur l' "association" entre certaines Astéries (Echinodermata) et des poissons Carapidae. *Revue de Zoologie Africaine* **88**(4): 789–796.

Jangoux, M. (1978). Biological results of the Snellius Expeditions XXIX. Echinodermata, Asteroidea. *Zoologische Mededlingen* **52**(25): 287–300, pls 1–3.

Jangoux, M. (1980). Le genre *Leiaster* Peters (Echinodermata, Asteroidea: Ophidiasteridae). *Revue Zoologie Africaine* **94**(1): 87–108, pls 5–8.

Jangoux, M. (1984). Les Astérides (Echinodermes) des Terres Australes ramenés par l'expédition Baudin (1800–1804): catalogue commente des dessins inédits de Charles-Alexander Lesueur conserves au Muséum d'Histoire Naturelle du Havre. *Bulletin trimestriel de la Société Géologique de Normandie et des Amis du Muséum du Havre* **71**(4): 25–56, pl. 1.

Jangoux, M. (1985). *Catalogue commenté des types d'Echinodermes actuels, conservés dans les collections nationales suisses, suivi d'une notice sur la contribution de Louis Agassiz à la connaissance des Echinodermes actuels.* Muséum d'Histoire Naturelle, Geneva.

John, D.D. (1948). Notes on Asteroids in the British Museum (Natural History) 1. The species of *Astropecten. Novitates Zoologicae* **42**: 485–508.

Keesing, J.K., Graham, F. and Irvine, T.R. (2011). Synchronous aggregated pseudo-copulation of the sea star *Archaster angulatus* Müller and Troschel, 1842 (Echinodermata : Asteroidea) and its reproductive cycle in south-western Australia. *Marine Biology* **158**: 1163–1173.

Kendrick, G.W. Wyrwoll, K.H. and Szabo, B.J. (1991). Pliocene — Pleistocene coastal events and history along the western margin of Australia. *Quaternary Science Review* **10**: 419–439.

Kenny, R. 1969. Growth and Asexual Reproduction of the starfish *Nepanthia belcheri* (Perrier). *Pacific Science* **23**(1): 51–55.

Keough, M.J. and Dartnall, A.J. (1978). A new species of viviparous asterinid asteroid from Eyre Peninsula, South Australia. *Records of the South Australian Museum* **17**(28): 407–416.

Koehler, R. (1910a). Astéries et Ophiures des îles Aru et Kei. *Senckenbergische Naturforschende Gesellschaft Abhandlungen* **33**: 265–295, pls 15–17.

Koehler, R. (1910b). An account of the shallow-water Asteroidea. *Echinoderma of the Indian Museum. Part VI.* pp. 1–192.

Komatsu, M. (1973). A preliminary report on the development of the sea-star, *Leiaster leachii*. *Proceedings Japanese Society of Systematic Zoology* **9**: 55–58, pl. 1.

Komatsu, M. (1983). Development of the sea-star, *Archaster typicus* with a note on male on female superposition. *Annotationes Zoologiscae Japonenses* **56**: 187–195.

Komatsu, M. Kawai, M. Nojima, S. and Oguro, C. (1994). Development of the multiarmed seastar, *Luidia maculata* Müller and Troschel. *In:* B. David, A. Guille, J-P. Féral and M. Roux (eds), *Echinoderms Through Time*. Balkema, Rotterdam. pp. 327–333.

Lamarck, J.B.P.A. de M. (1816). Stellerides. *In:* J.B. Baillière (ed.), *Histoire Naturelle des Animaux sans Vertèbres.* pp: 522–568.

Lane, D.J.W. and Hu, J.M.L. (1994). Abbreviated development in *Iconaster longimanus* (Möbius); Planktonic lecithotrophy in a tropical goniasterid sea star. *In:* B. David, A. Guille, J-P. Féral and M. Roux (eds), *Echinoderms Through Time*. Balkema, Rotterdam. pp. 343–346.

Lane, D.J.W. and Rowe, F.W.E. (2009). A new species of *Asterodiscides* (Echinodermata, Asteroidea, Asterodiscididae) from the tropical southwest Pacific, and the biogeography of the genus revisited. *Zoosytema* **31**(3): 419–429,

Lane, D.J.W., Marsh, L.M., VandenSpiegel, D. and Rowe, F.W.E. (2000). Echinoderm fauna of the South China Sea: an inventory and analysis of distribution patterns. *The Raffles Bulletin of Zoology. Supplement* **8**: 459–493.

Lawrence, J.M. Keesing, J.K. and Irvine, T.R. (2010). Population characteristics and biology of two populations of *Archaster angulatus* (Echinodermata: Asteroidea) in different habitats of the central-western Australian coast. *Journal of the Marine Biological Association of the United Kingdom* **91**(8): 1577–1585.

de Loriol, P. (1885). Catalogue raisonné des Echinodermes recueillis par M.V. Robillard à lîle Maurice II Stellérides. *Memoires de la Société de physique et d'histoire naturelle de Genève* **29**(4): 1–84, pls 7–22.

Lemmens, J.W.T.J. Arnold, P.W. and Birtles, R.A. (1995). Distribution patterns and selective feeding in two *Astropecten* species (Asteroidea (Echinodermata)) from Cleveland Bay, Northern Queensland. *Marine and Freshwater Research* **46**: 447–455.

Liao, Y. and Clark, A.M. (eds.) (1995). *The Echinoderms of Southern China*. Science Press, Beijing.

Livingstone, A.A. (1930). On some new and little-known Australian asteroids. *Records of the Australian Museum* **18**(1): 15–24, pls 4–8.

Livingstone, A.A. (1931). On a new asteroid from Queensland. *Records of the Australian Museum* **18**(4): 135–137, pls 17–19.

Livingstone, A.A. (1932a). Asteroidea. *Scientific Reports / Great Barrier Reef Expedition 1928–29* **4**: 241–265, pls 1–12.

Livingstone, A.A. (1932b). The Australian species of *Tosia* (Asteroidea). *Records of the Australian Museum* **18**(7): 373–382.

Livingstone, A.A. (1933). Some genera and species of the Asterinidae. *Records of the Australian Museum* **19**(1): 1–20, pls 1–5.

Livingstone, A.A. (1934). Two new asteroids from Australia. *Records of the Australian Museum* **19**(3): 177–180, pl. 18.

Livingstone, A.A. (1936). Description of new Asteroidea from the Pacific. *Records of the Australian Museum* **19**(6): 383–387, pls 27–28.

Lucas, J.S. and Jones, M.M. (1976). Hybrid crown-of-thorns starfish (*Acanthaster planci* x *A. brevispinus*) reared to maturity in the laboratory. *Nature* **263**: 409–412.

McKnight, D.G. (1977). Classification of Recent paxillosid sea-stars (Asterozoa: Echinodermata). *New Zealand Oceanographic Institution Records* **3**(12): 113–119.

MacNeil, M.A. Chong-Seng, K.M. Pratchett, D.J. Thompson, C.A. Messmer, V. and Pratchett, M.S. (2017). Age and Growth of an outbreaking *Acanthaster* cf. *solaris* population within the Great Barrier Reef. *Diversity* **9**(1): article 18. [online] doi: 10.3390/d9010018

Mah, C.L. (2005). 'Cladistic analysis of the Goniasteridae (Asteroidea: Valvatoidea): phylogeny, evolution, and biodiversity'. Unpublished PhD dissertation. University of Illinois.

Mah, C.L. (2007). Systematics, phylogeny and historical biogeography of the *Pentagonaster* clade (Asteroidea: Valvatida: Goniasteridae). *Invertebrate Systematics* **21**: 311–339.

Mah, C.L. (2017). Overview of the *Ferdina*-like Goniasteridae (Echinodermata: Asteroidea) including a new subfamily, three new genera and fourteen new species. *Zootaxa* **4271**(1): 1–72.

Mah, C.L. (2019). World Asteroidea database. [online] available at: www.marinespecies.org/asteroidea

Mah, C.L. and Blake, D.B. (2012). Global Diversity and Phylogeny of the Asteroidea (Echinodermata). *PLoS ONE* **7**(4): e35644. [online] doi:10.1371/journal.pone.0035644

Mah, C.L. and Foltz, D. (2011a). Molecular phylogeny of the Valvatacea (Asteroidea: Echinodermata). *Zoological Journal of the Linnean Society* **161**: 769–788.

Mah, C.L. and Foltz, D. (2011b). Molecular phylogeny of the Forcipulatacea (Asteroidea: Echinodermata): Systematics and Biogeography. *Zoological Journal of the Linnean Society* **162**: 646–660.

Mah, C.L., McKnight, D.G., Eagle, M.K., Pawson, D.L., Améziane, N., Vance, D.J., Baker, A.N., Clark, H.E.S., Davey, N. (2009). Phylum Echinodermata: sea stars, brittle stars, sea urchins, sea cucumbers, sea lilies. *In:* D.P. Gordon (ed.), *New Zealand inventory of biodiversity 1. Kingdom Animalia: Radiata, Lophotrochozoa, Deuterostomia*. Canterbury University Press, Christchurch. pp. 371–400.

Maluf, L.Y. (1988). *Composition and distribution of the central eastern Pacific echinoderms.* Technical Reports No. 2. Natural History Museum of Los Angeles County.

Marsh, L.M. (1974). Shallow-water Asterozoans of south-eastern Polynesia 1. Asteroidea. *Micronesica* **10**(1): 65–104.

Marsh, L.M. (1976). Western Australian Asteroidea since H.L. Clark. *Thalassia jugoslavica* **12**(1): 213–225.

Marsh, L.M. (1977). Coral Reef Asteroids of Palau, Caroline Islands. *Micronesica* **13**(2): 251–281.

Marsh, L.M. (1988). Spawning of coral reef asterozoans coincident with mass spawning of tropical reef corals. *In:* R. D. Burke, P.V. Mladenov, P. Lambert and R.L. Parsley (eds), *Echinoderm biology*. Balkema, Rotterdam. pp. 187–192.

Marsh, L.M. (1991). A revision of the echinoderm genus *Bunaster* (Asteroidea: Ophidiasteridae). *Records of the Western Australian Museum* **15**(2): 419–433.

Marsh, L.M. (2000). Echinoderms of Christmas Island. *Records of the Western Australian Museum* Supplement **59**: 97–101.

Marsh, L.M. (2009). A new species of *Thromidia* (Echinodermata: Asteroidea) from Western Australia. *Records of the Western Australian Museum* **25**(2): 145–151.

Marsh, L.M. and Marshall, J.I. (1983). Some aspects of the zoogeography of northwestern Australian echinoderms (other than holothurians). *Bulletin of Marine Science* **33**(3): 671–687.

Marsh, L.M. and Pawson, D.L. (1993). Echinoderms of Rottnest Island. *In:* F.E. Wells, D.I. Walker, H. Kirkman and R. Lethbridge (eds), *The Marine Flora and Fauna of Rottnest Island, Western Australia.* Western Australian Museum, Perth. pp. 279–304.

Marsh, L.M. and Slack-Smith, S.M. (2010). *Field Guide to Sea Stingers and other Venomous and Poisonous Marine Invertebrates of Western Australia.* Western Australian Museum, Perth.

Martin, R.B. (1970). *Asteroid Feeding Biology.* MSc Thesis, University of Auckland, New Zealand.

Migazawa, K. Higashiyama, Hori, K. Noguchi K. and Hashimoto, K. (1987). Distribution of tetrodotoxin in the different organs of starfish *Astropecten polyacanthus. Marine Biology* **96**: 385–390.

Moran, P.J. (1986). The *Acanthaster* phenomenon. *Oceanography and Marine Biology Annual Review* **24**: 379–480.

Moran, P.F. (1990). *Acanthaster planci* (L.): biographical data. *Coral Reefs* **9**(3): 95–96.

Mortensen, T. (1925). Papers from Dr. Th. Mortensen's Pacific Expedition 1914–16. XXIX. Echinoderms of New Zealand and the Auckland-Campbell Islands. III–V. Asteroidea, Holothuroidea and Crinoidea. *Videnskabelige Meddelelser fra Dansk naturhistorisk Forening i København* **79**(15): 261–420.

Mortensen, T. (1931). Contributions to the study of the development and larval forms of echinoderms. I–II. *Kingelige Danske Videnskabernes Selskabs Skrifter Naturvidenskabelig og Mathematisk Afdeling 4, Raekke* **4**(1): 1–39, pls 1–7.

Mortensen, T. (1937). Contributions to the study of the development and larval forms of echinoderms. III. *Kingelige Danske Videnskabernes Selskabs Skrifter Naturvidenskabelig og Mathematisk Afdeling 9, Raekke* 7(1): 1–65, pls 1–15.

Mortensen, T. (1938). Contributions to the study of development and larval forms of echinoderms. IV. *Kingelige Danske Videnskabernes Selskabs Skrifter Naturvidenskabelig og Mathematisk Afdeling 9, Raekke* 7(3): 1–59, pls 1–12.

Mukai, H., Nishihira, M., Kamisato, H. and Fujimoto, Y. (1986). Distribution and Abundance of the Sea-Star *Archaster typicus* in Kabira Cove, Ishigaki Island, Okinawa. *Bulletin of Marine Science* **38**(2): 366–383.

Müller, J. and Troschel, F.H. (1840). über die Gattungen der Asteriden. *Archiv für Naturgeschichte* **6**(1): 318–326.

Müller, J. and Troschel, F.H. (1842). *System der Asteriden.* F. Vieweg und Sohn, Brunswick.

Müller, J. and Troschel, F.H. (1844). Beschreibung neuer Asteriden. *Archiv für Naturgeschichte* **10**(1): 178–185.

Naughton, K.M. and O'Hara, T.D. (2009). A new brooding species of the biscuit star *Tosia* (Echinodermata: Asteroidea: Goniasteridae) distinguished by molecular, morphological and larval characters. *Invertebrate Systematics* **23**: 348–366.

Oguro, C. (1983). Supplementary notes on the Sea-stars From the Palau and Yap Islands. I. *Annotationes Zoologicae Japonenses* **56**(3): 221–226.

O'Hara, T.D. and Poore, G.C.B. (2000). Patterns of distribution for southern Australian marine echinoderms and decapods. *Journal of Biogeography* **27**: 1321–1335.

O'Hara, T.D. Mah, C.L. Hipsley, C.A. Bribiesca-Contreras, G. Barrett, N.S. (2019). The Derwent River seastar: re-evaluation of a critically endangered marine invertebrate. *Zoological Journal of the Linnean Society* **186**(2): 483–490.

O'Loughlin, P.M. (2002). New genus and species of southern Australian and Pacific Asterinidae (Echinodermata, Asteroidea). *Memoirs of Museum Victoria* **59**(2): 277–296.

O'Loughlin, P.M. (2009). New asterinid species from Africa and Australia (Echinodermata: Asteroidea: Asterinidae). *Memoirs of Museum Victoria* **66**: 203–213.

O'Loughlin, P.M. and Bribiesca-Contreras, G. (2015). New asterinid seastars from northwest Australia, with a revised key to *Aquilonastra* species. *Memoirs of Museum Victoria* **73**: 27–40.

O'Loughlin, P.M. and O'Hara, T.D. (1990). A Review of the genus *Smilasterias* (Echinodermata, Asteroidea), with descriptions of two new species from south-eastern Australia, one a gastric brooder, and a new species from Macquarie Island. *Memoirs of Museum Victoria* **50**(2): 307–323.

O'Loughlin, P.M. and Rowe, F.W.E. (2005). A new asterinid genus from the Indo-West Pacific region, including five new species (Echinodermata: Asteroidea: Asterinidae). *Memoirs of Museum Victoria* **62**(2): 181–189.

O'Loughlin, P.M. and Rowe, F.W.E. (2006). A systematic revision of the asterinid genus *Aquilonastra* O'Loughlin, 2004 (Echinodermata: Asteroidea). *Memoirs of Museum Victoria* **63**(2): 257–287.

O'Loughlin, P.M. and Waters, J.M. (2004). A molecular and morphological systematic revision of genera of Asterinidae (Echinodermata: Asteroidea). *Memoirs of Museum Victoria* **61**(1): 1–40.

O'Loughlin, P.M. Waters, J.M. and Roy, M.S. (2002). Description of a new species of *Patiriella* from New Zealand and a review of *Patiriella regularis* (Echinodermata, Asteroidea) based on morphological and molecular data. *Journal of the Royal Society of New Zealand* **32**(4): 697–711.

O'Loughlin, P.M. Waters, J.M. and Roy, M.S. (2003). A molecular and morphological review of the asterinid *Patiriella gunnii* (Gray) (Echinodermata: Asteroidea) based on molecular data. *Memoirs of Museum Victoria* **60**(2): 181–195.

Ottesen, P.O. and Lucas J.S. (1982). Divide or broadcast: interrelation of asexual and sexual reproduction in a population of the fissiparous hermaphroditic sea star *Nepanthia belcheri* (Asteroidea: Asterinidae) *Marine Biology* **69**: 223–233.

Pattiarachi, C. and Woo, M. (2009). The mean state of the Leeuwin Current system between North West Cape and Cape Leeuwin. *Journal of the Royal Society of Western Australia* **92**: 221–241.

Pearce, A. and Pattiarachi, C. (1999). The Capes Current: A summer countercurrent flowing past Cape Leeuwin and Cape Naturaliste, Western Australia. *Continental Shelf Research* **19**(3): 401–420.

Péron, F. (1807–1816). *Voyage de découvertes aux terres australes, exécuté par ordre de S.M. l'Empereur et Roi, sur les corvettes le Géographe, le Naturaliste et la goëlette le Casuarina pendant les années 1800, 1801, 1802, 1803, 1804.* Partie historique 1. Imprimerie Impériale, Paris. pp. 1–495, pls 1–11.

Perrier, E. (1875). Révision de la collection de Stellérides du Muséum. *Archives de Zoologie Experimentale et Generale* **4**: 265–450.

Pope, E.C. and Rowe, F.W.E. (1977). A new genus and two new species in the family Mithrodiidae (Echinodermata: Asteroidea) with comments on the status of the species of *Mithrodia* Gray 1840. *Australian Zoologist* **19**(2): 201–216.

Potts, D.C. 1981. Crown-of-thorns starfish: man-induced pest or natural phenomenon? *In*: R.C. Kitching and R.E. Jones (eds), *The Ecology of Pests*. CSIRO Publishing, Melbourne. pp. 55–86.

Pratchett, M.S. (1999). An infectious disease in crown-of-thorns starfish on the Great Barrier Reef. *Coral Reefs* **18**: 272.

Pratchett, M., Vytopil, E. and Parks, P. (2000). Coral crabs influence the feeding patterns of crown-of-thorns starfish. *Coral Reefs* **19**: 36.

Prestege, G.K. (1998). The distribution and biology of *Patiriella vivipara* (Echinodermata: Asteroidea: Asterinidae) a sea star endemic to southeast Tasmania. *Records of the Australian Museum* **50**: 161–170.

Price, A.R.G. (1982). Western Arabian Gulf Echinoderms in high salinity waters and the occurrence of dwarfism. *Journal of Natural History* **16**: 519–527.

Purwati, P. and Lane, D.J.W. (2004). Asteroidea of the Anambas Expedition 2002. *The Raffles Bulletin of Zoology* **11**: 89–102.

Rideout, R.S. (1978). Asexual reproduction as a means of population maintenace in the coral reef asteroid *Linckia multifora* on Guam. *Marine Biology* **47**: 287–295.

Roediger, L.A. and Bolton, T.F. (2008). Abundance and distribution of South Australia's endemic sea star, *Parvulastra parvivipara* (Asteroidea: Asterinidae). *Marine and Freshwater Research* **59**: 205–213.

Rowe, F.W.E. (1976). The occurrence of the genus *Heteronardoa* (Asteroidea: Ophidiasteridae) in the Indian Ocean, with the description of a new species. *Records of the Western Australian Museum* **4**(1): 85–100.

Rowe, F.W.E. (1977). A new family of Asteroidea (Echinodermata), with the description of five new species and one new subspecies of *Asterodiscides. Records of the Australian Museum* **31**(5): 187–233.

Rowe, F.W.E. (1985). Six new species of *Asterodiscides* A.M. Clark (Echinodermata, Asteroidea), with a discussion of the origin and distribution of the Asterodiscididae and other 'amphi-Pacific' echinoderms. *Bulletin Muséum National d'Histoire Naturelle, Section A* **7**(3): 531–577.

Rowe, F.W.E. (1989). Nine new deep-water species of Echinodermata from Norfolk Island and Wanganella Bank, northeastern Tasman Sea, with a checklist of the echinoderm fauna. *Proceedings of the Linnean Society of New South Wales* **111**(4): 257–291.

Rowe, F.W.E. and Albertson, E.L. (1987). A new species in the echinasterid genus *Echinaster* Müller and Troschel, 1840 (Echinodermata: Asteroidea) from southeastern Australia and Norfolk Island. *Proceedings of the Linnean Society of New South Wales* **109**(3): 195–202.

Rowe, F.W.E. and Albertson, E.L. (1988). A new genus and four new species in the family Echinasteridae (Echinodermata: Asteroidea). *Proceedings of the Linnean Society of New South Wales* **110**(1): 83–100.

Rowe, F.W.E. and Gates, J. (1995). Echinodermata. *Zoological Catalogue of Australia* **33**: 1–510.

Rowe, F.W.E. and Marsh, L.M. (1982). A revision of the asterinid genus *Nepanthia* Gray, 1840 (Echinodermata: Asteroidea), with the description of three new species. *Memoirs of the Australian Museum* **16**: 89–120.

Run, J.Q. Chen, C.P. Chang, K.H. and Chia, F.S. (1988). Mating behavior and reproductive cycle of *Archaster typicus* (Echinodermata: Asteroidea). *Marine Biology* **99**: 247–253.

Saito, T. and Kishimoto, H. (2003). Tetrodotoxin attracts toxic starfish *Astropecten polyacanthus. Bulletin of the Institute of Oceanic Research and Development Tokai University* **24**: 45–49.

Sastry, D.R.K. (2007). Echinodermata of India: An Annotated List. *Records of the Zoological Survey of India, Occasional Paper* **271**: 1–387.

Saville-Kent, W. (1897). The Naturalist in Australia. Chapman and Hall, London.

Shepherd, S.A. (1967a). A review of the starfish genus *Nectria* (Asteroidea: Goniasteridae). *Records of the South Australian Museum* **15**(3): 463–482.

Shepherd, S.A. (1967b). A revision of the starfish genus *Uniophora* (Asteroidea: Asteriidae). *Transactions of the Royal Society of South Australia* **91**: 3–14, pl. 1.

Shepherd, S.A. (1968). The shallow water echinoderm fauna of South Australia 1: The asteroids. *Records of the South Australian Museum* **15**(4): 729–756.

Shepherd, S.A. (2014). Echinoderms. *In:* S.A. Shepherd and G.J. Edgar (eds), *Ecology of Australian Temperate Reefs: The Unique South.* CSIRO Publishing, Melbourne. pp. 233–258.

Shepherd, S.A. and Hodgkin, E.P. (1965). A new species of *Nectria* (Asteroidea, Goniasteridae) from Western Australia. *Journal of the Royal Society of Western Australia* **48**(4): 119–121.

Sladen, W.P. (1882). *Challenger* Pterasteridae. *Journal of the Linnean Society of London (Zoology)* **16**: 189–246.

Sladen, W.P. (1883). The Asteroidea of H.M.S. *Challenger* Expedition (Preliminary Notices). 2. Astropectinidae. *Journal of the Linnean Society of London (Zoology)* **17**: 214–269.

Sladen, W.P. (1889). Asteroidea. *In: Report on the scientific results of the voyage of the H.M.S. Challenger during the years 1873–76, Volume 30.* H.M. Stationary Office, London. pp. 1–893, pls 1–118.

Sloan, N.A. (1980). Aspects of the feeding biology of Asteroids. *Oceanography and Marine Biology Annual Review* **18**: pp. 57–124.

Spencer, W.K. and Wright, C.W. (1966). Asterozoans. *In:* R.C. Moore (ed.), *Treatise on Invertebrate Paleontology. Part U. Echinodermata 3.* Geological Society of America and University of Kansas Press. pp. V4–V107.

Steene, R. (2014). *Colours of the Reef, Volume 2.* Self published, Cairns. pp. 485, 491.

Stickle, W.B. and Diehl, W.J. (1987). Effects of salinity on echinoderms. *In*: M. Jangoux and J.M. Lawrence (eds), *Echinoderm Studies 2*. Balkema, Rotterdam. pp. 235–285.

Studer, T. (1883). Über der Asteriden, welche während der Reise SMS Gazelle um die Erde gesammelt wurden. *Sitzungsberichte der Gesellschaft Naturforschender Freunde zu Berlin* **8**: 128–132.

Studer, T. (1884). Verzeichnifs der während der Reise S.M.S. *Gazelle* um die Erde 1874–76 gesammelten Asteriden und Euryaliden. *Abhandlungen der Preussischen Akademie der Wissenschaften. Gelehrier* **2**: 1–64, pls 1–5.

Sukarno P., and Jangoux, M. (1977). Révision du genre *Archaster* Müller et Troschel (Echinodermata, Asteroidea: Archasteridae). *Revue de Zoologie Africaine* **91**(4): 815–844, pls 4–6.

Sweatman, H. (2008). No-take reserves protect coral reefs from predatory starfish. *Current Biology* **18**: R598–R599.

Taylor, J.G. and Pearce, A.F. (1999). Ningaloo Reef currents: Implications for coral spawn dispersal, zooplankton and whale shark abundance. *Journal of the Royal Society of Western Australia* **82**: 57–65.

Teidemann, R., Sarnthein, M. and Shackleton, N.J. (1994). Astronomic timescale for the Pliocene Atlantic $\delta^{18}O$ and dust flux record of Ocean Drilling Program site 659. *Palaeoceanography* **9**: 619–638.

Tortonese, E. (1937). Descrizione di una nuova stella di mare (*Goniodicaster australiae* n. sp.). *Bollettino dei musei di zoologia ed anatomia comparata della R. Università di Torino* **45**(69): 293–297, pl. 1.

Town, J.C. (1979). Distribution and Dispersal of the genus *Astrostole* Fisher, 1923 (Echinodermata: Asteroidea) *Journal of the Royal Society of New Zealand* **9**(4): 385–395.

Town, J.C. (1980). Movement, morphology, reproductive periodicity, and some factors affecting gonad production in the seastar *Astrostole scabra* (Hutton). *Journal of Experimental Marine Biology* and *Ecology* **44**: 111–132.

Town, J.C. (1981). Prey characteristics and dietary composition in intertidal *Astrotole scabra* (Echinodermata: Asteroidea). *New Zealand Journal of Marine and Freshwater Research* **15**(1): 69–80.

VandenSpiegel, D., Lane, D.J.W., Stampanato, S. and Jangoux, M. (1998). The asteroid fauna (Echinodermata) of Singapore, with a distribution table and an illustrated identification to the species. *The Reffles Bulletin of Zoology* **46**(2): 431–470.

Verrill, A.E. (1867). Notes on the Radiata in the Museum of Yale College, with descriptions of new genera and species. *Transactions of the Connecticut Academy of Arts and Sciences* **1**(2): 247–351.

Vogler, C., Benzie, J., Lessios, H., Barber, P. and Wörheide, G. (2008). A threat to coral reefs multiplied? Four species of crown-of-thorns starfish. *Biology Letters* **4**: 696–699.

Vogler, C., Benzie, J., Barber, P.H., Erdmann, M.V., Ambariyanto, S.C., Tenggardjaja, K., Gérard, K. and Wörheide, G. (2012). Phylogeography of the crown-of-thorns starfish in the Indian Ocean. *PLoS ONE* **7**(8): 1–10. [online] doi: 10.1371/journal.pone.0043499

Waters, J.M. and Roy, M.S. (2003). Global phylogeography of the fissiparous sea-star genus *Coscinasterias*. *Marine Biology* **142**: 185–191.

Wells, F.E. and Lalli, C.M. (2003). *Astropecten sumbawanus* (Echinodermata: Asteroidea) in Withnell Bay, northwestern Australia. *In:* F.E. Wells, D.I. Walker and D.S. Jones (eds), *The Marine Flora and Fauna of Dampier, Western Australia, Volume 1.* Western Australian Museum, Perth. pp 209–216.

Wells, P. and Okado, H. (1996). Holocene and Pleistocene glacial palaeoceanography off southeastern Australia, based on foraminifera and nannofossils in Vema cored hole V18–222. *Australian Journal of Earth Science* **43**: 509–523.

Wilkinson, C.R. (ed.) (1990). *Acanthaster planci. Coral Reefs* (special issue) **9**(3): 93–172.

Wilkinson, C.R. and Macintyre, I.G. (eds) (1992). The *Acanthaster* Debate. *Coral Reefs* (special issue) **11**(2): 51–122.

Wilson, B.R. and Allen, G. (1987). Major components and distributions of marine fauna: *In:* G.R. Dyne and D.W. Walton (eds), *Fauna of Australia*, General Articles, Volume 1A. Australian Government Publishing Service, Canberra. pp. 43–68, figs 3.1–13.

Wood, G.L. (1982). *Guiness Book of Animal Facts and Feats* (3rd ed.). Guiness World Records Limited, London.

Wyrwoll, K.-H. Greenstein, B.J. Kendrick, G.W. and Chen, G-S. (2009). The palaeoceanography of the Leeuwin Current: implications for a future world. *Journal of the Royal Society of Western Australia* **92**(2): 37–51.

Yamaguchi, M. (1973). Early life histories of coral reef asteroids with special reference to *Acanthaster planci* (L.). *Biology and Geology of Coral Reefs* **2**(1): 369–387.

Yamaguchi, M. (1974). Effect of elevated temperature on the metabolic activity of the coral reef asteroid *Acanthaster planci* (L.). *Pacific Science* **28**: 139–146.

Yamaguchi, M. (1975). Coral-reef asteroids of Guam. *Biotropica* **7**(1): 12–23.

Yamaguchi, M. (1976). Estimating the length of the exponential growth phase: Growth increment observations on the coral-reef asteroid *Culcita novaeguineae. Marine Biology* **39**(1): 57–59.

Yamaguchi, M. (1977a). Population structure, spawning and growth of the coral reef asteroid *Linckia laevigata* (Linnaeus). *Pacific Science* **31**(1): 13–30.

Yamaguchi, M. (1977b). Larval behavior and geographic distribution of coral reef asteroids in the Indo-West Pacific. *Micronesica* **13**(2): 283–296.

Yamaguchi, M. and Lucas, J.S. (1984). Natural parthenogenesis, larval and juvenile development, and geographical distribution of the coral reef asteroid *Ophidiaster granifer. Marine Biology* **83**: 33–42.

Yuasa, H., Higashimura, Y., Nomura, K. and Yasuda, N. (2017). Diet of *Acanthaster brevispinus*, sibling species of the coral-eating crown-of-thorns starfish, *Acanthaster planci* sensu lato. *Bulletin of Marine Science* **93**(4): 1009–1010.

Zeidler, W. (1992). Introduced starfish pose threat to scallops. *Australian Fisheries* **51**(10): 28–29.

Zeidler, W. and Rowe, F.W.E. (1986). A revision of the southern Australian starfish genus *Nectria* (Asteroidea: Oreasteridae), with the description of a new species. *Records of the South Australian Museum* **19**(9): 117–138.

Zeidler, W. and Shepherd, S.A. (1982). Sea-stars (Class Asteroidea). *In:* S.A. Shepherd and I.M. Thomas (eds), *Marine Invertebrates of Southern Australia Part 1. Handbook of the Flora and Fauna of South Australia*. South Australian Research and Development Institute, Adelaide. pp. 400–418.

Ziedler W. (1995) Case 2951. *Nectria* Gray, 1840 (Echinodermata, Asteroidea): proprosed designation of *Nectria ocellata* Perrier, 1875 as the type species. *Bulletin of Zoological Nomenclature* **53**(2): 164–165.

INDEX

TABLE OF SPECIES

GENERAL

D

E

F

G

T

NOTES ON PHOTOGRAPHY

Photographs of collection specimens and those sourced from digital and film photography archives remain copyright of the respective institutions and individuals. Without access to these invaluable resources, publication of this book would not have been possible.

Australian Museum

Collection specimens 145, 183–184, 190, 196–198, 204, 208, 219, 225, 245–246, 265, 266, 308, 356, 376–377, 385–386, 413–414, 473, 483, 497, 501

Anne Hoggett 29, 234, 286, 303, 309, 313, 327, 395

Commonwealth Scientific and Industrial Research Organisation

John Keesing 64–65, 72, 75, 94, 125, 254, 301, 338, 339, 357, 380, 430–431

Coral Reef Research Foundation

Patrick Colin 456

Florida Museum of Natural History, United States of America

Robert Lasley 438

François Michonneau 233, 259, 335

Institut de Recherche pour le Développement, France

Pierre Laboute 140–141, 218, 352

James Cook University

John Lucas 97

Marine Life Society of South Australia

David Muirhead 505

Muséum National d'Histoire Naturelle, France

Collection specimens 97, 212

Frédéric Ducarme 298

Museums Victoria

Collection specimens 175, 254, 410–411, 488

Leon Altoff 132–133, 489

Julian Finn 191–192, 216, 252, 284–285

Mark Norman 23, 479, 496

David Paul 174, 491

Reef Life Survey

Graham Edgar 40, 61, 137, 148, 153, 156, 177, 179, 181–182, 250, 267, 274, 373, 406–407, 450, 474, 487

Andrew Green 209, 475, 498–499, 503, 505

South Australian Museum

Collection specimens 55, 172, 188, 215–216, 250–251, 253, 255–256, 264, 280, 285, 362, 464, 469, 490, 492, 494, 504

Queensland Museum

Neville Coleman 29, 111, 121, 127, 130–131, 146, 151–154, 158–159, 162–163, 173, 186, 189, 193–194, 223–224, 232, 251, 257, 282–283, 296, 312, 334, 338, 342, 345, 349, 351, 357, 440, 454, 462, 472, 474, 477, 494, 501–502

Merrick Ekins 90

Tasmanian Museum and Art Gallery

Simon Grove 78

University of Singapore

David Lane 247, 252, 442

University of Sydney

Maria Byrne 113, 117, 123, 135

University of Western Australia

Anne Brearley 372, 443

Ernest Hodgkin 420

Western Australian Museum

Collection specimens 54–59, 63, 65, 69–70, 73–75, 77–79, 83–85, 87, 89, 96, 98–99, 105, 107, 116, 119, 126, 128, 141, 143–144, 155, 166, 167, 169, 170, 192, 202–203, 205–206, 229, 231, 239, 272, 275, 276–277, 283, 289, 295–297, 317–318, 321, 325–326, 329, 344, 346, 348, 350, 353–355, 368–369, 373, 375, 379, 382–

Individual contributors

Alamy Stock Photography

Published 2020 by the
Western Australian Museum
Locked Bag 49, Welshpool DC,
Western Australia 6986
www.museum.wa.gov.au

Layout and design by Tim Cumming.
Illustrations by Jill Ruse and Tim Cumming.
Printed in Australia by Scott Print, Perth.

ISBN: 978-1-925040-39-5 (paperback).
A catalogue record for this book is available from the National Library of Australia.

Front cover: *Tosia australis*, Gulf St Vincent, South Australia (F. Bavendam).

Back cover (from top): *Echinaster superbus*, Dampier Archipelago, Western Australia (Z. Richards); *Nepanthia crassa*, Geographe Bay, Western Australia (C. Bryce); *Asterodiscides truncatus*, Shark Point, New South Wales (J. Turnbull).

Frontispiece: *Neoferdina cumingi*, Ashmore Reef, Timor Sea (G. Allen).

The Western Australian Museum acknowledges and respects the Traditional Owners of their ancestral lands, waters and skies.

This publication would not have been possible without the generous support of:

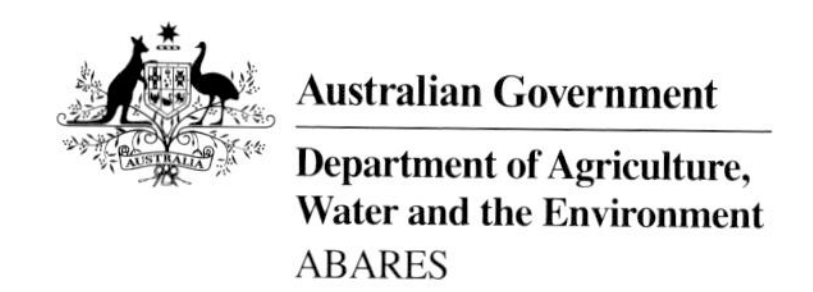